Public Welfare-Oriented Production of Food

Albert Sundrum

Public Welfare-Oriented Production of Food

Substantially New Impulses for a Sustainable Agro-Food Sector

Springer

Albert Sundrum
Department of Animal Nutrition and Animal
Health University of Kassel
Witzenhausen, Germany

ISBN 978-3-662-69039-0 ISBN 978-3-662-69040-6 (eBook)
https://doi.org/10.1007/978-3-662-69040-6

This Springer imprint is published by the registered company Springer-Verlag GmbH, DE, part of Springer Nature.
The registered company address is: Heidelberger Platz 3, 14197 Berlin, Germany

Paper in this product is recyclable.

We have inherited from our forefathers the keen longing for unified, all-embracing knowledge. … But the spread, both in width and depth, of the multifarious branches of knowledge during the last hundred odd years has confronted us with a queer dilemma. We feel clearly that we are only now beginning to acquire reliable material for welding together the sum total of all that is known into a whole; but, on the other hand, it has become next to impossible for a single mind fully to command more than a small specialized portion of it.

I can see no other escape from this dilemma (lest our true aim be lost forever) than that some of us should venture to embark on a synthesis of facts and theories, albeit with second-hand and incomplete knowledge of some of them—and at the risk of making fools of ourselves. So much for my apology.

Erwin Schrödinger in the preface of "What is life",1944

The burden of proof lies with those who support a thesis which is difficult to reconcile with life-experience realities and convictions.

Julian Nida-Rümelin, German philosopher

Dedicated to all the victims of a misguided agricultural development, which has focussed for decades on the increase in productivity while widely ignoring the undesirable side effects deriving therefrom,and thus missing the context-depending point where the positive marginal profit of a diminishing marginal utility turns into negative impacts for the majority of the stakeholders involved.

Preface

Every day, we resort to food in the most diverse forms and preparations to supply our bodies with vital nutrients. Most people in the Western world have long since become accustomed to these being available in abundance and can be purchased at comparably low prices. What appears to be a matter of fact does not usually require any reflection on the background of its production and the systems it derives from. After all, we don't need to know how a smartphone was made to use it. Therefore, how food is produced is unsurprising, and the associated consequences have largely disappeared from the public consciousness. Only now and then are we startled by scandalous reports in the media. These give the impression that food production is not always at its best. However, very few people can interpret the morsels of information from a largely unknown world or even place them adequately in an overall context.

Food production, processing, and trade cannot be understood by googling individual terms. The processes involved in cultivating plants, keeping livestock, and processing and distributing food are too complex and diverse to be reduced to a few explanations. This applies all the more to the consequences of these processes. What interacts with each other on farms and in the further processes up to the consumer's plate are factors of very different dimensions. These include the biological processes of the growth of microorganisms, plants, and animals and technical, economic, sociological, cultural, and political factors. In addition, many people intervene in the processes at the most diverse points and try to influence them according to their interests. We are dealing with a sector interwoven with diverse, often conflicting interests. The result is a supposedly opaque conglomerate of diverse individual aspects.

It seems rather presumptuous to unravel this conglomerate and make it understandable in an enlightened sense. On the other hand, despite all the shortcomings of such an approach, there is no way around a comprehensive analysis of the food production system if it is to be changed concerning the public welfare interests. There is widespread agreement among stakeholders that change is urgently needed. The extent of undesirable side-effects and harmful effects is too great, the waste of food and financial resources too large, the economic hardship of many primary producers too bad, and consumers too distrustful for 'business as usual' to be an option. But what has to be changed to come closer to a public welfare-oriented food production? This book deals with this core question by considering various sub-aspects and integrating them into a superordinate perspective. It does not claim to do justice to all aspects comprehensively. Nevertheless, the aim is not only to present a system analysis that is as profound as possible but also to break away from the previous, often one-dimensional models of thinking. A coherent counter-concept is also propagated via a multi-perspective and goal-oriented approach to the complexity of food production.

The remarks begin with a brief historical outline of the development of agriculture from its beginnings to the present day. The current state of the agri-food sector is too closely linked to its prehistory to understand the present situation without reflecting on the history of its development. Furthermore, it is, above all, the agro-industrial and technological developments of modern times that have revolutionized production structures and left their mark on them. These structures were decisively promoted by the economic framework conditions of a so-called "free" market economy. This exposed food production not only to local but also to global competition. Agricultural commodities such as meat, milk, and grain compete on the world market for the lowest production costs. While competition initially acted as a powerful growth engine, bringing a wide range of productive forces to bear, undesirable knock-on effects soon emerged in many sub-sectors. However, as long as one's interests are not directly affected, these can be and often are ignored. The less the polluter-pays principle takes effect and gives way to organized non-responsibility, the easier this becomes. Due to factual and mental path dependencies, many farmers do not see themselves in a position to deviate from the growth course they have once embarked upon and from ongoing intensification. This is true even though this course has already led the majority of farms to economic ruin.

With their considerable gain in detailed knowledge, the agricultural sciences have played a significant role in increasing the productivity of production processes. Now, however, when undesirable side effects and adverse effects are gaining the upper hand, they lack the competence to solve problems that arise from the complexity of the overall interrelationships. This is the price that the agricultural sciences are paying for decades of differentiation and specialization. The resulting inability to reorient itself in an interdisciplinary and transdisciplinary way makes the agricultural sciences in their current state part of the problem rather than part of the solution. Agricultural economics also appears trapped in its explanatory approaches. "*It's the economy, stupid.*" This slogan, which has been making waves since Bill Clinton's 1992 election campaign, has also been the prevailing motto in agricultural policy. Who would doubt that economics is also essential in the agricultural sector? Only slowly is the realization gaining ground that economics is not everything and is by no means decisive for all areas. Ecological requirements are of far greater relevance for the viability and survival of living systems.

The book devotes a separate chapter to the differences between economic and ecological approaches. After weighing up various reasons, it is concluded that the economy must no longer hold the scepter in its own hands but must be made to place itself at the service of ecological and qualitative objectives. This "subordination" is one of the book's several impositions for the reader. One of these is that ecological objectives are not to be understood in the sense of organic farming. Despite the pioneering achievements of the past, in its current orientation, it is only of limited use as a model for sustainable agriculture.

According to consumers' wishes, food should be affordable for all, of high quality, available everywhere and at all times, and produced, processed, and marketed in an animal- and environmentally friendly, sustainable, and socially responsible way. Marketing responds to these wishes with images on the packaging that evoke rural agriculture, traditional craftsmanship, and an intact natural environment, giving customers a good feeling. As many consumers do not want to know the production conditions and consequences in detail, they are satisfied with references to compliance with higher minimum standards, which marketing offers them via labels of origin and trademarks. This brand strategy reduces the complexity of food quality to individual aspects while at the same time excluding elementary areas.

This approach contrasts with what constitutes quality at its core: the totality and interaction of all properties of a product, process, or system. Product and process qualities do not differentiate based on different minimum standards that relate only to a few partial aspects but rank in the totality of relevant properties on a scale from very low to very high. The equation of standards with quality (**„Quality standard“**) and the use of false, because explanation-poor, frames of reference is based on an elementary inductive fallacy that ignores the interactions in living systems between the parts and the whole. At the same time, this way of thinking stands in the way of the urgently needed increase in value creation in primary production. For the latter, the only way out of existing dependency relationships is to turn away from striving for cost leadership and change towards quality leadership, which realizes value creation through the proof of quality and public welfare services rendered. Through various examples, it is shown how the production of increased product qualities, as well as individual animal and environmental protection services, can succeed.

Therefore, biological, technical, economic, and sociological interactions must be illuminated from different perspectives and brought into an overall context. The only way out of the confusion of one-dimensional, short-sighted, and self-referential approaches is a clear overarching objective. In this context, the adequate use of terms is of particular importance. Central terms are defined in this book in a glossary to remove them from an interest-driven interpretation. Finally, it is also a matter of countering the dominance of particular interests and reorienting them towards overriding public welfare interests.

To show ways out of a situation in agriculture and the agricultural and food industry that is perceived to be rather complex, I felt compelled to take the readers on a broad journey of thought through very different subject areas. Repeatedly, familiar paths of thought are left behind to open up new perspectives on the complexity of causal relationships and to point out new approaches and options for action for the necessary transformation. The book blends different strands of thought that have engaged me in the past into what I hope is a coherent whole. For better readability, I have taken the liberty of using the plural or the masculine form for persons, whereby all gender identities are included.

I want to thank all current and former department members who, over the past years, have worked with me in various discussions and on numerous occasions to penetrate the complex issues intellectually. Special thanks go to Dr. Susanne Hoischen-Taubner for her critical review of the manuscript, her valuable suggestions, and her support in the editorial work.

The author would like to see an active engagement with the book's contents that goes beyond the subject-specific context. Reflections and discussions are of central importance for the necessary transformation process. These should not only take place in the agricultural and food industry, agricultural sciences, and agricultural policy or among NGOs. After all, food production concerns us all. We are all both the recipients of the vital services of the economic sector and sometimes more or less affected by its harmful effects.

Göttingen
January 2024

Albert Sundrum

Contents

Abbreviations

AHW	Animal health and welfare
ASF	Animal-sourced food
BMEL	Federal Ministry of Food and Agriculture
BMELV	Federal Ministry of Food, Agriculture and Consumer Protection
BUND	Association for the Environment and Nature Conservation Germany e. V.
DBV	German Farmers' Federation
DLG	German Agricultural Society
EC	European Commission
ECJ	European Court of Justice
EEA	European Environment Agency
EEG	Renewable Energy Sources Act
EFSA	European Food Safety Authority
EWG	European Economic Community
FAO	Food and Agriculture Organization of the United Nations
GAP	Common Agricultural Policy of the European Community
GATT	General Agreement on Tariffs and Trade
GMO	Genetically modified organisms
IMF	Intramuscular fat (marbling)
N	Nitrogen
NEC	National Emission Ceilings Directive for certain air pollutants
NGO	Non-Governmental Organization
NOx	Nitrogen oxides
Nr	Reactive nitrogen
NRL	Nitrates Directive
OIE	World Organization for Animal Health (Office International des Epizooties)
P	Phosphorus
QS	Quality and Safety GmbH—Testing system for food safety

r	Correlation coefficient
SCC	Somatic cell counts
WHO	World Health Organization
WTO	World Trade Organization

List of Figures

List of Tables

Introduction

1

Managing agricultural land by cultivating crops and keeping livestock is not a matter that concerns only farmers. Farmers provide the raw materials processed into food and vital nutrients to the population. Moreover, agricultural land is part of a landscape, and farms are part of a communal community structure and economic and social systems. It is also part of agriculture that various harmful effects emanate from the production processes. These few indications alone justify the statement: agriculture concerns us all! To what extent this is the case is what this book is intended to answer. This cannot be satisfied with descriptions of the current situation. It is no less essential to trace the complexities of the interests and driving forces involved and to shed light on their share in the developments. The same applies to reflections on embedding agriculture in diverse contexts and causal relationships. After all, agriculture is not an independent sector but is highly dependent on the framework conditions within which farms operate. According to Langthaler (2010), agriculture can be understood as a temporally and spatially varying interrelationship of ecological, economic, political, social, and cultural elements—in short, as a **complex system**:

> *"Essential here is the external and internal differentiation of the agricultural system: externally, the agricultural system is embedded in natural and social environments—and is regulated through coupling with the respective ecological and social systems. Internally, it comprises material and immaterial resource flows that are self-regulated to some extent by the actors working and living within it."*

The outlined relationships are illustrated in Fig. 1.1.

The complexity of the areas involved results in different perspectives from which agricultural systems can be viewed and assessed. Langthaler names some of these perspectives: agriculture as a part of nature "colonized" by human societies, as that area of labor and family relations in which the majority of the world's population produced and

A. Sundrum, *Public Welfare-Oriented Production of Food*,
https://doi.org/10.1007/978-3-662-69040-6_1

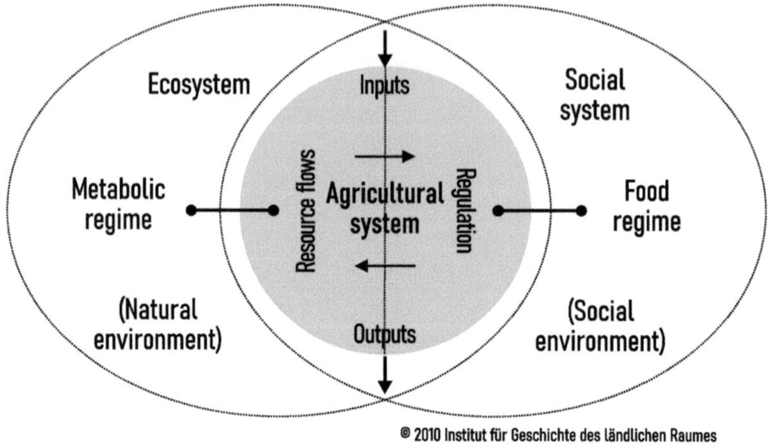

Fig. 1.1 Agriculture as an agricultural system (Langthaler 2010)

reproduced until the 1970s, and as it continues to do in many countries; as a relevant economic sector alongside industry and services; as an object of political regulation at (supra-)national and global level; as a projection surface for (urban) notions of 'rurality,' and many perspectives more. Of particular relevance are the perspectives of primary producers, upstream and downstream economic sectors, agricultural sciences, agricultural policy, and many other positions taken by different non-governmental organizations (NGOs), which significantly influence the public perception of agriculture. No less important are the views of those affected to a greater or lesser extent by the processes in and in connection with agriculture. The mixture of different perspectives from which agricultural processes are examined and evaluated results in ambiguities and confusion that call for classification and structuring.

Conflicts thus stands at the beginning of a cultural-historical development of human communities and represents a terrain on which diverse interests have always been directed and more or less pronounced conflict situations have been fought out. Accordingly, a reflection on the development of agriculture cannot avoid looking at the background of the emergence of conflicts, how they are dealt with, and the various ways disputes can be resolved. Agricultural systems do not only have to cope with the conflicts of interest that arise internally or through interactions with the outside world. They are also confronted with many requirements that arise from the various sub-areas and are probably not to be found in this variety in any other industry. Not only physical and chemical laws but, above all, biological laws and regularities that determine the processes in agriculture must be observed. Even the term multi-disciplinarity does not adequately describe the all-rounder skills required. Mastering the interplay of the various areas, each following its laws is decisive for production performance. If some of the requirements are only insufficiently considered, this can result in serious malfunctions. Furthermore, farms are economic systems that are subject to economic laws. Farmers cannot avoid considering economic requirements if they want to make a profit to keep the farm viable in the long term. Furthermore, farms are part of a social system that provides the infrastructure and acts as a buyer of the products sold. In the long run, farms cannot afford to disregard social concerns and societal interests if they do not want to lose societal support. Farms are part of a social system that provides the infrastructure and acts as a buyer of the products sold. In the long run, farms cannot afford to disregard social concerns and societal interests if they do not want to lose societal support.

Throughout their history, agricultural systems have gone through different phases. These are closely related to developments in the environment of farming systems. For a long time, **subsistence** was predominant for long periods, with farms being closely dependent on natural conditions on the one hand but, on the other hand, having a high degree of self-sufficiency. In the course of further developments, dependence on the natural environment decreased. On the other hand, dependence on the markets in which farmers sold their products increased. In all phases, agricultural systems were demanded to be highly adaptable. This is no less true for the prevailing phase of industrialized (modern) agricultural systems. Analogous to the evolutionary processes in biology, the adaptation processes of agricultural systems secure existence via the creation of value from utilizing biological growth processes. Whereas for a long time, the main concern was to extract sufficient nutrients from natural conditions to enable one's clan to survive, today the aim is to produce large quantities of products for sale as cheaply as possible and to achieve sufficient revenues on the markets to cover the livelihood of one's own family, ongoing operating costs and investments.

If the current economic conditions make it difficult for farmers to compete for cost leadership, they are increasingly confronted with further demands from society. Many social groups are taking increasing offense at the undesirable side-effects and harmful effects associated with agricultural production processes. On the one hand, these concern internal areas such as the welfare of farm animals. At the same time, as the public becomes increasingly aware of the need for environmental and climate protection, the climate-relevant emissions from agriculture and the **external effects** are also coming under criticism. (Mankiw and Wagner 2004). As a rule, they are not included in the decision-making calculations of those who cause them. In economic terms, they constitute a form of market failure and can—according to theory—make state intervention necessary. Criticism of the internal and external effects of agriculture reveals the extent to which value creation in agriculture has become detached from any appreciation by society.

In anticipation of rising global demand, especially for products of animal origin, intensive expansion of agricultural production capacities was pursued in many countries. However, hopes of increasing prices and profits have not been fulfilled. Instead, developments have led to overproduction. Significant uncertainties and volatile price developments in products for sale and means of production characterize regional, national, and global agricultural markets. Unstable weather conditions and competition for land, including from energy crops, influence the availability and, thus, the prices of feedstuffs, which account for the largest share of production costs in livestock production. In addition, endemic animal diseases cause turbulence in the global and, thus, regional markets. They restrict free trade and change the relations between supply and demand in the remaining trading areas. On the other hand, many farmers have become path-dependent in reducing production costs by specializing in individual production segments. This means they hardly have any options left to adapt to changing requirements.

Squeezed between in-house requirements, volatile market prices, and societal expectations, primary producers' room for maneuver is severely limited. This causes considerable uncertainty. Additional stress is caused by the intensification of unequal income developments in other economic sectors and the lack of appreciation. Many farmers feel an inevitable disorientation and a sense of powerlessness. Without **subsidies**, many farms would not be able to gain economic viability, and many farms would not survive even with them. At the same time, voices are becoming louder and louder, calling for subsidy payments to be linked to services that benefit the community. Here, fault lines within the agricultural systems and conflict areas with the social and economic systems become visible, urgently requiring relief.

From the perspective of **public welfare interests**, it will be essential to establish production methods that are associated with the lowest possible internal and external harmful effects. This requires nothing less from the actors in agriculture than a transformation from the various forms of industrialized agricultural systems to **agroecosystems**. However, the greening of production methods must not be equated with a conversion to organic farming **organic farming**. Recent developments in organic agriculture, particularly increased **minimum requirements**, are not a blueprint for coping with the manifold internal challenges and integrating external demands. On the contrary, it illustrates how not to do it if one wants to meet the internal challenges of the farm and simultaneously public welfare interests. Organic farming is subject to the same market conditions that have left their mark on agro-industrial agriculture. As a result, the supposedly "organic" farms lack the necessary resources to provide ecological services and the incentives and clear targets that must be regarded as an essential prerequisite for reorientation and evidence-based greening of production processes.

There are no easy solutions to the **agricultural crisis**. While there should be little disagreement about the necessity and urgency of change, it cannot be brought about by changing attitudes or behavior alone. It is a misconception to conclude from the factual and normative necessity of change that change is possible and enforceable. (Nassehi 2020). Dealing with the complexity of agricultural, biological, and social systems requires careful consideration of the advantages and disadvantages, the costs and benefits of implementing measures, and fair negotiation processes with the interest groups involved. Because of various conflicting goals, questions arise about the ability to control complex systems. Steering can only be successful if there is clarity about the objective. Regulatory measures must not be directed against the logic of the individual functional systems.

The modern form of agriculture is permeated by a scientific understanding of reality and efforts to break down the interrelationships into individual parts and illuminate them down to the last micro-area through analytical procedures. In this way, they are more amenable to manipulation. They can also be better integrated into standardized procedures and more efficiently used for one's interests. A postmodern form of agriculture will have to find other ways of approaching the complexity of the causal structures and different strategies for **reducing complexity** if it is to internalize the external effects and generate product and process qualities. These are not based on individual characteristics but on the totality of the properties of a product, process, or system. In generating quality, it is crucial to bring together the sub-areas to create synergies; thus, the whole is more than the sum of the parts. This requires comprehensive knowledge of how biological systems function and interact with the outside world. The mental comprehensibility of the interrelationships is also an elementary prerequisite for solving the problems that arise from them. Since production processes in agriculture are based on the functioning of living systems, an understanding of systems-oriented towards the functional and teleological orientation of living systems is required.

Living systems are imbued with the desire to maintain themselves and to ensure the continuation of the species through reproduction. Life carries its purpose within itself; thus, it is an end. All functional areas of living systems are subordinated to this end. For this purpose, they use diverse means, above all in the form of nutrients, which are metabolized and thereby create growth and self-preservation. The living systems (microbes, plants, and animals) are provided with appropriate nutrients and growth conditions to produce agricultural commodities. Their sale is the central means of promoting the economic existence of farming systems. The processing industry processes raw commodities to make them suitable as nutrients in food for consumers and their quest for self-sustainability. In this way, the raw commodities produced by agriculture are **means to life** and, therefore, **food**, provided that they do not contain harmful substances that impair the life processes of consumers. For Regulation (EC) No 178/2002, food is any substance or product intended to be, or reasonably expected, ingested by humans in a processed, partially processed, or unprocessed state. This anthropocentric perspective does not make a fundamental distinction between food and foodstuffs, which is why the two terms are often used interchangeably.

However, suppose the reference system for the assessment of food is not limited to its suitability for consumers but is extended to the upstream and downstream areas. In that case, this opens up the possibility—initially in broad terms—of differentiating between food and foodstuffs and including the interests of the common good. In this sense, food as a means of life differs from aliments in that the production process also considers the life processes of many other living beings in the upstream and downstream production contexts. In other words, food is only food if the life processes of other living beings are not massively impaired in the methods of its production. This includes the fact that the output is not excessively detrimental to flora and fauna in the environment or to farm animals and also not harmful to primary producers and other people, i.e., does not run counter to the public welfare interests. Aliments are limited to the immediate context of consumers, whereas foods, provided that the above conditions are met, also include the higher-level context of the production processes.

Nutrients do not exist in isolation but are integrated into **material cycles** integrated. The substances in foods were previously in other living things and will be part of other living things again after passing through consumers. They are, therefore, part of a much larger context than they are assigned as food for humans. In this context, as the Nobel laureate in physics, Erwin (Schrödinger 1992), already formulated in 1944 in his book What is Life?, it is not only about the pure exchange of substances. Material means for life consist of micro- and macronutrients, whereby the latter, in addition to the material content, are also endowed with negative **entropy**, which is the essential fuel of life. The living organism escapes decay and thus increases entropy by exchanging material (**metabolism**) and absorbing negative entropy. Entropy is a fundamental thermodynamic state variable with the SI unit Joule per Kelvin (J/K). Just like the addition of heat or matter, all processes that occur within a living system cause an increase in entropy. The entropy of a system can only decrease through the release of heat or matter. A living system in which processes occur must be coupled with its environment, which absorbs the increase in entropy and changes its state. By absorbing nutrients charged with negative entropy, physical life processes are fueled. The excess entropy that living things constantly generate is released into the environment by releasing heat and undigested residual matter. This happens equally on a micro and macro scale and, as global warming shows us, on a global scale.

In addition to the physical, chemical, and biological premises of the processes of life, in the search for a definition of life, one also comes across a definition by the Nobel Peace Prize winner Albert Schweitzer, from whom the following quotation comes: "*Every life is a life that wants life amid lives that want life.*" In his guiding principle, "Reverence for life," Schweitzer clarifies with this sentence that human beings are not the only living beings who want life. We are part of a community of living beings, and as part of a larger whole, we should also show respect for other living beings. On the one hand, this can be interpreted as a moral appeal, leaving it up to the individual to decide whether and to what extent the appeal elicits a particular response. Schweitzer is said to have been "unable to harm a fly." Consumers who consider themselves to be vegans or vegetarians will interpret the appeal as a call to do without products of animal origin to a large extent. Flexitarians will infer that the farm animals we keep and eventually kill to produce products of animal origin should have lived good lives. Others will argue that animal products significantly provide essential nutrients to humans. Moreover, farm animals can utilize nutrient resources and convert them into foods unsuitable for human consumption. From an ecological perspective, the use of animals for food production can also be seen as highly relevant in maintaining higher-level material cycles. Still, other groups of people will emphasize that farm animals are sentient beings; they are considered fellow creatures and are thus to be classified as part of the community. Farm animal husbandry, aligned with the overriding contexts, is just as much in the interest of the common good as the concern that farm animals are protected as far as possible from pain, suffering, and harm throughout their life cycle. If this cannot be guaranteed, the products of animal origin produced in this way are not food but, at best, foodstuffs, which can only be attested to a reduced quality and accorded a low value.

What consumers choose when buying the products offered on the markets is mainly up to them. Legal minimum standards ensure that extremely harmful excesses in production conditions are avoided and that recognizably defective products are not sold on the markets. From the perspective of the common good, however, it would be urgently necessary to develop yardsticks to distinguish between food for human consumption and aliments, by which the production processes are accompanied by excessive destruction, undermining other living beings' ability to exist. The transitions between food-supporting life and food-supporting life are fluid. They, therefore, require differentiation based on meaningful and robust assessment criteria. Notwithstanding the methodological challenges discussed in more detail in the book, it should be agreed to focus first on those production processes associated with particularly conspicuous harmful effects.

The complexity of food production can only be penetrated mentally if the topic is approached from different perspectives. A change of perspective facilitates access and makes the extensive interrelationships more comprehensible. The following explanations do not claim to consider all potentially possible perspectives similarly. Essentially, the question is how a transformation of current agricultural systems towards the design of agroecosystems can be realized. In particular, the challenges are addressed regarding how existing internal resources can be harnessed without causing excessive adverse internal and external effects to the detriment of public welfare interests. This will not succeed without subjecting the economic framework within which primary producers operate to a fundamental transformation.

While the challenges are significant, the potential for positive change is equally immense. To be able to give a well-founded answer to the challenges outlined, readers can first expect a brief treatise on the historical development from the beginnings of arable farming and animal husbandry to today's forms of industrialized agriculture. Today's agricultural conditions and the various forms of agricultural systems are the result of decisions taken far back in the past. Tracing the evolutionary process that has made agricultural systems what they are today helps to sound out the options for further developing the remaining agricultural systems to remain viable. This will only succeed if agricultural production processes are no longer at the expense of the common good interests but are in harmony with them. An orientation towards common good interests requires conceptual considerations and theories. The focus here is on the possibilities of operationalizing theoretical considerations to translate them into practical action, inspiring a future of sustainable agriculture.

Agriculture is based on the utilization of cultivable land and livestock. While the availability of land could hardly be expanded, it was only possible to intensify its cultivation and expand and intensify livestock farming. To understand the extent of the changes, the primary factors and driving forces behind the biological, technical, and economic successes in livestock farming are analyzed in detail. For long periods, the focus was on productivity successes, but today, questions about the welfare of farm animals are on the agenda. Reflections on past strategies to improve farm animal welfare are not only significant for correcting undesirable developments. They are also instructive for developing comprehensive concepts with which the adverse external effects of agriculture on nature and the environment, climate, and consumer protection can be reduced.

The protection of common goods collides in many ways with the particular interests of diverse groups. Thus, it is necessary to identify the driving forces to understand what drives primary producers to prioritize production services and ignore undesirable side effects and harmful impacts. It is also required to explore the room for maneuvering available to the responsible actors to minimize the damaging effects. The adverse consequential effects emanating from agriculture affect very different areas. However, in the public perception influenced by the media, only individual aspects are generally perceived. However, focusing on particular elements is more likely to obscure the true extent of the **harmful effects** than to make it visible to outsiders. Last but not least, the diversity of harmful effects and the significant variation between farms contribute to a lack of clarity. This will be countered by structuring and classifying the diverse impacts in a separate chapter.

It is not only the primary producers who are responsible for the internal and external damaging effects of agriculture to the detriment of the common good. The interest groups located in the upstream and downstream sectors of agriculture exert considerable influence on the economic and social framework conditions with their respective particular interests. Primary producers must align themselves with these if they are to survive economically. Therefore, the question is examined to what extent the interests brought to the farms from outside run counter to the interests of the common good and hinder the primary producers in expanding their contributions to promoting the common good.

A reflection on the overall topic cannot avoid illuminating the role of agricultural science. Particularly after the Second World War, it had a decisive influence on the direction in which production processes developed. Now that undesirable severe developments can no longer be ignored, the question arises as to what extent an institution that has played a significant role in promoting the processes that are today partly responsible for the occurrence of harmful effects is also suited to contribute to the containment of negative consequences.

Agricultural economic maxims and economic liberal principles permeate the development of modern agriculture. The conflict between economy and ecology often cited and referring to various economic sectors is particularly effective in agriculture. This makes the question of how to reconcile economic and ecological concerns in the agricultural sector all the more urgent. It will be necessary to analyze the opposing objectives and common intersections to determine the potential of conflict management. Furthermore, proposals must be worked out to expand the commonalities at the expense of the conflict areas.

It is concluded from the analyses and reflections that maintaining the orientation of production processes towards cost leadership is unsuitable for reconciling the various particular interests with the interests of the common good. The watchword is not a further reduction in production costs but an increase in market prices to make more resources available to primary producers, which they need to be able to take better account of the interests of the common good. Higher prices can only be justified and sustained if higher-quality services match the higher prices. The potentials of evidence-based quality production are explored using various examples relating both to product quality characteristics and to the process qualities of animal welfare and environmental protection services. The concluding chapter discusses how this can be operationalized and what obstacles must be overcome.

References

Langthaler E (2010) Landwirtschaft vor und in der Globalisierung. In: Sieder R, Langthaler E (eds) Globalgeschichte: 1800–2010. Böhlau Verlag; Vandenhoeck & Ruprecht GmbH & Co, KG, Wien, Köln, Weimar, pp 135–170

Mankiw NG, Wagner A (2004) Grundzüge der Volkswirtschaftslehre, 3rd edn. Schäffer-Poeschel, Stuttgart

Nassehi A (2020) Komplexitätsprobleme: Der Klimawandel aus soziologischer Perspektive. Forschung & Lehre 27:908–909

Schrödinger E (1992) What is life?: The physical aspect of the living cell. Cambridge Univ. Press, Cambridge

A Brief History of Agriculture 2

Abstract

The history of humankind is inextricably linked with the agricultural exploitation of natural resources. This still forms the basis of human life today. What began locally has expanded into a global system of agribusiness and international markets in agricultural products. Food procurement processes have shaped not only diets but also social interaction. While significant progress has been made in ongoing efforts to increase productivity, these advances have always been accompanied by undesirable side effects. Looking at the fast-motion developments helps to understand an agricultural economic system whose driving forces have perfected the exploitative use of natural resources. At the same time, the historical retrospective reveals the challenges facing current decision-makers when steering the agricultural economy away from a course that threatens to destroy its livelihoods.

2.1 To the Beginnings of Agriculture and Animal Husbandry

The first evidence of human activity suggesting the beginnings of agriculture and animal husbandry is dated back about 11,500 years. Strategies that changed from gathering and hunting for food appeared independently in different regions, including southeastern Turkey, western Iran, and Central and South America. The earliest agricultural settlements have been found in the Levant, an area bordering the eastern Mediterranean, including present-day Israel, Palestine, Jordan, Lebanon, Syria, and southern Turkey (Hands 2015). Concerning today's perception of time, development and spread were very slow. The transition from a nomadic way of life to sedentary communities was connected with a fundamental change in food procurement. Development was set in motion,

A. Sundrum, *Public Welfare-Oriented Production of Food*,
https://doi.org/10.1007/978-3-662-69040-6_2

which can be interpreted today as an expansive and continuously increasing exploitation of natural resources. Humans have now appropriated most of all agriculturally usable land with their food procurement strategy.

Ideas that hunter-gatherers lived peacefully and in harmony with nature, as in the Garden of Eden, are more likely to belong to the realm of myth. Prehistoric finds show that raids and massacres occurred frequently in prehistoric times and sometimes claimed many victims (Hill and Hurtado 1996). According to the prehistoric findings, about a quarter of adult males died by physical violence. Deaths by violence also included children and older people. Infanticide of orphans and killing of the old and sick were probably widespread and can be interpreted as part of the group's struggle for survival. In short, life was hard for gatherers and hunters. Periods of food in abundance alternated with periods when people were predominantly hungry. People were directly dependent on what the respective living conditions offered them for their ability to survive. Sudden natural phenomena or slow changes occurred repeatedly without people being able to assign a cause to these changes. Consequently, their only option was to adapt to the given living conditions. This was only possible through group cohesion. On the other hand, those who could not contribute to the nomadic search for food could not be supported permanently at the expense of group interests.

With hunter-gatherers, the evolutionary principle for preserving the species functioned no differently than with other animals. Only a portion of the offspring survived the first critical phases of life. All offspring were exposed to natural selection pressures caused by limited food availability and threats from visible or invisible (e.g., microbial) enemies. Availability of food and protection from threats determined how many of the offspring produced survived. At the same time, the demand for food determined the extent of efforts to improve supply. In times of scarcity, interdependence implied efforts to bring in as much food as possible so that as many could survive. Keeping oneself and one's offspring alive was then (and still is today) the most significant driving force of human endeavor. Survival chances were only offered by belonging to a group. These were generally better in a large group than in a small group.

The remains of bones, crops, tools, pottery, and buildings preserved from the Neolithic phase provide evidence of the transition to sedentarism. From today's point of view, we can only speculate about the manifold backgrounds that promoted a changed food procurement strategy. Among other things, it is estimated that a substantial increase in the carbon dioxide content of the atmosphere from 180 to 280 ppm in a few thousand years during the transition from the Pleistocene to the Holocene may have acted as a trigger because it had a positive effect on photosynthesis and plant productivity (Schaefer 2007). The domestication of various animal species, starting with sheep and goats and then extending to cattle, pigs, horses, and camels, improved the supply of protein, especially when crops were meager (Alvard and Kuznar 2001). With sedentarism, foresight, stewardship, cooperation, and problem-solving skills became increasingly important. Coordinated action in larger groups offered improved possibilities for obtaining food, which was accompanied by advantages for survival. Cultivation, care, harvesting, and protecting the harvested crop from human and animal enemies required the coordinated cooperation of people in the group. This was even more true for the construction of permanent dwellings. Rather than chasing after wandering animals, capturing some animals and letting them live eventually proved more beneficial. They could then be killed if they were needed for their food. Thus, the fixing of wild animals amounted to a certain stockpiling. The specific practices of obtaining food and people's behavior towards each other were shaped by the conditions of life. The better people understood how to adapt to the conditions, the greater the chance of improving individual and community survivability. Not everywhere did agriculture and animal husbandry promise a survival advantage. In some places, the transition to agriculture and livestock was never made. Survival International estimates that even today, an almost inconceivable number of about 150 million people live as gatherers and hunters in small groups in more than 60 countries world wide. This points to the significant variation in living conditions and the remarkable adaptability of the human species.

People began to settle primarily in those places that offered good conditions for the growth of plants suitable for consumption. The use of plants and animals was improved over long periods, changing wild plants into cultivated plants and wild animals into domestic animals. Influencing the growth of selected plants was done according to the principle of trial and error. Accordingly, the methods of action that proved successful under the respective conditions prevailed. This way, the available food increased, which could feed more people. This by no means meant better nutrition (Harari 2015). Instead, it can be assumed that the diet was much more one-sided compared to the life of hunter-gatherers. Above all, however, it entailed various consequential effects, which led the author to assess that "not men domesticated wheat, but wheat domesticated men." The cultivation of crops entailed, among other things, the need to sow, tend, harvest, and process the plants. In addition, measures had to be taken to protect the plants from the access of wild animals and other humans.

Adaptation strategies included stockpiling, which helped through periods of sparse food availability. As with any new development, stockpiling was associated not only with advantages but also with disadvantages. Stockpiles not only improved the food supply but also created covetousness. Fortifications had to be built to protect against living beings who tried to access the stores without permission. Above all, protection became necessary against other tribes forced to abandon their territory due to agricultural overexploitation, drought, or other natural disasters and who felt compelled to seize possessions from different communities to survive. Accordingly, as the settlements grew, so did the fortifications. About 1000 years after the first settlements, the settlements in the Levant grew to gather several hundred to 2000 people (Hill and Hurtado 1996). The settlements were extended more and more and finally enclosed by stone walls. However, even stone walls could not prevent conquests. Conquest and destruction were followed by reconstruction, usually in the form of an even larger fortification.

The initial form of agriculture is called a natural economy. The extended family or horde produced all food, articles of daily use, and weapons and implements (Bergmann

1962). The natural economy was associated with multiple social interactions. It welded people into communities of fate, without which survival could not succeed. Whereas hunter-gatherers depended on traveling great distances in small groups, in the case of the localization associated with land cultivation, they squatted together in one place. The need to work together, secure food resources, and fend off threats from enemies gave rise to new ways of life. At the same time, they created new ways of trying things out and passing on ideas using signs and language. For example, people invented techniques to make farmland easier to cultivate, expand, or improve tilling, harvesting, and storage. The development of techniques was accompanied by a growing awareness of how one's actions could influence the environment to one's advantage. One's experiences were passed on to the next generation through communicative forms. Accordingly, the descendants could build materially and intellectually on what had been developed and proven generations before. The main driving force may have been the concern that, in the event of an unforeseen shortage, the supply of food for oneself or one's clan might not be sufficient to sustain everyone. Expanding communal activities required coordination among people in a group and some planning. Experience mixed with projections into the future induced measures of preparedness far beyond what would have been necessary or possible for hunter-gatherers. With the successive expansion of food procurement, more people could be fed. More food also meant more births and children surviving the particularly critical first years of life. Consequently, by making efforts to obtain food, the survivability of the members of one's clan could be improved. Food procurement was a communal task in which more or less all clan members were involved. Violations by individuals against the higher interests of the clan were punished. The maximum punishment that could befall individual members was exclusion from the community, which meant certain death.

From the individual's concern for the preservation of his own life and that of the members of his clan, a general trend towards an increase in efforts to secure food becomes comprehensible. In the hunter-gatherer way of life, people had to make do with whatever food sources nature offered them. With land cultivation, they could influence the availability themselves, even if they were not immune to unforeseen setbacks. As people intervened in the growth and life processes of plants and animals and gained experience, the degree of their effectiveness gradually expanded. Nevertheless, this was very limited compared to today and was repeatedly questioned by events interpreted as interventions by superhuman powers. People tried to contain the feeling of control- and helplessness- by giving names to these unknown powers, personifying them as gods, and attempting to communicate with them.

It took time for the strategies to become more efficient and the yields that could be extracted from the soil to become more significant, providing food for the survival of an increasing number of members of one's clan. Food production was laborious and, at the same time, constantly exposed to the danger of insufficient food to feed people in need or even being lost. Over time, improvements due to increased harvests, which made life easier, alternated with changes threatening survival. Changes were caused not only by the forces of nature in the form of rain, storms, drought, heat, and cold but also by the living environment in the form of hostile humans, wild animals, or parasitic microbes. Some of the changes were also likely caused by poor planning and decisions by the people themselves. With a growing number of people, the demand for food increased, causing shortages and thus threat scenarios. In situations of limited availability, competitive behavior, and conflicts grew, turning other people into victims.

The "struggle for daily bread" was not a metaphor but a reality of life. It was linked to a struggle for dominance, i.e., for control over agricultural and other resources. Battles were fought both within and between settlements. Social hierarchies developed early on to contain competitive behavior. Individuals placed themselves in leadership positions to guide the fortunes of communities. Those who ruled over others developed their dynamics to maintain and, if possible, further consolidate or expand their respective positions. More and more, rulers and elites emerged who created structures of order that gave them access to resources. For the "ruled," this could result in advantages such as protection on the one hand and the obligation to pay taxes on the other. Hierarchical structures led to centralization and not only promoted forms of cooperation but also enforced them. Elites of rulers, priests, soldiers, and merchants developed, who shared in the surpluses produced by the peasants—partly with the support of slaves. Thanks to the surpluses produced, more and more people could settle in settlements and cities and engage in activities beyond food production. At the same time, elites ensured that they did not miss out on the distribution of available food by creating the conditions to enforce priority consideration of their claims, even in the face of opposition. Hierarchical structures ordered the balance of power and the interpretative sovereignty over the manifold phenomena for which the people of the time had no explanations. Faced with a constant confrontation with unpredictable phenomena classified as supernatural, various forms of superstition arose from a combination of imagination, lack of understanding of natural phenomena, and fear of the unknown. Organized religions developed from superstitions, reflecting the increasing size, complexity, structure, and specialization of human societies. The rulers and priests provided the people with explanatory worldviews and, thus, orientation based on a simple, vertical hierarchy.

Rules of conduct were needed to contain conflicts within groups. The earliest legal text we know of is the Code of Hammurabi, from 1776 BC. It regulated the coexistence of hundreds of thousands of people in the Babylonian Empire, one of the largest empires of its time. This code is a catalog of judicial decisions "to bring justice to bear in the land, to destroy the wicked and the evil, lest the strong harm the weak." (Harari 2015). In the Code of Hammurabi, about 300 examples are listed, which always followed the same basic scheme: if someone does such and such, such and such judgment takes place. These conditional sentences not only showed the community members what consequences and undesirable actions entailed but also set limits to the arbitrariness of the powerful. The code preceded punishment with a review, in which it first had to be assessed whether the conditions for punishment apply.

Derivations

The transition from gatherers and hunters to sedentary farmers and herders is a process that took place over very long periods. People who had to adapt to the environment to survive became people who began to adapt to the environment to their needs. The driver of this development was the struggle for survival. These people necessarily took on as a group or community. Accordingly, the beginnings and further development of land management cannot be understood without the interdependence with the development of communities. The evolution of communities of living organisms is characterized by an interplay of **competition** and **cooperation** (Nowak and Highfield 2011). At the beginning of the transition, cooperation within agricultural communities may have predominated. With increasing differentiation of activities and a hierarchization of groups within communities, competitive behavior increased, requiring measures to counteract the disintegration of communities. In addition, from the beginning, community growth was momentum, leading to further measures to meet the growing need for food. The growth dynamic was accompanied by ignorance about the consequences of growth and the inability to contain undesirable consequences effectively.

The pattern of socio-cultural development of the human species is determined from the beginning by a growing power of disposal over the external conditions of existence for self-preservation and securing the existence of the descendants. Habermas formulates it in his way:

"Purposeful action represents the form of active adaptation which distinguishes the collective self-preservation of socialized subjects from the species-preservation of animal species. Humans have learned how to bring relevant living conditions under control and to adapt environments culturally to human needs, rather than, like animals, merely adapting to external nature" (Habermas 1969).

2.2 Subsistence Farming

From the beginnings of agriculture and animal husbandry, a way of working and living gradually developed in which commercial production and private reproduction (choice of partner, reproduction, education) were borne by the entire clan in one house. In many regions, this also included slaves and unfree people. The functioning of the community determined whether sufficient quantities of food were available for each individual to survive. Equally essential to survival was the protection that the community provided against many—though not all—life-threatening dangers. Individuals contributed to the functioning of the community in different ways depending on their respective capabilities. This led to a more significant division of tasks and necessitated an organizational structure through which the various requirements were coordinated. Inevitably, this also led to a comparison between one's situation and the situation of the community in relation to other communities. Other people and communities had things one needed or wanted to possess. With the growth of needs, on the one hand, desires increased; on the other hand, there was also an increased exchange between different communities. Not only food and other materials were exchanged, but also ideas. Goods covered the central part of the food supply the people grew or acquired through a network of barter relationships (Illich and Lindquist 1983). The goal of farming was primarily self-sufficiency to ensure a family's or community's livelihood. The technical term is subsistence farming. Except for weapons, utensils, and jewelry made of metal and precious stones, almost all objects of daily use were produced within the fundamental economic and sociological unit of the farm or village. This by no means excluded the fact that several particular goods were made by a division of labor and offered or purchased on local markets. However, the proportion of bought-in food tended to be a measure of bad times. The less a community had to purchase goods from outside, the more successfully it valued its achievements.

With the communal forms of land cultivation and food production, more and more people could be fed. At the same time, this was the basis for the first processes of civilization and the beginning of an increasing environmental change by man. From the centers of their origin, agriculture and animal husbandry spread worldwide. In the process, hunter-gatherers were displaced by numerically superior arable farmers. Forests were cleared so that agriculture could be practiced on the cleared land. The natural site conditions essentially shaped the forms of life and the size of the communities. Climate and soil conditions largely determined yields and what surpluses could be produced. The surpluses allowed for more extensive stockpiling and the exchange of natural goods for other services. Other craft and food processing skills developed through their activities could contribute to the better provisioning of communities. The expansions of food production not only led to the settlements becoming larger. There was a need for counting, writing, and bookkeeping to manage the surpluses. What we now understand as human cultural achievements originated in agrarian cultural achievements, which created freedom for people to engage in activities outside of immediate subsistence.

The selection of plants and domestic animals that were particularly suitable for one's utilization improved the yield situation and, at the same time, expanded the genetic variation. The exchange between the settlements did the rest so that the growth capacities of plants and domestic animals were successively increased through a targeted selection of the best-adapted plants and animals. Thus, in the Middle Ages, several new crops were introduced into cultivation, corresponding to the expanded needs of a growing population. For example, beer hops were added in the thirteenth century. Accordingly, hop cultivation developed in many regions. This also happened when the climatic conditions were unfavorable. Besides the nobility, a bourgeoisie developed with the estates, which became increasingly demanding concerning food. The ecclesiastical cult of the Eucharist also led to the cultivation of wine in many regions, although certainly with varying degrees of success. In addition, fish farming was introduced, which offered a welcome change, especially during the ecclesiastically prescribed Lent, and was also an essential source of protein. In many regions, it took much time to get to the next small town on market day by horse and cart to sell one's products to the few townspeople there. Given the poor transport facilities, there was little exchange of food goods between settlements. Various crops were cultivated on site even when the site conditions were unfavorable.

It was not until the Middle Ages that an independent craft developed, specializing in the manufacture of clothing and footwear and the production of utensils and tools. Tanners, shoemakers, tailors, and clothiers produced items for personal use. Blacksmiths, wheelwrights, carpenters, and all the building trades were devoted to manufacturing household goods, utensils, tools, and buildings. However, apart from blacksmiths and wheelwrights, their activities were confined to the towns, while the rural population still strove to produce most of their daily necessities. They could purchase implements, clothing, and services only to the extent that they sold agricultural products in the towns. As the roads at that time were consistently poor and consequently the transport costs relatively high, agriculture only enjoyed the advantages of a particular division of labor in the urban and peri-urban areas. In the more remote regions, even on most miniature farms, flax had to be grown in addition to grain and vegetables so that the spinning wheel and loom did not come to a standstill during the winter and the families could be supplied with all the necessities. In addition, different kinds of animals were usually kept. In addition to cattle or goats and pigs, there were sheep and at least two types of poultry.

For the time towards the end of the Middle Ages, various traditions describe an increasingly diverse arrangement of both plant and animal production. Different cereals and other crops were cultivated simultaneously on many farm estates, and various types of animals were kept. Available figures show that the number of livestock hardly increased over the centuries (Backhaus 1894). This suggests that the cattle were often sparsely fed. There are no indications of an idyllic coexistence between humans and animals, although conditions may have varied regionally and locally depending on vegetation. Until the beginning of the nineteenth century, cattle and sheep were fed with hardly anything else in winter than straw and possibly hay. Pasture and fallow land were used in summer. Pigs were kept alive almost entirely on scraps. The reasons for keeping cattle consisted primarily of supplying manure for the fields and gardens ("dung cattle"), performing farm and transport work, providing households with animal products, and occasionally supplying markets with milk, meat, fat, hides, and bones (Zorn and Sommer 1970). The animals had to make do with what the farmers left them in the form of fodder that was not suitable for human consumption. Milk and meat yields were correspondingly low. With the introduction of improved three-field farming (Albrecht von Thaer, 1752–1828), it was possible in many regions to improve the fodder basis by cultivating the fallow field. In many areas, the "Norfolk four-field system" became the model for crop rotation, in which nitrogen-consuming summer and winter cereal crops alternated with nitrogen-enriching clover and fodder beet crops (Mazoyer and Roudart 2006). With the additional yield of field fodder, it was possible to improve the supply of animals. At the same time, more nitrogen-rich manure was obtained and fed into the fields' nutrient supply. The additional effort in developing the fodder basis, the selection of the breeding animals, and the husbandry conditions were accompanied by an expansion of meat and milk production.

The diversity of agricultural activities suggested that, depending on the size of the rural families, there were different numbers of people who could or had to be entrusted with the various tasks over the year so that all members of the multiple communities contributed their share to the livelihood. Consequently, food security depended not only on local conditions but also on the availability of labor. Drawing on the traditions of their ancestors and their own experience, those crops were selected from the available varieties that were best adapted to the particular climate or soil conditions, to their own needs, or the farm's work processes. Thus, various crops were grown in different combinations in the garden or the field. The specific sequence of cultivation of crops (crop rotation) revealed interactions that gave rise to further improvements. Livestock keeping was also characterized on many farms by the presence of different domestic animal species. This versatility provided a certain degree of security of supply. In addition, the other production directions also offered more variety on the menu, a better distribution of the necessary work during the day and the year, and improved use of available fodder.

On the other hand, three-field farming, compulsory cultivation, and the cultivation of common land resulted in fundamental restrictions that stood in the way of effective agriculture. Three-field farming meant that after two years of grain being mainly grown, the land remained fallow in the third year. Cattle were allowed to graze on the fallow land, leaving valuable manure on the land. In many regions, owners of farmland and the lord of the manor ordered the cultivation of individual pieces of agricultural land in the community in areas using three-field farming. The aim was to prevent a farmer from gaining an advantage by harvesting earlier or cultivating other produce from what had previously been agreed upon. The agreements were also supposed to avoid damage to the land due to people walking or even driving onto the fields due to a lack of roadways. In addition, they were to prevent the theft of vegetables or other produce.

The **commons** gave everyone the right to graze their animals (cattle, pigs, sheep, or geese) on the communal meadows and fallows at any time. As poverty increased, the benevolent principle of the commons gave way to an ever-increasing lack of rules. In 1801, the New Hanoverian Magazine, in an article entitled "Of the little benefit of common pastures," enumerated seven points that set out how much this practice damaged the community itself:

> *"Everywhere moles would spill the grass, puddles, and hollows would not be drained, so that only bad or no grass would grow, all kinds of thistles rushes and reeds would find their way in, which would spread further and further by their seeds and would always be left standing by the grazing cattle, while "the good, nutritious herbs" would be greedily gnawed off even before sowing and therefore could not reproduce. It also complained that the pigs of the village rooted up the meadow uncontrollably, geese tore out the grass by the roots and dropped their corrosive dung everywhere, and sheep additionally spoiled the whole thing by their stench and their biting down to the roots. Moreover, the cattle and horses were "disgusted by their dung," and they could no longer be grazed there"* (quoted from Ruge 2020).

Derivations

The operational forms of land cultivation and livestock farming developed in interaction with very different site and ownership conditions. On the one hand, people were exposed to very different local conditions and changeable alterations in living conditions. On the other hand, they developed strategies to better cope with the alterations. In this context, the division of labor within and between communities was among the relevant rationalization principles at an early stage. The internal division of labor was geared toward exploiting the available resources of cultivable land, arable crops, accessible fodder bases, and usable husbandry conditions for livestock, and not least to the labor capacities available in the course of the year. The overriding aim of farming was to ensure self-sufficiency and to produce surpluses to build up reserves for poorer years in good harvest years or to exchange the produce for other goods. The inter-farm division of labor was limited to those objects whose production required particular prerequisites in the form of specific materials and skills. In addition, the products had to be valuable and easy to transport. Thus, the division of labor remained low as long as there were few marketplaces and opportunities for exchange.

2.3 From Subsistence to Market Economy

With the increasing expansion of settlements and cities and the beginning of industrialization, there was a growing need to grasp the developments intellectually and to classify them in a superordinate scheme, i.e., a **theory**. Adam Smith was one of the most influential thinkers who tried to bring the many different forms of the economy into an overall context. He was born in Scotland and is considered the founder of classical national economics. In his famous work "The Wealth of Nations," published in 1776, he also dealt with the increasing division of labor between population groups as one of the decisive prerequisites for growing productivity and the resulting prosperity. Today, in economics, the division of labor is understood as any form of division of the social production of goods into different sub-processes, which are then carried out by different economic units (producers, production sites, regions). Every division of labor presupposes a corresponding form of organized coordination of the various sub-processes simultaneously. In addition to increased effectiveness, mutual dependencies are promoted simultaneously, a further characteristic of the division of labor.

According to Smith, three reasons, in particular, are responsible for an increase in the productivity of labor associated with the division of labor: specialization and thus the promotion of the more excellent skill of each worker, the time saved, and, last but not least, technical progress, which occurs more quickly in the course of the specialization. The effect of specialization is that actors concentrate on the part of the overall production process in which they have comparative advantages. This applies to individuals as well as to companies, and in the same way to a national as well as an international division of labor. Each actor could now devote his time and effort solely to other sub-activities in which he is particularly productive. However, Smith also pointed out some dangers of the division of labor. For example, permanently repetitive activities do not exactly contribute to people's professional satisfaction and mental challenges, as with comprehensive activities in overall contexts.

The division of tasks and labor was already a decisive factor for successful food security among hunter-gatherers and in cultivating land and livestock farming. Adam Smith, however, saw specific differences between the emerging industrial production processes and the agricultural production processes, which we will deal with in more detail in the following:

> *"The nature of agriculture indeed does not admit of so many subdivisions of labour, nor of so complete a separation of one business from another, as manufactures."* (Smith 1776)

Albrecht Thaer (1752–1828), considered the founder of agricultural science, attempted to make Adam Smith's teachings on the division of labor applicable to agriculture. He recommended the division of labor by referring to various practical and technical examples but without elaborating on its economic significance. In the autumn of 1809, the first of four volumes on the principles of rational agriculture appeared. As a "systematic, not fragmentary work in a subject which had not yet been treated scientifically," it became one of the significant works of agricultural science. The following quotation is found in the first volume:

> *"Agriculture is a trade which aims to generate a profit contribution or acquire money through the production—sometimes also through the distant processing—of vegetable and animal substances. The higher this profit contribution is sustained, the more completely this purpose is fulfilled. The perfect agriculture, therefore, draws the highest sustainable contribution to profit from its operation, according to the ratio of the forces' assets and the circumstances."* (Thaer 1809)

At the beginning of the nineteenth century, the mixed economy of arable farming and large-scale livestock farming that had existed since the Middle Ages was largely confined to Europe. The agriculture business served primarily self-sufficiency but increasingly also interests that went beyond this, combined with attaining positions of power. These are reflected, among other things, in the high lordly attitudes and status conceit of large farmers. Around 1900, a farmer with 50 hectares of land had about 20 farmhands and maids (Bergmann 1962). The shady sides were far-reaching dependencies and powers of disposal of a lord over so-called "serfs," as they were widespread until modern times. Serfs were obliged to perform personal services for their landlords and were not allowed to move away from the lord's manor. They were also only allowed to marry with the lord's permission and were subject to his jurisdiction. In many cases, serfs were also landowners, i.e., they farmed the land of their landlord and owed him in kind and manual labor in return (Ullmann 2007).

In the development from a subsistence economy to a trade, the number of people whom the proceeds of rising productivity could feed increased. At the same time, labor forces were released, and they tried to make a living by working in the wake of emerging industrialization. In contrast to industrial development, large parts of agriculture in the nineteenth century were still far removed from a pronounced division of labor and market orientation. However, depending on regional conditions, the differentiation of work steps gradually increased without losing sight of self-sufficiency. This also applies to people who were already engaged in other occupations.

Significant increases in productivity in connection with an expansion of the division of labor between agriculture and trade were not realized until the beginning of the twentieth century. This was mainly due to three developments (Bergmann 1962):

1. non-agricultural energy sources in the form of the steam engine and the electric and internal combustion engine,
2. large-scale machines and automatic machines, which allowed mass production and processing far cheaper than the original manual work in small businesses,
3. efficient transport networks using railways, land and waterways.

Of the three, the development of the transport network and a significant reduction in transport costs are probably the most important. This has given many companies access to markets that were inaccessible in the past. This assessment can be understood by looking at developments in different countries. In his treatise published in 1894, Backhaus proved to be a good connoisseur of German, English, and American agriculture at that time. Unlike in Germany, in Great Britain, the development towards increased specialization and division of labor was advanced much earlier. This may also have had something to do with the fact that the English were exposed to overseas competition much earlier. For example, the cultivation of oil crops was discontinued early on because they could be imported so cheaply from other parts of the world that Britain was no longer competitive in growing them. One of the main focuses of farming was fodder production and, thus, the keeping and breeding of domestic animals. Often, only one type of animal was kept, so pure sheep farms, cattle farms, or dairy farms were to be found in England early on. Dairy farms differentiated between those producing milk for fresh sale and those specializing in processing milk into butter and cheese.

A much more pronounced division of labor in agriculture was already established in North America. For Backhaus, the reason for this was, among other things, the natural conditions of the climate and the large-scale cultivation possibilities. In addition, the transport possibilities on waterways and railways had been developed very early. What seemed particularly relevant to him, however, was that

> *"the American is an excellent speculator and calculating merchant, who therefore set up the agricultural enterprise less according to the principles of agriculture than according to the prospects of the quickest possible return of money."*

As a result of the pronounced division of labor, agriculture in North America was characterized by a high degree of one-sidedness. This went furthest in the pure grain farms, where only one crop, e.g., wheat, was often cultivated for decades. This had the immense cost advantage that no permanent staff had to be employed, but additional labor needed for sowing and harvesting was hired only for a limited time. At the same time, there was an incredible drive to further reduce the number of workers through machinery. Specialization forced the technical development of machines, which made it possible to replace human and animal labor. Consequently, very early on in America, some entrepreneurs moved from farm to farm with harvesters to carry out the mowing of grain for smaller farms. As early as the nineteenth century, self-binders were common in North America for grain harvesting and threshing machines that ginned the ears directly after cutting them from the stalk in a threshing drum while the stalks remained on the land as straw. The harvested grain was rarely stored on the farm but in specialized storage facilities, where the grain was also cleaned. From here, the grain was transported by rail to its destinations at a comparatively low cost. Backhaus does not conceal the fact that the one-sidedness of the cultivation was also connected with disadvantages:

> *"There is usually no livestock husbandry for the utilization of arable products on these farms; thus, complete overexploitation is practiced."*

The cultivation of individual crops resulted in a one-sided utilization of the soil, so the compensation for these disadvantages was complex or associated with considerable costs. Organic manure was suitable for the return of the extracted soil nutrients. This, however, would have required livestock and fodder farming, depriving the economy of its specialized character. Another disadvantage was the proliferation of pests, which found good conditions when monocultures were grown. Moreover, one-sided cultivation caused more significant fluctuations in yields and, as a result, financial uncertainties. Furthermore, the one-sidedness of the cultivation required a high capital requirement for using machines, which could only be utilized minimally.

Not only in crop production but also in livestock-oriented farms in North America, a one-sidedness prevailed. As a rule, not only one type of animal was kept, but only one specific production direction was pursued. Mixed farms, on the other hand, were rare. Slaughtering was increasingly done in purpose-built slaughterhouses. In her book "In Meat We Trust—An Unexpected History of Carnivore America," Maureen Ogle gives an apparent overview of the historical development from the beginnings of colonization to the present day from the perspective of meat production. She writes about conditions at the start of the nineteenth century:

> *"Over time, carnivorous paradise begot lethal legacy. The abundance of meat spawned waste and fostered indifference bordering on cruelty. "The Cattle of Carolina are very fat in Summer," charged one critic, "but bone bags in winter because their owners refused to protect them from cold rains, frosts, and snows." Settlers dismissed such criticisms, claiming they could spare neither time nor labor to build animal shelters of fencing, occupied as they were with "too many other affairs." As a result, cattle and hogs scattered their droppings*

hither and yon, left uncollected because no one could spare the labor to gather and spread them on corn and tobacco fields. Thus developed a cycle of destructive extravagance that Americans passed from one generation to the next. Abundance of land nurtured an abundance of livestock that enabled settlers to eat well and accumulate tangible wealth with a minimal labor investment." (Ogle 2013)

In Germany then, livestock farming primarily utilized the farm's fodder resources. Versatile livestock farming was advantageous for their utilization. Pig **farming** was not asked everywhere. After all, employing the available family members and, if necessary, other workers throughout the year and having them carry out the most diverse work involved in a varied crop and livestock production was also a question. Accordingly, the urge to mechanize and the availability of capital to purchase machinery was comparatively low. Due to the diversity of farm structures that prevailed in Germany, the expansion of the technical division of labor was prolonged. In an agricultural inquiry report of Baden from 1883, it is stated:

"The surveys have shown in both respects, or rather have confirmed the previously known fact, that the more varied the agricultural enterprise and the more varied the use of the nature of the soil and the climate, the more satisfactory conditions tend to develop for the peasant population and that the most precarious and, under certain circumstances, most critical conditions very easily arise where everything is, so to speak, put on one card." (Anonym 1883)

Accordingly, in the agricultural business theory of the time, the recommendation was to strive for the most significant balance of agricultural work, always able to employ the available labor force profitably for economic reasons.

From the knowledge of both agricultural worlds, Backhaus (1894) described the German and American forms of economic activity by two different stereotypes:

"The German farmer is a careful cultivator, a good stock-raiser, and an industrious landlord, but the American is more of a calculating, speculative merchant, and the latter in particular should be highly recommended to the German farmer for imitation. ... How useful it is to make the transition from simple landlord to agricultural industrialist is proved by the fact that the American farmer can compete with our farmers, with wages twice as high as in Germany, with by no means better soils, much lower yields, and very considerably lower prices for the products, which often reach only half the prices usual with us. ... To be sure, ruthless overexploitation was and still is practiced in very many parts of America, but more and more there, too, people are turning to a substitute economy."

While the industry in Germany was already making good use of the expansion and progress of transportation and the division of labor, its utilization in agriculture progressed only slowly. The division of labor was most feasible in cultivating commercial crops (market fruits) that could be transported over longer distances. These included crops processed in factories, such as sugar beet, potatoes, oil fruits, and vegetables. This also included the processing of grain into spirits. For example, between 1836 and 1886, the number of distilleries in Saxony's agricultural sector fell from 1684 to 629. At the same time, the amount of product produced more than tripled. This was probably due to good storage and transportability of the raw materials and end products, a higher degree of mechanization in processing, and an increasing demand for the end products. In general, however, the market economy was still in its infancy, in which individual goods and services were exchanged for money on the market to obtain other goods or services with the resulting income.

An elementary prerequisite for the division of labor was a receptive market in which a lively exchange could occur between individual enterprises and buyers of the products. What was comparatively easy to realize in small-scale sectors required, above all, in sparsely populated regions, a functioning transport system, which brought the products from the places of production to the areas of further processing and consumption, thus connecting the needs of the primary producers with the needs of the consumers. This also required a well-organized trade.

Most arable crops were advantageous if grown in larger areas, especially when cultivating fruits that required less manual labor or allowed the use of machines. However, this presupposed that labor was available in sufficient numbers or machinery when needed. Where this was not the case, smallholder structures had advantages in the availability of labor, which could be employed throughout the year because of the versatility of the work. Smallholders were also ahead of large farms in livestock production, especially in those branches where success depended on careful husbandry. This also applied to fodder production, which had to ensure livestock supply. If the animals were not adequately fed, significant economic losses were likely.

The different farm structures had various advantages and disadvantages. However, the division of labor between differently structured farms has only occurred to a limited extent. In the long term, however, the joint procurement of means of production, such as seeds, has established itself as a form of **cooperation** in the rural cooperative system. Among the pioneers were Friedrich Wilhelm Raiffeisen and Wilhelm Haas, who founded the first rural cooperatives in the second half of the nineteenth century. The pooling of farmers' purchases and sales through cooperatives helped them compete with large farms abroad. The cooperatives aimed to meet the suppliers of required means of production, such as fertilizer suppliers, on an equal footing, organize a joint utilization of agricultural products, and open up new sales markets for the produced and processed products.

Regardless of place and time, farm organization was a priority in the tension between integrating and differentiating forces that worked in opposite directions (Brinkmann 1922). The forces oriented towards integration and balance within the farm strove for high self-sufficiency for diverse fields of activity with which the available labor resources could be utilized. In addition, the preservation of soil fertility and largely loss-free utilization and refinement of all accumulating and unsaleable fodder materials in livestock farming and security and balancing risks were in the foreground. The integrating forces acted on the farm structure from within, meaning farms were generally organized very diversely. The differentiating forces included the differences in natural production conditions (soil, climate, and terrain), the transport situation and the related level of transport costs, the state of mechanization, and the differences in the developing market conditions connected with the respective price levels. These differentiating forces had an external effect on the farms. Consequently, the farms oriented their organization and production differently to the changing framework conditions. Gradually, the structure of the agricultural economy was changed by mechanization. The individual farms became more extensive, and production was broken down into sub-processes. In the course of specialization, only raw goods were produced on the farms, while further processing was increasingly shifted to a developing food industry.

During the rationalization process, those branches of the farm that were of no great significance for the economic result were initially hived off. These included, above all, those branches of land use and livestock farming that essentially operated for reasons of self-sufficiency or were too small in scale to be mechanized. From farm simplification, further specialization steps developed, accompanied by changes in farm organization. Increased differentiating forces lead to pronounced regional differences in farm structures and production potentials. Depending on the location, farms in favorable locations could exploit the advantages of high-yield potential. Farms in more or less disadvantaged regions, where hardly any or no arable farming could be practiced, were left empty-handed. The different forms of inheritance law, which promoted or counteracted the cohesion of farm capacities and the spatial proximity or distance to the cities and, thus, to the markets, also had a considerable influence on farm structures. Finally, the ability of the farm managers to coordinate the various resources and the diverse processes of agricultural food production had a significant impact on the development and prosperity of the farms.

Derivations

From the beginning of agricultural development, one of the most significant challenges was dealing with the vicissitudes of foreseeable and especially unforeseeable events and changes in a way that guaranteed the supply of the clan and the maintenance of production capacities. By efficiently utilizing the available production and labor resources in the form of a multifaceted subsistence economy, the goals of self-preservation and the provision of the community were generally guaranteed. As developments progressed, surpluses increased, feeding more people outside the communities. The expansion of production served an increased demand for food by the population and the need for a growing industry for raw materials for further processing. Barterable and saleable products were offered for sale in the developing markets. On the one hand, this increased the financial resources of the farms, which could be invested in the development of the business. On the other hand, however, dependency ratios and risks increased, as producers could only adapt to the differentiating forces and the development of the market to a limited extent, depending on the initial conditions.

While the productivity of production increased on many farms due to the increasing division of labor, at the same time, many synergistic effects, depending on an integrated approach, were lost. The division of labor places increased demands on coordinating internal and external requirements. However, the biological processes, which can only be predicted and thus planned to a limited extent, stand in the way of effective coordination. Technical developments have made it possible for farms in various cultivation situations to use machinery to free themselves from their significant dependence on the availability of labor. However, competition between individual farms is increasing with the advent of the market economy. Everyone tries to gain an advantage in the competition in their own way. Unlike in industry, where appropriate administrative procedures are developing, these are hardly used in agriculture. Exploiting the potential of inter-farm cooperation would have required appropriate coordination structures and quantifying positive and negative effects to counteract actual or perceived overreaching. However, authorities that can oversee the various options for cooperation to generate a win–win situation for all participants, monitor the agreements, and intervene effectively in the event of misconduct are missing.

2.4 Modern Agriculture

The advances in the mechanization of work processes and the expansion of transport networks benefited the industry and agricultural processes. In addition, there were advances in knowledge about the functional relationships in various areas, which could be used to increase the productivity of farming processes. For example, the developing science of animal nutrition and feeding created the basis for achieving higher performance by improving the supply of nutrients to animals. The considerable differences between the numerous landraces and country breeds kept on the farms provided indications of genetically determined differences in the animals' performance. This led to targeted mating and the establishment of the first breeding associations.

The increasing demand for food by a growing population was the decisive driver of development and stimulated production increases. Because the emerging agricultural sciences significantly increased the knowledge of functional relationships, striking increases in output could be realized (Sundrum 1996, 1996). New techniques rationalized work steps and increased labor productivity in sowing, weed control, harvesting, and, last but not least, further processing the products. These efforts increased the average annual output of a dairy cow from about 1200 kg in 1860 to an average of 3400 kg in 1960 (Zorn and Sommer 1970). However, the significant increases in the production of animal products should not obscure the fact that performance varied significantly between farms and was influenced to a large extent by the respective site conditions and farm structures, such as the herd size, the available pasture, and arable land and their respective yield potential.

Incentives to increase production output also arose from the need to feed one's family with the proceeds from sales and to ensure the economic viability of the farm. Despite significant increases in output, farm profits did not increase proportionately. On the contrary, food prices fell, and farm workers' wages rose. A price gap began to open up, causing farm profits to melt away and creating deficits for many farms. As industrialization progressed, the demand for labor increased, as did the need for skilled regular workers. This intensified competition for labor between the agricultural and industrial sectors. In addition, the development of railways and steamships reduced transport costs and created world markets for grain. The expansion of transport opportunities meant that farms outside Europe competed with European producers and put downward pressure on market prices. Imports from overseas were used to supply a growing industrial workforce and were met in return with exports of manufactured goods. At home, more efficient means of transport and a denser network of roads and railways ensured that industrial centers of consumption were linked to agricultural production areas. Under these conditions, companies emerged near the cities that processed agricultural raw products into foodstuffs and luxury foods, such as sugar, spirits, or dairy products.

Depending on the endowment of resources in the form of agricultural land and labor, industrialization in the farming sector replaced the scarcer, and consequently more expensive, factor of production with capital by creating a cheap supply of labor-saving technologies. Falling market prices and a consequent decline in profitability fueled an intensification of land management and increased use of capital. Conversely, agricultural productivity gains increased demand for industrial goods and freed up labor forces for the industry. To a certain extent, industrial and agricultural development influenced each other, even if neither benefited in the same way.

After the Second World War ended, there was a clear break in farming and livestock keeping. Under the impression of hunger and supply shortages during and after the war years, which had left deep marks on the consciousness and attitude to life of the population, in the post-war years, politicians across party lines saw it as the primary objective to increase agricultural yields to improve the supply situation significantly. The industry had demonstrated how to mass-produce goods through measures of specialization and standardization of production processes. The strategies that had been successful in industry—the obvious conclusion at the time—had to be transferred to agriculture. In 1957, the founders of a united Europe immediately enshrined the concept in Article 39 of the Rom-Treaty. It states:

The objectives of the common agricultural policy shall be:

(a) to increase agricultural productivity by promoting technical progress and by ensuring the rational development of agricultural production and the optimum utilization of the factors of production, in particular, labor;
(b) to ensure a fair standard of living for the agricultural community, in particular by increasing the individual earnings of persons engaged in agriculture;
(c) to stabilize markets;

(d) to assure the availability of supplies;
(e) to ensure that supplies reach consumers at reasonable prices.

Even though these objectives were pursued with vigor, only three of the five goals of the EU's Common Agricultural Policy (CAP) have been achieved to date. As predicted, the instruments of industrialization led to a significant increase in agricultural productivity in the following decades. It also improved the supply of food to the population in a short period, at market prices that continued to fall in relation to other products. However, this did not lead to per capita income in agriculture keeping pace with wage increases in different sectors of the economy and to farmers being able to secure a decent standard of living and economic livelihood in the long run. It has also stabilized markets for more extended periods, but not permanently. Above all, however, the CAP has had several knock-on effects, the costs of which are still being borne heavily by the Community today. Section 5.2 discusses these costs in more detail. First, it will be explained how the successes of EU policy, in particular the productivity gains, came about. Four factors, in particular, were decisive in this respect (Sundrum 1996):

1. To modernize agriculture as quickly as possible and to overcome the skepticism of farmers, who initially had reservations about industrially organized farming techniques, governments everywhere fell for the same trick: they excluded the agricultural sector from the market economy. They offered farmers buy-up prices for products that were higher than market prices. The incentive to produce was immense. Within a few years, agricultural production picked up tremendously. A golden age for agriculture was dawning. Farmers only needed to produce larger quantities to earn more. The self-regulating market mechanism, whereby prices fall when the supply of goods exceeds demand, was suspended by guaranteed prices. Investing, expanding, and specializing were now the overriding motto.
2. However, not only did the guaranteed prices break the boundaries of previous production processes. With regard to the availability of means of production, which were the fuel for growth, the Haber–Bosch process virtually initiated a revolution. This chemical process, patented in 1908, made it possible to produce ammonia synthetically from atmospheric nitrogen and hydrogen, using large amounts of energy. Ammonia is the starting substance for producing various reactive nitrogen compounds, which are used on the one hand as explosives and, on the other hand, as nitrogenous chemicals and thus as fertilizers. While the process was soon used to produce explosives, it took another half century, during which two world wars with enormous amounts of explosives were fought, before mineral nitrogen fertilizers became widespread in agriculture. To fully appreciate the importance of reactive nitrogen compounds in agriculture, it is necessary to know that they are the first limiting factor of biological growth processes for all living organisms, i.e., for microbes, plants, and animals. Overcoming the previously existing limitations in the availability of reactive nitrogen compounds and thus in exploiting the growth potentials, an enormous explosive

force and fundamentally changing production conditions developed. Since these changes were slow to unfold their effects, their continuing explosive power was and is not perceived by most people. The quantities used rose worldwide from 1–2 megatons (Mt = 10^9 kg) around 1950 to 11 Mt around 2000 (Bouwman et al. 2013). The undesirable consequences are discussed elsewhere (see Chap. 4). In this context, it is essential that crop farmers' dependence on the availability of nitrogen fertilizers, which were available from the cultivation of legumes through the fixation of atmospheric nitrogen by microbes and via the excrements of animals, was essentially eliminated. As a result of significantly increasing grain yields per hectare, an increasing proportion of grain could be used as animal feed. Extensive imports of protein- and energy-rich feed from third countries enabled a considerable expansion of pig and poultry farming and increased availability of the products of animal origin derived from them. Since the intensification processes were similar to those in other industries, the terms "plant production" and "animal production" became common after World War 2 and are still used today. The import of high-quality feed was a prerequisite for the operation of large "animal production plants," which could thus be mainly operated independently of the availability of forage land. Furthermore, developing new husbandry techniques, in which many animals could be kept in a confined space without straw, using sophisticated ventilation techniques contributed to a significant expansion of production capacities and increased labor productivity.

3. A change in tax law made it possible for non-agricultural enterprises to invest in agriculture and to claim the investment funds as depreciation for tax purposes. This made borrowed capital generated outside agriculture available for the expansion of agriculture and gave a starting signal for establishing "agricultural factories."
4. However, the enormous increases in performance and restructuring only became possible because an expansion of agricultural sciences provided the necessary *know-how*. State research and educational institutions were institutionalized on a large scale, and a state-federal research and advisory apparatus was established to develop and disseminate technology and **knowledge transfer** and lobby groups to harness and exploit agricultural resources. In the process, the quest for knowledge was initially directed, above all, toward the possibilities of increasing productivity.

In addition, agricultural production was supported in a variety of ways through subsidies (Hartenstein et al. 1997). However, these did not solve the fundamental problem of farmers' insufficient earnings while maintaining the compulsion for continuous increases in productivity. For their part, farmers reacted to the changes in the general conditions of agriculture with far-reaching changes in their internal farm structures. These included, above all, the intensification of production processes through the increased use of means of production, the mechanization of work processes, and the specialization of partial aspects of the production chain. Focusing on a few and expanding individual farm branches to increase productivity through improvements in labor technology was only possible on most farms if other farm branches were significantly restricted or abandoned.

The internal connection between cereal cultivation and livestock farming was primarily broken down. Neither was the extent of livestock dependent on cereal production, nor was the extent of cereal cultivation based on the concentrated feed requirements of cattle, pigs, and poultry. Instead, the feed grains and other supplementary feeds were purchased primarily. One of the advantages of this was that the nutrient composition of the feed rations was more balanced and supplemented with minerals, trace elements, and vitamins. It thus gave rise to higher performance expectations than using the farm's feed mixtures.

The process of differentiation and specialization progressed rapidly, especially in poultry farming, where breeders, hatchers, egg producers, and fatteners focused on individual sub-processes. Analogous structures developed in pig and cattle farming, although less pronounced. The possibility of purchasing fodder allowed farms with little land to expand their livestock production. A large number of smaller farms saw no other way of increasing their income than to increase the size of the farm from within, i.e., by buying in feed without requiring funds to expand the farm area. Internal cost reduction was replaced by striving for the most significant possible increase in the production of sales products per worker. However, investment had to be made in barn construction. Since converted space was expensive, animals were kept in confined spaces. The resulting increase in germ pressure from pathogenic and facultative pathogenic germs was kept in check by hygiene concepts using disinfectants and with the help of antibiotics for the animals.

While the production of stable manure, e.g., for the supply of root crops, was the focus of farm organization issues for a long time, many growers of cash crops refrained from keeping livestock and buying organic fertilizers. The supply of mineral fertilizers could compensate for nutrient deficiencies in the soils. Implementing an extended crop rotation to conduct weed and pest control was no longer necessary, as has been the case in the past. Synthetic chemical agents were now available to combat pests and weeds, seemingly eliminating the need for extensive crop rotation, such as root crops for weed control. Specialization fostered the expertise required for diverse production processes. It was hardly possible anymore for individuals to be sufficiently expert in all areas of agricultural production and utilization to achieve the highest possible level of productivity according to the knowledge available at the time. It also became increasingly difficult for farm managers to keep track of the markets for all products and inputs and the expected development of their respective prices. More and more farms were seeking to simplify the organization of their operations. The advantage of saving person-hours per animal while increasing the number of animals was joined by the benefit of mechanization as a further driving force for specialization. The work cost could be reduced if the technical facilities were fully used. In addition, mechanization made work easier and reduced physical strain. The fewer person-hours that had to be spent per product unit, the more units could be produced.

While value added to the rest of the economy grew strongly in the post-war years, value added to agriculture could not keep pace with this development. Accordingly, the share of agriculture in the economy's value-added steadily declined. This shrinking process was the inevitable consequence of increased incomes, as occurred in all highly industrialized countries. For example, the wage bill per average employed worker in West Germany rose from DM 2911 in 1950 to DM 6100 in 1960 (Statistisches Bundesamt 1961). While technical progress in industry resulted in a steady increase in incomes, the share of total expenditure on food fell as incomes rose. Rising incomes of employees went hand in hand with higher food consumption, but only to a limited extent. There was only a shift in consumption towards so-called "processed products." Per capita consumption of products of animal origin increased by 50% in the EU between 1960 and 2007, doubling in relative terms (Westhoek et al. 2011), and doubled in relation to 1900 (Smil 2002).

Even the consumption of processed and animal-sourced products could not rise to the same extent as the incomes of those employed. It was not only in this respect that agriculture differed from many other sectors of the economy, which could count on a substantial expansion of their markets as national income rose. The technical progress, which occurred, among other things, due to the availability of cheap means of production, labor-technical facilitations, and manifold advances in knowledge, caused a disproportion between supply and demand of food at the latest when the saturation level was reached. Accordingly, the markets for agricultural products came under increasing price pressure. Since neither an unlimited increase in sales nor a sweeping rise in prices was possible, farms had little choice but to increase productivity per worker further and to outsource labor for which a wage was payable.

However, the decline did not narrow the income gap between industry and agriculture. Instead, the industrial comparative wage continued to increase and widen the gap. Accordingly, many workers sought and found their income and livelihood outside agriculture. For the remaining farmers, the possibilities of increasing their income continued to be limited. Bergmann (1962). At the beginning of the 1960s, the farms were faced with an unavoidable choice between four options: (1) a fundamental change in the farm size structure and farm organization, (2) phasing out and abandoning further adjustments, (3) farming as a side-line or (4) abandoning the farm. In the further development, for most farms, neither the renunciation of adaptation and thus the decoupling from the income development nor the maintenance of the farm as a side-line business were sustainable alternatives, so options for most farms soon condensed to the short formula: "Growing or giving way."

However, this short formula only inadequately reflects the development. Before giving up production on a farm, many farm managers first tried to maintain economic viability through further investment in growth and specialization processes, only to have to leave the race and give way after a time lag. This process continues to this day, in which the closure and failure of the business is preceded by efforts to preserve the business for the next generation through adaptation processes and investment. In some cases, considerable debts were accepted and easily granted to farms due to the available (scarce) production factor "land." Agricultural policy and economics have used the term "structural change" to describe this development. However, this term is more suited to envelop the rapid upheavals and their consequences and conceal their extent than to describe them adequately. For example, almost 1.8 million farms were counted in the territory of the Federal Republic of Germany in 1949; in 1960, there were still about 1.5 million. From then on, the development corresponded less to a change than to a rapid upheaval, combined with a drastic change in production conditions. By the end of 2017, the number of farms had fallen to around 270,000 (German Farmers' Federation 2020). By 2040, the number of agricultural enterprises could drop to around 100,000 (Niegsch and Stappel 2020). At the same time, this would bring German agriculture closer to the standard farm sizes among family farms in the US (2018: 179 ha on average). After the Second World War, parallel developments occurred in American agriculture, described in detail by Cochrane (1993).

The family farm model, which prevailed for a long time and was characterized by independent farmers, small farm units, and family members helping out, has rapidly lost importance. However, this does not mean that farmers themselves or various groups of the population and consumers have abandoned the old images; more on this elsewhere (see Sect. 6.2).

Part of holding on to outdated ideas is that the economic situation of the agricultural sector is often misjudged. The farm sector's sales potential in developed countries has been exhausted for quite some time. Accordingly, the industry can only grow assuming positive price trends and increasing foreign demand. However, the increasing liberalization of agriculture in the European Union led to a growing influence of European and global agricultural markets on the member states' agriculture without a significant increase in exports of farm products. One of the consequential effects was that the extent of price fluctuations increased significantly. In contrast, the development of market prices for foodstuffs lagged noticeably behind the development of other consumer prices. According to the Federal Statistical Office, agricultural producer prices have increased marginally at best from 1981 to the present, while general consumer prices have increased by more than 200% over the same period. When consumers buy products from domestic agricultural production, on average, only 23 cents of every euro go to primary producers. In the 1970s, the share was still 48%. At just under 4%, products from bread cereals have the lowest producer share. Regarding meat and sausages, 22.5 cents of every euro still reached primary producers, surpassing milk at almost 40 cents and eggs. Here, about 60 cents of every euro remains with the farmers (Anonym 2019). The differences are also related to the decreasing degree of further processing. While **eggs** are merely sorted and packaged, the path from grain to bread roll and from carcass to sausage is considerably longer and involves various stages of manual processing.

In parallel, the capital input in agriculture increased. For instance, in 1991, an agricultural job in Germany cost, on average, about €176 thousand per person employed (German Farmers' Federation 2020). Capital input increased to € 552 thousand in 2017. To make matters worse, only 41% of agricultural land is owned, i.e., 59% of the land used is rented. Leasing prices have to be served for the right to cultivate, which on average doubled from € 141 to € 288 per ha per year between 1991 and 2016 while at the same time showing considerable regional differences. About one-third of the tangible capital is financed with borrowed funds. Per hectare of agricultural land, the borrowed capital is around 2500 €. The more industrial growth outstripped the growth of the rural economy, the wider the gap between non-farm and farm incomes widened in the post-war decades. Beginning in the 1990s, government subsidies enabled many farmers to produce raw goods and energy to feed them into the public grid based on long-term fixed purchase prices. As a result of the incentives provided by the state through the Renewable Energy Sources Act (EEG), many farmers became producers of renewable energies and thus tapped into new sources of income. At the same time, they intensified the competition for the availability of land. In many regions, the expansion of bioenergy production led to a shortage of arable feed land and a drastic increase in rental prices. Since then, this has been particularly hard on dairy farmers who depend on forage land.

Large capital-intensive and managerially organized companies using state-of-the-art technologies will shape the agricultural sector in the long term. However, the new technologies are associated with enormously high investments. **Digitalization** is expected to enable a further increase in production efficiency. At the same time, it is assumed that "the classic farm will increasingly develop into a modern production site for agricultural commodities, just as more and more farmers have become modern, technically and economically well-trained entrepreneurs."

Derivations

The industrialization of agriculture was accelerated mainly by the interconnections with upstream and downstream industries. Limits in the availability of farmland and labor capacity were sought to be compensated by capital input as a factor of production, mainly in the form of technology input. Fossil fuels directly (e.g., fuel for motorized agricultural machinery) and indirectly (e.g., chemical-industrially produced mineral fertilizer) replaced solar energy (via the detour of plant photosynthesis) as the primary source of energy. The production factors and the products went through capital-intensive processing steps before they reached consumers via supermarkets. Many sub-processes in the production chain were outsourced. Farms were reduced to the function of raw goods producers and buyers of means of production from upstream and downstream industries. With the replacement of endogenous resources by exogenous ones, farms got caught in an "agricultural treadmill" in which intensification, specialization, and farm concentration set in motion the self-reinforcing spiral of economies of scale according to the logic of "growing or giving way."

The interlocking of breeding and scientific and technological developments has immensely increased the productivity of crop and livestock production. Yield-increasing production techniques could be better mastered on specialized farms than on versatile mixed farms. However, high mechanization was only economically viable if sufficient utilization of machinery and productive use of the person-hours freed up was ensured. Also, only those who were the first to use technical progress could generate above-average income advantages (pioneer profits). As soon as many enterprises used the new production methods, prices, and costs were settled at a new level. This essentially eliminated the additional profit from technical progress. This way, the urge to increase productivity using technical developments became system-immanent (positive co-coupling). Reciprocal conditions created path dependencies on the chosen growth path. On the other hand, increasing specialization meant that price fluctuations in the markets could be absorbed less and less by different sources of income. This increased the dependence on demand trends in the respective market segment. At the same time, agriculture became increasingly capital-intensive and thus even more dependent on outside capital. Expanding production capacities and rising product volumes resulted in sustained pressure on producer prices. This was met with further efforts to reduce production costs. The products that could not be sold domestically were sold on global markets with the help of (export) **subsidies** and caused further problems there.

In the past, it was the rule to run the farm as a family business over a long succession of generations, but this changed fundamentally after the Second World War. To keep up with the development of income outside agriculture, the production productivity had to be further increased. Intensification, mechanization, and specialization were the key instruments to achieve this. However, even the enormous productivity increases could not make up for what was denied to the farms by the economic framework conditions set by the politicians, namely an adequate rise in food prices and, thus, in farm incomes. The stimulation of volume production and the resulting oversupply of food prevented price increases in the food sector that were natural for other industrial product lines. Consequently, farmers couldn't keep pace with non-farm earnings regarding income growth. Even with the help of government subsidies, the majority of farms succeeded only temporarily in generating a viable income. In the EU, the linking of direct subsidy payments to the land area increased rental prices, whereas the subsidies go primarily to those who own the land and not to those who work the land and have to earn the increased rental prices. Without the direct subsidy payments and the additional national support measures, the income from sales revenues is insufficient for most farms to survive economically. More and more farmers have to admit defeat in the competition for the highest productivity and withdraw from production. The remaining farms use the freed-up farmland to expand their production units to achieve economies of scale. In this way, they can hold their own in the cut-throat competition, at least temporarily, but without being able to foresee over what periods this will succeed.

2.5 Globalized Agriculture

The term "**globalization** goes back to developments in the 1960s (Chanda 2007). It refers to the increase in worldwide interconnections in many areas (environment, economy, politics, culture, communication) between individuals, companies, societies, and institutions. What is disseminated or exchanged, and at what speed, is inextricably linked to the nature of the connections. Without the connections created and established by humans over great distances, there is merely an uncontrolled spread in space and time of those substances that are not locally bound. Before humans, birds, fish, and whales had already traveled the globe. However, the spread of dead matter and living organisms dependent on a specific transport vehicle only began to expand with the travel activities of humans. Travel has always been associated not only with the transport of desirable but also with undesirable matters. For example, the connection between human travel and the spread of microorganisms has been evident with the arrival of the returning Columbus ships in Naples in 1493, which had syphilis pathogens in their luggage. However, many people are slowly becoming aware that every form of travel and every exchange of dead or living matter across borders is ambivalent. The ambiguity results not only from the different perspectives when matter moves from locality A to locality B but is accompanied by various advantages and disadvantages at both the origin and destination and on the way there. In addition, every form of transport has a CO_2 backpack in its luggage. This would not require any particular emphasis if not because the spread of people and goods of living and dead matter across the globe along established routes has increased exponentially over the centuries. This is particularly true of agricultural products.

In the context of agriculture, international interdependencies initially focused on imports of goods that were not available locally. Early in human history, easily transportable spices and tropical fruits were shipped to Europe "from all corners of the earth." As early as the sixteenth century, cultivated plants such as potatoes and maize arrived in Europe from South America; their cultivation enriched and changed the local diet. Later, sugar, rum, cotton, tobacco, and indigo were among the most important nineteenth-century imports from the global economic periphery for the industrializing countries of Europe. The "goods" exchanged also included people who were forced to work and enslaved against their will as laborers on the cotton plantations in the American southern states, among other places. Importing from near and far regions of the world made it possible to pursue one's interests in utilizing goods. According to today's terminology, this was a win–win situation for importers and exporters and not least for the middleman, provided that an appropriate return was made for the acquired goods. It is indisputable that, in many cases, trade was and still is practiced at the expense of others. This is without prejudice to the fact that many negative consequences for others are not even acknowledged as such because they occur outside one's horizon of perception. Or they are ignored or argumentatively rationalized away to be able to continue to profit from the advantages. This is not the place to comprehensively discuss the history of globalization and its manifestations in the context of agricultural production. However, it is relevant for further discussion that in the direct exchange of goods between people and living systems and exchange over long distances, the question always resonates: ***who benefits from the exchange, and who may be harmed by it?***

Using several examples, we will first identify the driving forces behind the growing global exchange of agricultural products to address the above question in greater detail at another point. The exchange of matter over long distances had an early influence on the structural developments of agriculture. Then, it developed to an extent that shapes and dominates today's agriculture. In 1806, Alexander von Humboldt brought the first guano samples (excrement of seabirds) from the coasts of South America to Europe, and chemists analyzed them. They found nitrogen and phosphorus highly enriched and organically bound in the excrement. Although with a considerable time lag, this led to an excellent guano boom, which peaked between 1845 and 1880 and lasted until the end of the nineteenth century (Rott 2016). The many uses of guano as a fertilizer in agriculture and as a raw material in manufacturing explosives contributed significantly to this boom. This only changed when the German chemist Fritz Haber succeeded in synthetically producing ammonia from hydrogen and nitrogen in 1908.

While the synthetic production of ammonia essentially shut down guano imports, another **reactive nitrogen compound** became an import or export hit. The scale has already surpassed that of guano many times over. We are talking about soybeans. As an oil crop with an oil content of around 20%, they are essential for human nutrition. After the oil has been extracted, what remains is a very protein-rich residue that, as animal feed, is now the indispensable fuel for the industrialization of livestock farming in Europe (Bund für Umwelt und Naturschutz Deutschland e. V. 2019). Soybeans are mainly processed into animal feed in the form of extraction meal. Worldwide, soybeans are the most important protein supplier for pigs and poultry, and their growth rates depend on the supply of high-quality protein components. The protein content of soybeans is four times higher than that of cereals, at around 40% in dry matter. The world's largest producer of soybeans is the USA, closely followed by Brazil and Argentina. In 2018/19, the world soybean crop was estimated at 367.5 million tons; China proved to be the leading importer with about 90 million tons, followed by the EU with about 16 million tons. In importing countries, soybean extraction meal is an ideal feed component to supplement domestic grain and other feed resources. Soybean extraction meal is used in the rations of pigs and poultry and in feeding beef cattle and dairy cows. The latter is done not so much because of the lack of domestic alternatives but mainly because of cost advantages. Without the international trade in cheap feed from North and South America, the scale of intensification and expansion of livestock farming in Europe could not have taken place.

The expansion of global trade in goods called for rules created after World War II with the international agreement 'General Agreement on Tariffs and Trade' (GATT) of 1947. The agreement primarily excluded agriculture from free trade and legitimized the protectionism that had been institutionalized to deal with the Great Depression and World War II in the 1930s and 1940s. Later, under the auspices of the International

Monetary Fund (IMF) and the World Trade Organization (WTO), the successor to GATT, a global set of rules for agricultural trade emerged. The United States accepted the European Community's Common Agricultural Policy (CAP) protectionism against grain imports. In return, soy-based feed imports from the US were exempted. New sales markets were sought with state support for the production surpluses that could not be sold within the European member countries. As early as the beginning of the 1990s, the EU decided to orient its agriculture towards the world market and its prices. The markets in Africa and Asia, in particular, were to be supplied. Subsidizing exports with considerable sums of money and infrastructure initially gave the EU various advantages over other countries already active on the world markets and seeking to expand their **exports**.

At the so-called Uruguay Round of the World Trade Organization (WTO) in 1992, it was agreed that agricultural products would also be subject to the rules of international trade in goods. Accordingly, price support and market regulations for rural markets were gradually abandoned. In return, farmers receive financial support (**subsidies** in the EU have been completely decoupled from volume production and are granted direct payments via the so-called 1st pillar on an area basis. In addition, the second pillar finances investments in agriculture and rural areas to improve competitiveness and cushion structural change. Compared to the 1st pillar, which is financed 100% by the EU, the EU funds of the 2nd pillar usually have to be co-financed with national public funds.

After a phase in the world markets of the agricultural industry characterized, according to Sieder and Langthaler (2010), primarily by "productivist" strategies, the development of a "post-productivist" food regime emerged from the 1990s onwards, comprising several partly contradictory processes. They highlight four strands of development:

1. Production processes intensified competition for food producers, processors, and traders in developed countries who had operated under the protective cloak of protectionism for decades. At the same time, new producer countries emerged (including Brazil and Argentina), where subsistence agriculture was displaced by commercialized export production. The main driving force behind the rapid growth in production in South American countries was the continuing demand for protein-rich soybean meal as animal feed, initially mainly in the EU and later also in emerging Asian countries. As a result, the US lost the dominance it had gained in the feed and food trade since 1945.
2. Within the neoliberal regulatory framework, agro-industrial companies gradually gained more leeway and, as "global players," replaced the nation-states as central regulators of the agricultural sector. The growing concentration of market power in the hands of a few corporations was the consequence and, at the same time, the driver of further developments in globalization. Today, five corporations, including Bayer, Syngenta, and Dupont, control more than half of the global seed market. The global market for pesticides and fertilizers is also controlled by a few large corporations, as is the international grain trade, 70% of which is handled by five large companies. Based on a flexible use of capital, they can react quickly to changes in production and

consumption conditions. In doing so, they are no longer subject to the influence of the nation states, which are losing importance, but act primarily in competition.

3. The consequences of technological developments are penetrating more and more food production areas. Technical innovations are fueled by the investments provided by nationally and internationally operating companies and the state. They are forcing concentration processes in the agricultural economy. At the same time, demand is growing for "organic" food, i.e., food produced without chemicals. These opposing trends are promoting the segmentation of the food market. This is roughly divided into a market for agro-industrially produced mass goods geared to consumer groups with greater purchasing power in emerging and transition countries. In addition, there is a market for regionally produced products and organic products, which serves the 'post-materialistic' lifestyles of the middle and upper classes in the industrialized countries of the North, who still have greater purchasing power.
4. During the WTO's liberalization efforts, price supports were dismantled in most industrialized countries, and production was decoupled from income. State transfer payments replaced these, some linked to specific conditions, to buffer farmers' income losses. Under the changed conditions, entrepreneurial and farmer-oriented family farms developed different economic styles.

Even though export subsidies were successively reduced, exports of agricultural products nevertheless increased. Between 2002 and 2019, EU agricultural trade more than doubled. Meanwhile, trade agreements comprise one-third of the EU's total trade in goods (2018); this also applies to agricultural trade.

World feed grain production increased by 64% between the 2000/01 and 2016/17 business years, from 863 million tons to 1414 million tons (German Farmers' Federation 2020). Over the same period, wheat production for food increased by only 31%, from 581 to 763 million tons. In the EU, of the 372 million tons of cereals available in 2017/18 based on annual production, stocks, and imports from third countries, only 18% was used for food and about 47% for animal feed. 9% of this was used in industry, including for bioethanol production; the self-sufficiency rate was 109%.

Between 1998 and 2018, EU meat production increased by around 25% to 48.1 million tons. Production was dominated by **pork** (approx. 50%), followed by **poultry meat**) (around 31%) and **beef and veal** (about 17%), while sheep and goat meat accounted for only about 2% of total production. For an extended period, **meat consumption** remained relatively unchanged in Europe at an average of 69.3 kg per capita. However, the summit on meat consumption seems to have reached Western countries. Here, scandals and media reports about the negative impacts of industrialized meat production, such as a high level of animal health and welfare problems, are responsible for the decreased demand for animal-sourced foods in many European countries (European Union 2017). In the long run, this development threatens the business model of livestock farming. The negative image decreases the demand for animal-sourced food and raises questions about societal support and the continuance of generous public subsidies. Livestock farms characterized by an inferior animal health and welfare level might even lose the social licenses to produce animal-sourced food (Hampton et al. 2020). For further details, see Sect. 4.4.

With the outbreak of African swine fever (a pandemic disease that spreads among pigs), exports to third countries were halted, combined with a dramatic collapse in market prices. Over the years, the pathogen has spread from Africa via Russia and Eastern European countries to Western European countries. The 2018 outbreak in China, by far the world's largest producer and consumer of pork, has led to an immense loss of animals there (up to 40% of the total pig population in China). As a result, market prices for pork rose significantly worldwide. Whereas pig farmers from countries spared from the pandemic disease could enjoy significant price increases, the market price in export countries where disease outbreaks occurred sank to new lows due to the vast surplus quantities produced for export.

The EU also plays a significant role in milk production. In 2016, the EU was still the largest milk producer in a global comparison. Because the effects of a prolonged drought in northern Europe reduced growth, India displaced the EU from first place in 2017. The third largest milk producer in the world is the USA. Germany is the largest milk producer in the EU and ranks fourth in the global rankings (Food and Agriculture Organization 2020). The leading importers of dairy products are the countries of Asia, which account for around 60% of global milk imports. The USA, Mexico, New Zealand, Argentina, Uruguay, and Australia show strong export growth. Nevertheless, the EU is the largest exporter of milk and dairy products, accounting for 27% of the world milk trade. Around 12% of the milk delivered to dairies by European farmers was exported to third countries in 2018, and the trend is rising. The largest share of German milk production goes into cheese processing (about 45% in 2017). In addition, the use of milk in the form of drinking milk or fresh products significantly influences the revenue generated by the milk trade (approx. 25%).

Product quantities and productivity, i.e., the quantity ratio between what is produced (output) and the means of production used (input), were significantly increased by the intensification processes forced after the Second World War. However, increased productivity should not be equated with increased profitability, where costs and expenses are related to yield or revenue. For most primary producers, increased productivity has not paid off; too often, market prices were not high enough to cover costs. Due to an oversupply of agricultural products in national and global markets, market prices usually remained low. Producers are forced to make ongoing efforts to reduce production costs even further to survive economically. The large number of farms that give up every year because they have exhausted themselves in competition shows the extent of the futile efforts to secure long-term economic viability under these conditions. The primary beneficiaries of these developments have been the processors (including slaughterhouses and dairies) and the agricultural and food retail sectors, which have seen enormous increases in turnover and profits. **Consumers** have also benefited. In 1970, consumers had to work an average of 96 min for a kilogram of pork chop, but in 2017, this was only

22 min (DBV 2018). The share of private household expenditure on food, beverages, and tobacco products in consumer spending was still 61% per person in 1850; it fell to 25% in 1970 and moved to 13.4% in 2017 (Statistisches Bundesamt 2020).

The **subsidies** have contributed significantly to the fact that market prices do not reflect the true production costs and remain at a low, often not cost-covering level. Even without linking subsidies to the quantities of products produced and irrespective of the respective national adjustment responses, the agricultural subsidies of the industrialized countries distort world market prices for food and rob developing countries, in particular, of urgently needed development opportunities. Instead, developing countries with high proportions of rural populations increasingly depend on food imports from developed countries to feed their rapidly growing populations. Imports are paid for, among other things, with industrial crops (e.g., cotton) or other exportable products such as tropical fruits. To generate foreign exchange earnings, the cultivation of export crops is pushed. This adversely affects traditional, subsistence-oriented agricultural systems and increases dependence on food imports. Thus, consumption by industrialized countries significantly affects land use in developing countries, leading to various land-use competitions there (Paulitsch et al. 2004).

Globalization mainly affects farmers in poor countries. This is not just about exporting poultry meat and dairy products, which are produced in developed countries and account for more than 50% of supply in much of Africa. In many places, these supplies have displaced local producers. "The children of African farmers are now becoming migrants rather than farmers," says Jürgen Mauer, executive director of the Forum Environment and Development, describing the consequences. Unlike in Europe, family farms in developing countries are not subsidized. Technically and financially, they are at the mercy of unconquerable competition. Who can still produce at a cost-covering level when market prices fall due to imported goods? Many farms no longer profit and can no longer take out investment loans. Production capacities and productivity fall accordingly.

Since the Second World War, the world population has tripled. According to the FAO, food production has increased 3.25 times worldwide. However, this did not mean that poorly-fed people were now well-fed. Three out of seven billion people—more than a third of the world's population!—live below the poverty line. They have no money to feed themselves adequately. According to a report by the Food and Agriculture Organization of September (2018), 11% of the world's population suffers from hunger; that's about 821 million people and 17 million more people than the previous year. If the figure alone is depressing, it is irritating that around 75% of them live in rural areas. People in Asia (515 million), Africa (257 million), South America, and the Caribbean (39 million) are particularly affected. According to FAO data, hunger is most prevalent in areas where armed conflicts are being fought. Approximately 60% of the world's hungry live in such areas. Added to this are the effects of **climate change**. According to the FAO, 9 million people die of hunger every year. 80% of the hungry people live in rural areas; 70% are farmers. The media draws the world's attention, particularly to urban hunger. The international aid initiated as a result arrives mainly in the cities, while the people in the countryside suffer the most tremendous hunger. These are then forced to flee to the cities to try to stay alive there.

Even if interested parties justify the necessity of further increases in productivity with the argument of fighting hunger, the production outputs achieved so far have not been able to alleviate the **hunger problem** in the world. If many primary producers in the world, currently achieving only low productivity, were supported to increase their output, there would be the possibility of doubling the production, even without overusing agrochemical inputs. This, however, would require changes in the economic framework and the distribution of resources to improve the existing production potentials significantly, for example, through greater trade justice for developing countries. In this context, it should also be mentioned that the 821 million undernourished people in the world are compared with about 1.9 billion people who are considered overweight; of these, about 672 million people are obese. For further details, see Sect. 5.2.

Derivations

The agricultural market has finally developed into a world market by abolishing the last protectionist measures under the 1992 Uruguay Round of the World Trade Organization (WTO). The long-lasting history of the production and procurement of nutrients, called agriculture, has developed from a common task of a community based primarily on cooperation to a competition between single actors on the global level. The common goal is not to produce sufficient and appropriate foodstuffs according to the community's requirements to enable food security but to ensure individual earning opportunities. The global market is influenced by powerful interests, which are not only manifested in the form of a few large global players. Political forces are also trying to influence the markets, in some cases massively, in the supposed interests of domestic producers and many other stakeholders involved, as well as representatives of the agricultural lobby for their clientele. However, an alignment of workers' wages in agriculture and with the operating profits of comparable companies cannot be achieved in this way. The political measures of national governments or the European Community have, at best, a buffering effect to somewhat mitigate the drastic consequences of global predatory competition. The long time it takes for EU member states to agree on a Common Agricultural Policy and the long periods (7 years) that the decisions are then valid make the EU a pawn in decisions taken elsewhere.

Instead, the changes in the climate are coming to the fore as harbingers of a climate crisis of enormous proportions. These will affect food production in a particular way. It is already becoming clear how much events in other parts of the world will also affect the local agricultural market. If there are yield losses in other countries relevant to the world agricultural market, these have an impact on world market prices and, thus, also on local pricing. This can affect food prices and, hence, the demand situation or influence feed prices. On the one hand, the price in world markets is determined by the supply and demand of products. On the other hand, supply and demand are not God-given variables but are subject to market interests and the political influence of various stakeholders. The oversupply of products in the world markets is not the most minor consequence of the general economic conditions. Among other things, expanding production capacities and services results in ruinous cut-throat competition, which is fought out on a global scale. The crises caused by the spread of epidemics, as well as the crises caused by climate change, expose the weak points and the enormous negative consequences of a globally oriented agricultural economic system that is subject solely to market interests. Political decision-makers have not sufficiently acknowledged the collateral damage caused by this system, let alone taken it as a reason to change course.

References

Alvard MS, Kuznar L (2001) Deferred harvests: The transition from hunting to animal husbandry. Am Anthropol 103:295–311. https://doi.org/10.1525/aa.2001.103.2.295

Anonym (1883) Erhebungen über die Lage der Landwirtschaft des Großherzogtums Baden.: Band IV. Großherzogliches Ministerium des Innern, Karlsruhe

Anonym (2019) Nur 23 Cent gehen an die Landwirte. Allgemeine Fleischerzeitung

Backhaus A (1894) Die Arbeitsteilung in der Landwirtschaft. Jahrbücher Für Nationalökonomie und Statistik 63:321–374

Bergmann H (1962) Arbeitsteilung und Spezialisierung in der Landwirtschaft. Feld und Wald, Essen

Bouwman L, Goldewijk KK, van der Hoek KW, Beusen AHW, van Vuuren DP, Willems J, Rufino MC, Stehfest E (2013) Exploring global changes in nitrogen and phosphorus cycles in agriculture induced by livestock production over the 1900–2050 period. Proc Natl Acad Sci 110:20882–20887. https://doi.org/10.1073/pnas.1012878108

Brinkmann T (1922) Die Ökonomik des landwirtschaftlichen Betriebes. In: Brinkmann T (ed) Grundriss der Sozialökonomik: Bd. VII Land- und forstwirtschaftliche Produktion, Versicherungswesen, vol 7. Mohr, Tübingen, pp 27–124

BUND - Bund für Umwelt und Naturschutz Deutschland e. V. (2019) Soja-Report - Die Bedeutung eines EU-weiten Eiweißplans. https://www.bund.net/fileadmin/user_upload_bund/publikationen/landwirtschaft/landwirtschaft_sojareport.pdf

Chanda N (2007) Bound together: How traders, preachers, adventurers, and warriors shaped globalization. A Caravan book. Yale University Press, New Haven

Cochrane WW (1993) The development of American agriculture: A historical analysis, 2nd edn. University of Minnesota Press, Minneapolis

DBV – Deutscher Bauernverband (2018) Situationsbericht 2018/19 Trends und Fakten zur Landwirtschaft. https://www.bauernverband.de/fileadmin/user_upload/Kapitel1.pdf

EU – European Union (2017) EU agricultural outlook for the agricultural markets and income 2017-2030. https://ec.europa.eu/info/sites/default/files/food-farming-fisheries/farming/documents/agricultural-outlook-2017-30_en.pdf. Zugegriffen: 09. Januar 2022

German Farmers' Federation (ed) (2020) Situationsbericht 2020/21: Trends und Fakten zur Landwirtschaft. Situationsbericht, 2020/21. Deutscher Bauernverband e. V., Berlin

Food and Agriculture Organization (2018) The state of food and agriculture 2018: Migration, agriculture and rural development. The state of the world, Rome

Food and Agriculture Organization (2020) Milk, whole fresh cow. https://www.fao.org/faostat/. Accessed 5 May 2020

Habermas J (1969) Technik und Wissenschaft als "Ideologie". Edition Suhrkamp. Suhrkamp, Frankfurt am Main

Hampton JO, Jones B, McGreevy PD (2020) Social license and animal welfare: Developments from the past decade in Australia. Animals 10. https://doi.org/10.3390/ani10122237

Hands J (2015) Cosmosapiens: Human evolution from the origin of the universe. Gerald Duckworth & Co, London

Harari YN (2015) Sapiens: A brief history of humankind. Harper Collins Publishers, New York, NY

Hartenstein L, Priebe H, Köpke U (1997) Braucht Europa seine Bauern noch?: Über die Zukunft der Landwirtschaft, 1st edn. Nomos Verl.-Ges, Baden-Baden

Hill K, Hurtado AM (1996) Ache life history: The ecology and demography of a foraging people. Routledge; de Gruyter, New York, Aldine de Gruyter

Illich I, Lindquist NT (1983) Fortschrittsmythen: Schöpferische Arbeitslosigkeit; Energie und Gerechtigkeit; Wider die Verschulung. Rororo aktuell, 5131//5131 : rororo aktuell. Rowohlt, Reinbek bei Hamburg

Mazoyer M, Roudart L (2006) A history of world agriculture: From the neolithic age to the current crisis. Monthly Review Press, New York

Niegsch C, Stappel M (2020) Branchenanalysen – Deutsche Landwirtschaft unter Druck; Research-Publikation. DZ BANK AG. https://docplayer.org/178798133-Branchenanalysen-deutsche-landwirtschaft-unter-druck.html

Nowak MA, Highfield R (2011) Super cooperators, altruism, evolution, and why we need each other to succeed. Simon & Schuster Inc., New York

Ogle M (2013) In meat we trust: An unexpected history of Carnivore America. Houghton Mifflin Harcourt, Boston

Paulitsch K, Burdick B, Baedeker C (2004) Am Beispiel Baumwolle; Flächennutzungskonkurrenz durch exportorientierte Landwirtschaft. Wuppertal Institut für Kima, Umwelt, Energie. Wuppertal papers (148)

Rott B (2016) Alexander von Humboldt brachte Guano nach Europa – mit ungeahnten globalen Folgen. HiN 17:82–109. https://doi.org/10.18443/234

Ruge U (2020) Bauern, Land: Die Geschichte meines Dorfes im Weltzusammenhang. Verlag Antje Kunstmann, München

Schaefer KW (2007) Was löste die Domestizierung von Pflanzen durch den Menschen aus? https://www.geschichtsforum.de/thema/was-loeste-die-domestizierung-von-pflanzen-durch-den-menschen-aus.16621/,. Accessed 27 Mar 2020

Sieder R, Langthaler E (eds) (2010) Globalgeschichte: 1800–2010. Böhlau Verlag; Vandenhoeck & Ruprecht GmbH & Co, KG, Wien, Köln, Weimar

Smil V (2002) Eating meat: evolution, patterns, and consequences. Popul Dev Rev 28:599–639. https://doi.org/10.1111/j.1728-4457.2002.00599.x

Smith A (1776) An inquiry into the nature and causes of the wealth of nations, vol 1, 1st edn. W. Strahan and T. Cadell, London

Statistisches Bundesamt (1961) Statistisches Jahrbuch für die Bundesrepublik Deutschland. W. Kohlhammer, Stuttgart, Mainz

Statistisches Bundesamt (2020) Volkswirtschaftliche Gesamtrechnungen des Bundes: Konsumausgaben der privaten Haushalte im Inland (nominal/preisbereinigt). https://www.destatis.de/DE/Service/Datenbanken/_inhalt.html. Accessed 29 Sept 2020

Sundrum A (1996) Von der Viehwirtschaft zur Tierproduktion. In: Reimer W (ed) Die Wissenschaft und die Bauern, 1st edn. Bauernblatt Verlag; ABL Bauernblatt-Verl, Rheda-Wiedenbrück, pp 97–103

Survival International (2004) Stories and Lives: 21st Century Tribal Peoples. http://assets.survivalinternational.org/static/files/books/stories_and_lives.pdf. Accessed 6 Jan 2022

Thaer A (1809) Grundsätze der rationellen Landwirthschaft: Erster Band. Begründung der Lehre des Gewerbes. Oekonomie oder die Lehre von den landwirtschaftlichen Verhältnissen (Bd. 1 von 4). Realschulbuchhandlung, Berlin

Ullmann I (2007) Die rechtliche Behandlung holsteinischer Leibeigener um die Mitte des 18. Jahrhunderts: Dargestellt unter besonderer Berücksichtigung der Schmoeler Leibeigenschaftsprozesse von 1738 bis 1743 sowie von 1767 bis 1777. Zugl.: Kiel, Univ., Diss., 2006. Rechtshistorische Reihe, vol 346. Peter Lang, Frankfurt am Main

Westhoek H, Rood T, van den Berg M, Janse J, Nijdam D, Reudink M, Stehfest E (2011) The protein puzzle: The consumption and production of meat, dairy and fish in the European Union. PBL Netherlands Environmental Assessment Agency, The Hague

Zorn W, Sommer OA (1970) Rinderzucht, 5th edn. Tierzuchtbücherei. Ulmer, Stuttgart

Communities, Commons, and Common Good

3

Abstract

From the beginning, the production of foodstuffs was a task intended to benefit all community members. However, access to the harvest was also confronted with the self-interests of community members and the desires of other communities. What benefits some may be a disadvantage to others. The "dilemma of the commons" became a common word to describe the inevitability of conflict. These prevent it from following the approaches of an eco-social market economy and the approaches of an eco-social market economy.

3.1 The Cement of Communities

More than any other economic sector, agriculture is intimately linked to the interest of every human being. We depend on the daily supply of nutrients and energy through food and can only survive for minimal periods without food intake. Accordingly, obtaining food was central to communal life and shaped people's daily lives for thousands of years. Just as the individual depended on the community's activities to participate in the available resources, the community relied on each individual to contribute directly or indirectly to procuring food according to their abilities. The survival of the community and its members depended on the functioning of the network of the relationships between the individual community members.

According to today's knowledge, the relations of a hunter-gatherer society are characterized by extreme care in avoiding competition. According to Service (1966):

A. Sundrum, *Public Welfare-Oriented Production of Food*,
https://doi.org/10.1007/978-3-662-69040-6_3

"Neither is there any pecking order based on physical superiority, nor is there any superiority or subordination at all based on other sources of power, such as wealth, inherited membership of certain classes. ... The only coherent superiority is that of a person of greater age and wisdom who can preside over a ceremony."

It is conducive to thriving togetherness if individuals do not derive personal advantage from their superiority over others and avoid excessive rivalry. A taboo of a particular kind prevents those more successful than others because of manifold talents from gaining unrestricted power and using it only for their advantage. Among many hunter peoples, therefore, the hunter is not allowed to consume the animal which he has killed (Gronemeyer 2009). The reason given is that this meat could harm him in the sense that the animal could take revenge on its killer through the meat. It is a cultural achievement when strict taboos secure the avoidance of excessive rivalry and attaining particular personal advantages. Cooperative behavior and sharing have emerged as a predominant social pattern of behavior in hunter-gatherer societies. They enable economic activity without scarcity and ensure the immediate accessibility of all to the necessities of life. In this sense, **cooperation** and sharing are not moral decisions but an existential prerequisite of the conditions of existence of hunter-gatherers, no matter how different the concrete manifestations of the rules of coexistence may have been.

In this way, it is possible to organize coexistence within the community so that it does not disintegrate into winners and losers and that some do not gain an advantage at the expense of others. Individuals may indulge in their interests as long as the community benefits from them or as long as they do not endanger the benefit of the community. As soon as someone claims authority that cannot be legitimized by service to the community, he is exposed as pompous and exposed to ridicule. All it takes is ridicule to be at risk of compromising respect and position in the community. Social status is subject to standard criteria of appropriateness (Illich 1982). They are at the community's disposition and intelligible to every community member. Life in the communities of gatherers and hunters is oriented towards mobility. Only by moving around according to the changing availability of food throughout the year can they secure their livelihoods. As a result, nomads are protected from accumulating extensive possessions because great importance must be attached to freedom of movement and the carrying capacity of what has to be transported during migratory journeys.

In contrast, the seizure of land, the appropriation of where the necessities of life flourish, where they can be found or produced, is a seizure of possession that precedes all other acts of gaining power through possession. Gronemeyer described this as follows:

"With private ownership of land, the exclusive right of disposal is claimed over everything that nature produces, its entire 'living and dead inventory.' The original form of possessive power is land ownership, and the original act of taking possession is the fencing of land. At the same time, the privatization of land seals the end of communal use; it abrogates the "right to commons," abrogates the common property, the 'commons,' in which all

> *community members could freely participate."* ... "In place of the *communal use of the land and the cooperative production of subsistence comes rivalry for the rationed means of life, the struggle for distribution. The principle of sharing gives way to that of allocation. People who, to eke out a living, have to 'get' something, are 'warring'* people, *i.e. also warlike* people" (Gronemeyer 2009).

Taking possession of land for use and the sedentariness accompanying it requires a high degree of skill in coordinating the activities associated with food procurement, preparation, and distribution. It is necessary to build up stocks and to ensure that the stocks are kept out of the reach of unauthorized persons and other parasitically behaving creatures. The better the coordination succeeds, i.e., the more purposefully the different abilities the community members have are used, the more successfully the food availability for the community can be secured. The more food is available with the further development of agriculture and animal husbandry, the more free spaces open where people can detach themselves from their immediate preoccupation with food procurement and pursue other interests and activities. On the other hand, the division of labor is also the starting point for competitive behavior, among other things, in the competition for the more pleasant or glorious activities that can be expected to bring a higher-ranking position in the community. At the same time, the division of labor opens up increased opportunities for spoilsports (Nowak and Highfield 2011). Individuals or small groups that more or less openly refuse to cooperate tend to prefer a parasitic lifestyle, i.e., act at the expense of the community and its function. Consequently, food availability for the community was threatened not only by unfavorable climatic and weather conditions but also by the covetousness of other communities. Still, it was also exposed to threats from within through a prevalence of parasitic behavior.

Communities are always also communities of purpose, to the preservation of which each individual has to contribute their part in relation to the respective abilities and possibilities. The basis of sharing is that the activities of the individuals result in a benefit for the actors themselves and the community. The overall goal of the community is to ensure its survival. This also applies to the individuals but includes finding a way to ensure their survival within the community. The diverse activities of individuals are functionally aligned, albeit to varying degrees, with this common goal. In the early days, communities were economically self-sufficient (autarkic). Only a few items were obtained from strangers, i.e., persons outside their community. Gradually, there was an increased exchange of objects and services between people of different communities, initially through barter and, as it developed, through purchase using introduced money. Markets developed where food was available to all those who could afford the goods on offer at the exchange value or prices demanded. At the same time, power secured access to part of the surpluses and stocks produced by the rural population, not only through access to the markets but also through introducing taxes and duties.

Derivations

All living individuals have in common that they want to maintain themselves and survive as long as possible. The survival of individuals can only be ensured through participation in a community. The functioning of the community is the common intersection of the interests of all community members. Farming and animal husbandry and the generation of foodstuffs that can be achieved through them make an indispensable contribution. Agricultural activities require **cooperation** between the members of the community. Their activities are based on their capabilities to contribute to the community's welfare. At the same time, individuals within communities compete with each other. The competition may be for the most nutritious and tasty food, for the favor of the opposite sex, or for the higher-ranking positions within the communities. The greater the availability of food, the more options are available to the community and individuals for expanding other activities and areas of life. Higher rank positions promise higher prestige and improved opportunities to pursue one's interests. The existential necessity of cooperation hems competition among individuals. The mutual dependency relationship, i.e., of the individual on the community and the community on the contribution of the individuals, is the basis for establishing rules to which the individuals subordinate themselves out of their interest for the good of the community and the superior requirements. Cooperation and coordination are based on the given circumstances, i.e., on the possibilities of organizing food production in a purposeful way and on the needs of the members of the communities. They are, therefore, communities of purpose that adapt their activities to the circumstances, the necessities, and the changes that occur. Rules are developed that stand the test of time under the prevailing conditions. Cooperation, coordination, and regulations ensure food availability as efficiently as possible and keep the communities of purpose together through joint land management and livelihood security. Communities of purpose together through the benefits of joint land management and livelihood security.

3.2 Markets Create Needs

The more comprehensive the range of surplus food, the more fundamental human issues arose: why should one be content with the fruit from one's fields or garden when other growers had other and possibly tastier fruit to offer? An obvious solution is the bartering of foodstuffs. The prerequisite, however, is that those willing to barter have something to offer that is of mutual interest. With the advent of money as a universal medium of exchange, trade possibilities were created. Almost anything could be exchanged for cash, provided a monetary exchange price could be agreed upon. Thus, food was no longer reduced to its importance in ensuring the survival of a community; it also served the producers as an object of exchange or money-making. Consequently, the production of foodstuffs developed from a purely **subsistence economy** and became more and more a trade. The possibility of selling surpluses that were not directly needed for self-sufficiency opened up opportunities to generate added value. The surplus value that farmers had extracted from the soil and livestock by the sweat of their brow was, however, no longer measured solely in terms of their valuation but also in terms of what could be exchanged for it, either directly or in monetary terms, for other more or less crucial material goods or services. The exchange value was increasingly determined by the relations of supply and demand that settled outside the agricultural production system.

As markets developed, the value that buyers and sellers attached to the respective foodstuffs was not measured by the degree of neediness of the actors involved. Those who could afford it and were not directly dependent on the exchange or sale of the goods had various options for negotiating the value of the goods with potential buyers. Through the emergence of commodity trade, food was assigned a commercial value fundamentally different from the exchange value. Trade value depended primarily on the extent to which the supply of goods exceeded demand in markets accessible, depending on the state of transportation and whether demand exceeded the supply or whether the two were roughly balanced. With the increase of production surpluses, the trade of commerce began to flourish and grow into the structures that today assign commerce a dominant position in the food industry.

The relationship between the effort that had to be made to produce food and the benefit that could be obtained through intra-community use or sale was increasingly put into perspective. The value of foodstuffs in the immediate context of the communities (small, large families or estates) was overlaid by the extent of the demand for the goods on offer, which grew out of village or urban structures. If other regional producers offered the same products at the markets, the surplus of goods caused a significant decrease in value, even if the production effort was the same. This changed only in times of scarcity, such as crisis and war, when a very different value was placed on food than in times of satiety. In times of famine, food was even weighed with gold and silver if this could still be offered in exchange.

Outside of crises, a market economy developed, which, by providing opportunities for the sale of surpluses, created new incentives for the production of surpluses and contributed to an increase in food availability. New skills were required, providing new income opportunities for those with them. This included the ability to foresee future developments in the supply of and demand for goods and to match them with production, harvesting, and storage capacities. The more reliably the demand development could be estimated, the more purposefully the supply side could align itself with it and use this to its advantage. This meant that the reliability of information on expected trends was vital even in the early days of market developments. The earlier and more reliably one knew what the future would bring, the better one could adapt to it and hold one's own against competitors. Speculation and the danger of speculation are inherent phenomena of market development. Little has changed in this basic constellation over time, even if the time factor is different today. Today, agricultural products are also traded on commodity futures exchanges; fractions of seconds determine whether a profit can be made from a minimal lead in assessing trend development.

The market functions as a **system** of interactions where all participants seek to optimize their self-interests. The "natural law of economics," according to which supply and demand regulate prices, ensures that production processes beyond subsistence agriculture increasingly align with market events. If there is an increased demand for specific products, this forces competition between food producers who supply the same market. Those who produce more can sell more; those who offer their goods more cheaply can out-compete their competitors. Those with only products to offer that are in surplus must be content with the prices buyers are willing to pay. On the other hand, buyers acquire goods according to their judgment at a price that seems acceptable or even cheap from their perspective. The value of food is thus also measured by the relationship between availability and demand. As transport routes and means are developed, trade directs the flow of goods to places with high demand and willingness to pay. **Traders profit from the difference between supply and demand between different locations.** Little has changed in these operating principles of the market to this day, either.

For farmers, developing markets create various options for selling self-produced products and services to acquire things they cannot produce themselves or expand their holdings. Markets thus become the driver of increasing diversification of goods and further differentiation and specialization in terms of production. Those who can produce a good more efficiently and offer it at a lower price gain an advantage over competitors. Another market strategy consists of providing goods that, due to limited availability or exceptionally high quality, are held in higher esteem by buyers and for which sellers can demand a higher price. Markets create desires for goods that individuals are eager to acquire beyond immediate need. Those who are willing and able to do so exert themselves more to acquire possessions and the accompanying increases in prestige, power, and influence.

The value attached to the products in the internal context of the farm system is irrelevant to market events. Buyers are primarily eager to acquire the products at the lowest possible price, regardless of how much effort has been expended to wring the products produced from the respective conditions. Thus, the products offered on the markets are largely decoupled from the preceding production processes and the associated advantages and disadvantages for internal and external effects. Buyers are not interested in whether the cultivation of the market crops was to the detriment of the regenerative capacity of the soils and the yield of the subsequently cultivated crops or what status the animals sold had in the farm context. The products sold have different internal and external consequential effects without this being reflected in the market prices.

Derivations

The expansion of production processes is developing with the nutritional needs of a growing number of people and increasing demands on the available quantity and the properties of food. Gronemeyer (2009) summarized this as follows:

"The expansion of production processes is causing subsistence-oriented activities to disappear. The production pattern of division of labor is replacing the life pattern of cooperation. Production displaces self-preservation in every sector of life it taps."

People become dependent on wages; they do not produce what they need but work for money that allows them or forces them to spend it on being represented in almost all aspects of life. While an ever-smaller proportion of the population is dependent on producing food itself, the distance to the processes of agricultural production and to the services that farmers provide for a differentiating community is growing. Trade anonymizes the products and removes them from the context of the production processes. The decisive factor for the market value is not the value of the food for the community and farm system in which it was produced but for those who can afford to purchase it. If there is an oversupply of products, they lose their market value and reduce producers' options to obtain the expected revenues for their more or less costly work and investments in monetary exchange.

3.3 Dilemma of the Commons

During industrial development, there was a steadily increasing differentiation of professional and social groupings. Each of them put its interests on the line against the interests of others and tried to safeguard them. Analogous differentiations also took place in the context of agricultural food production. From the nineteenth century onwards, small and large families or estates were at most still functioning as communities oriented towards the common goal of supplying their members with the necessities of life. These were in a close relationship of dependence; their survival depended on sufficient food in the community for all its members. Even today, the families or clans are communities of provision that try to ensure the elementary needs of their members.

Common goods are not owned by individuals who can withhold their property from others but are owned by communities and are thus accessible to all members of a community, at least in theory. Common goods are provided by a community (e.g., village community, state or private providers). Goods such as nature or the environment are also perceived as freely accessible. The perception of something as a common good (also in the sense of something "good") often only occurs when the **quality** of a common good is impaired, e.g., pollution of the air or groundwater with pollutants, or when access is temporarily or permanently restricted or even blocked. Communities are defined not least by drawing boundaries, i.e., who is assigned to them and which persons are (or can be) excluded from belonging. This is comparatively clearly regulated in the case of nationality, which distinguishes nationals from foreigners. The latter can be denied entry or ordered to be expelled. Also, for example, the members of an association have access to events or to the association's premises, which can be denied to non-members. As a rule, all members of a defined community are allowed to share the use of common resources from foreigners. The latter can be denied entry or ordered to be expelled. Also, for example, the members of an association have access to events or to the association's premises, which can be denied to non-members. As a rule, all members of a defined community are allowed to share the use of common resources.

At a time when the vast majority of land was not yet privately owned, some of the arable land was owned by a village community. This mainly included pastures and forests but also water reservoirs, the use of which was open to the members of the community. Forms of communal land use as common property are still common today, for example, in alpine pastures. Since freely accessible and, above all, scarce opportunities for use arouse covetousness, there is a need for decision-making structures, and in particular for regulation, when there is a danger of persistent overuse of common property (see Sect. 2.2). This is the case, for example, when too many animals are driven onto the pastures, too much wood is cut in the forest, or too much water is taken from the reservoirs. This results in consequential damage to the commons that impairs the ability of others to use them. The literature describes numerous examples of successful regulation of common property where overuse has been avoided. However, numerous examples will be discussed below where this has not (yet) been achieved. These include, among other things, the overfishing of the world's oceans and their overuse as waste dumps. The world's oceans are seen as a "no man's land" with open access for everyone.

In the case of resources that are communally used but limited in availability, a fundamental conflict has become known as the common good dilemma. The Canadian economist H. Scott Gordon was among the first to integrate the conflicts associated with using common good property into a **theory**. (Gordon 1954). Using fisheries as an example, he wrote:

> *"No one attaches any value to a possession that is freely available to all because everyone who is foolhardy enough to wait his turn eventually finds that someone else has already taken his share away."*

The American Ecologist Hardin (1968) expanded the term in an essay for the journal Science entitled: "The Tragedy of the Commons." He argued that overuse and even destruction of resources is inevitable without limits on the use of commons. Once a resource is unrestricted, some people will try to secure as much yield as possible for themselves. This works as long as the common good is not exhausted. However, as soon as the use exceeds a certain level, the tragedy of the commons takes hold: Everyone still tries to maximize his yield. Now, however, the good is no longer sufficient for all.

The community bears the consequences of overexploitation. For the individual, the immediate gain is rated much higher than the consequential costs, which are only felt in the long term. "*Freedom in commons brings ruin to all*" is one of Hardin's core theses. Therefore, the problem should not be left to individuals but should be seen and tackled as a community problem. However, Hardin felt compelled to make corrections years later and titled the essay "Tragedy of the Unmanaged Commons" (Ostrom 1999). In his essay, he described the situation of unfettered access to land that belonged to no one and did not differentiate between common and no man's land.

There are many options for dealing with common property and no-man's land, and they involve different actors and process levels. The example of the overuse of agricultural land—as has happened in the past and still happens in many ways today—illustrates the complexity of the problem. In a village community, all ruminant keepers grazing the common land benefit from the plant growths because the livestock can use the growths for self-preservation or growth or milk production processes. The more animals that individual livestock owners allow to graze on the common land the higher the benefit for the individual. This is true as long as the forage availability is not limited. The limited nature of the available area and the quantity and quality of the forage in relation to the forage intake capacity of the animals reduces the availability and creates a scarcity. In the case of scarce resources, the additional benefit to the individual livestock farmer who nevertheless drives an additional animal onto the commons is greater than the harm caused to the community members by the relative reduction in the total availability of forage. The person who seeks to maximize his benefit does so at the expense of the other community members. At the same time, this approach relies on the other community members being humble and not following this strategy. Unlike creeping, drastic overuse through a massive increase in the number of animals in a limited area makes the consequential effects quickly visible. The sward would be damaged, and the livestock would not increase in live mass in the usual way but would become emaciated. An apparent crisis forces the community members to react if the regenerative capacity of the growing areas is to be regenerative. Thus, the utilization potential of the commons is not to be irreparably damaged. Slow changes, on the other hand, rarely trigger reactions and countermeasures.

In the real world, overuse of common goods usually occurs as a gradual process, the adverse effects of which are not immediately apparent. To perceive changes, it is necessary to have a trained eye or to carry out continuous measurements to assess undesirable developments. Depending on the degree of overexploitation, the unwanted effects often only become apparent with pronounced delays. We speak of a chronic event (Chronos, the god of passing time). Unlike an acute event, there is a tendency in chronic processes for people to come to terms with the event, to delay an appropriate response, and to postpone it into the future (procrastination). There are many reasons for postponing, ranging from the consequential effects, which are often only vaguely discernible and thus speculative at the outset, to the effort that would have to be expended to record the consequential effects more precisely to the need to defend oneself against overuse and, if necessary, to risk a dispute or even a court case. More noticeable is a conflict avoidance strategy, possibly supported by the hope of a positive turn of events or justified by the fact that it is impossible to foresee the consequences of settling a conflict. In contrast, the grey areas in chronic processes require efforts to enable a robust assessment.

To make matters worse, there are rarely clear procedural rules for overuse, as there are, for example, in public road traffic with traffic regulations. The undesirable consequences and dangers that unregulated traffic poses to the life and limb of fellow citizens have prompted the community to implement traffic regulations to protect citizens. Road traffic regulations stipulate who may use the roads, under what conditions and in what manner, and how violations of the rules are recorded and punished. Exceeding the speed limit is relatively easy to quantify using technical devices; the costs of exceeding the speed limit are known to traffic offenders in advance as a catalog of fines. In the event of a dispute, legal proceedings can bring about clarity in the legal sense. However, most overuse of common property by individuals is far removed from such clarification processes. In the case of the beneficial and harmful effects of the overuse of grazing land by and at the expense of community members, this is not easy to assess. This also applies to the regenerative capacity of rangelands and the growth potential of livestock. Also, considerable time lags that often occur between the causal initial constellation and the undesirable consequential effects may mean that the primary causes and the polluters responsible for the consequential effects cannot be attributed. Thus, the overexploitation of resources by individuals at the expense of others results from the interaction between the two, i.e., those who overexploit and those who allow this to go unchallenged. Spoilsports take advantage of the supposed convenience of the other members of the community and secure a personal advantage contrary to the general rules of the game of a fair distribution of the utilization potentials of common goods. In a biological sense, they behave parasitically towards the community (Nowak and Highfield 2011).

Overexploitation of a resource occurs when too many users have the privilege of using it, and no one, as in the case of no-man's-land, has the right to restrict its use. Whether common land or no man's land (oceans, atmosphere, etc.), a threatening overuse of resources requires regulation to the advantage of a few and the disadvantage of many. In the case of goods that can be assigned to defined communities, it is easier to regulate the use of resources than in the case of no man's land. However, the proliferation of multiple problems directly or indirectly related to forms of overuse in the general increase of users suggests how difficult it is to implement effective regulation, even in the case of shared property use. There is an understandable reason that the original (Greek) word for **law** and order, "nomos," was the "law of pasture use." Nomads had to abide by the law of sustainable use of pasture. Also on alpine pastures, there are regulations governing the use of alpine pastures, albeit in very different, locally negotiated forms. In the exploitation of maritime fish stocks, riparian states have also agreed on a course of action intended to counteract overfishing and, thus, destroy the regenerative capacity of fish stocks.

The list of common goods that have a relevant potential for use and are simultaneously subject to adverse impacts and damage in the event of overexploitation is very long.

Common goods include non-renewable resources such as fossil energy sources (oil, gas, etc.) and other mineral resources, as well as the availability of agricultural land, which is no longer available for food production due to overbuilding or other devastation. Soil fertility can also be classified as a partially renewable resource, as it can be damaged to a greater or lesser extent by leaching and erosion, but also irreparably. The same applies to forest and fish stocks. Examples of renewable resources owned by communities are water resources and the quality of groundwater, which may be contaminated by inputs, degraded in quality, and subject to extra treatment costs (e.g., nitrate in groundwater). Another but no less critical resource is the effectiveness of **antibiotics** against harmful microbial organisms. Excessive use, e.g., in livestock farming, can impair or cancel out the effectiveness of antibiotics by forcing the **development of resistance**. This happens to the detriment of humans and animals who depend on antibiotic efficacy to be treated in the case of a bacterial infection, which significantly increases their chances of survival.

Free access to limited common goods creates problems if framework conditions are not established simultaneously, enabling counter-regulation in the event of overuse. On the one hand, these can occur in the form of appropriation problems if freely available scarce resources are overused and then no longer available to all (e.g., water reserves in times of drought). On the other hand, there is a problem of provision when public goods cannot be made available to those in need or cannot be made available in sufficient quantities (e.g., hygiene masks, vaccines, or care places in nursing homes). Ultimately, both problems concern the relationship between availability, need, and protection against negative consequences. Even if not all community members are equally endangered by the consequences of an overexploitation of limited available resources, the situation should be fair as far as possible. The principle of fairness is, therefore, of great importance. The protection against disadvantages and injustice should not be related to the wealth of the individual but should be granted independently of the person's reputation. This does not exclude the fact that specific groups, such as children, are given special attention and rights. The special attention paid to children or older people is primarily due to their particular need for protection.

Just as there might be little disadvantage for individuals when others use the commons excessively, there can also be little advantage for investors. This is the case if they invest in measures to protect the commons and promote the community (e.g., through donations and taxes), while others refrain from doing so but benefit from the protective measures because they cannot be excluded from them. The latter thus behave as so-called "free riders" who profit from the contributions of others without themselves making an appropriate contribution. If all were to invest according to their means in measures to protect common goods, the common investments would pay off best. If only a few bear the costs of the investments but they derive little benefit from them, this is a bad deal for self-interested investors. Accordingly, it is unsurprising if only the altruists and not the selfish invest. If a community (e.g., a community of states) defines an overriding goal, such as the reduction of CO_2 emissions, one might expect that everyone is required to do their part to achieve the goal. The more effort one country makes, the less effort the other countries have to make to reach the target. In a sense, the investments of the climate altruists subsidize the contribution of the climate egoists (Ockenfels 2020). Although widespread agreement exists on the benefits of climate action, implementations are too often doomed to failure. Common goods can consequently also be eroded by those who do not adequately participate in the investments to protect them, even though they share in the positive effects.

Derivations

The dilemma of the commons has preoccupied people for many generations. It is a basic anthropological constellation that repeatedly challenges human communities in a variety of forms and manifestations. It causes people to vacillate back and forth between their very own interests and the interests of a superordinate community to which they belong and on which they depend to very different degrees. How individuals deal with this ambivalence is as heterogeneous as the number of characters involved and as the respective communities allow. Among the nomads of early times, the sick and weak who could no longer contribute to the survival of the clan were left behind when food was scarce. Exclusion from the supply community was tantamount to a death sentence for those affected; at the same time, it increased the chances of survival for the other members of the clan. Such existential weighing processes are still necessary today, even if only in exceptional situations. Today, it is primarily a matter of chronic conflict situations in which different interests clash without a solution being found that takes account of both particular interests and the interests of the common good in a way that is acceptable to both sides.

Even if the individual's and others' survival is rarely directly at stake in the Western world, the basic constellation of perspectives: I or they or he/she or we, persists. It can refer to seemingly minor aspects, such as the diverging interests of the football team members, or to such serious problems as how the world community deals with the climate crisis. At the global level, the struggle of different interest groups to invest fairly in climate protection and to protect themselves from the effects of the climate crisis that has already occurred is already a struggle for the bare survival of many people. The scarcer global capacities of the atmosphere for absorbing climate-relevant greenhouse gases become, the more these disputes will intensify.

3.4 Dealing with Limited Common Goods

Common goods are goods that are freely accessible to all potential consumers. They can be provided by the state or by private providers. A characteristic property is non-excludability. Goods, such as access to the Internet, whose use is not subject to any limitations apart from the technical requirements, do not require access regulation. The situation is different with common goods whose availability is limited. Here, questions arise as to how to deal with them in a just and fair manner, to prevent individuals from gaining an unruly advantage through skill or cunning at the expense of access by others. All those who belong to a community, which can be a family, a clan, an association, a collective, a municipality, or a nation-state, must agree on how they will use the limited available **resources**. Consequently, there is a need for forms of agreement and regulation which, if possible, are comprehensible to all and bring with them a certain degree of reliability to contain recurring conflicts if they cannot be avoided in principle. Experience shows that uncertainty and anger spread within communities when clarity and orientation are lacking. But this is easier demanded than implemented.

Regulation can take place in a variety of ways. One prominent advisor who has devoted herself to the topic of commons and was awarded the Nobel Prize for her work is the American economist Elinor Ostrom. In her major work, Governing the Commons, published in 1990, she set out the principles of economic management of the commons (Ostrom 1999). These are based on the insights that Elinor Ostrom and collaborators have gained in numerous studies in different countries and communities about success in using the commons. She presented them again in her Nobel Prize speech in 2009. The design principles synthesize core factors that affect the probability of long-term survival of an institution developed by the users of a resource. They are outlined below (Ostrom 2009):

1A. *User Boundaries: Clear and locally understood boundaries exist between legitimate users and nonusers.*
1B. *Resource Boundaries: Clear boundaries that separate a specific common-pool resource from a larger social-ecological system are present.*
2A. *Congruence with Local Conditions: Appropriation and provision rules are congruent with local social and environmental conditions.*
2B. *Appropriation and Provision: Appropriation rules are congruent with provision rules; costs are proportional to the distribution of benefits.*
3. *Collective-Choice Arrangements: Most individuals affected by a resource regime are authorized to participate in making and modifying its rules.*
4A. *Monitoring Users: Individuals who are accountable to or are the users monitor the appropriation and provision levels of the users.*
4B. *Monitoring the Resource: Individuals who are accountable to or are the users monitor the condition of the resource.*
5. *Graduated Sanctions: Sanctions for rule violations start very low but become stronger if a user repeatedly violates a rule.*
6. *Conflict-Resolution Mechanisms: Rapid, low-cost local arenas exist to resolve conflicts among users or officials.*
7. *Minimal Recognition of Rights: The government recognizes local users' rights to make their own rules.*
8. *Nested Enterprises: When a common-pool resource is closely connected to a larger social-ecological system, governance activities are organized in multiple nested layers.*

Above all, there is the maxim of exploring the principle of **cooperation** and giving it validity to bring it into play as a counterweight to the simultaneously existing competitive behavior and the dominating competitive mechanisms. If some want to cooperate, but others do not, exit possibilities must be offered for the latter. For Ostrom, it is crucial that the potential of common resources cannot be developed without involving the users. Cooperative structures are based on participatory approaches. Only these make it possible to adapt regulations to local cultural and ecological conditions and to modify them if necessary. In this way, she explicitly contradicts the frequently encountered attempts to impose detailed rules on heterogeneously structured communities from the outside. The aim should be to increase self-determination. As a rule, this cannot be reconciled with patent remedies and blanket regulations. With her research, Ostrom was able to show that attempts to impose central solutions to resource use problems top-down are to be regarded as a misguided research approach, Ostrom was able to show that attempts to impose central solutions to resource use problems top down are to be regarded as a misguided approach.

What is already a particular challenge in organized communities is far more difficult in cases of (still) "unmanaged commons", as already thematized by Hardin. This includes, above all, nature in the broadest sense, which generally encompasses what is not man-made, as opposed to man-made culture. From the biological perspective, humans originate in nature and live from nature. They are intimately connected with it in an ongoing process to be able to sustain themselves. From this perspective, human beings are a part of nature. If one follows the common understanding, natural capital comprises nature with its species diversity, biotic communities, and ecosystems. The various services provided by nature, which are often summarized by the term **"ecosystem services,"**, refer to direct and indirect contributions of ecosystems to human well-being, i.e., services and goods that provide direct or indirect economic, material, health, or psychological benefits to humans. In an economic sense, nature constitutes capital; its services can be understood as dividends that accrue to society. Preserving the natural capital stock makes it possible to provide these dividends for future generations in the long term. In this approach, however, it is important to remember that natural capital and ecosystem services are **anthropocentric** concepts of the "intrinsic value of nature".

Derivations

Dealing with limited common goods is different from different perspectives. From an economic perspective, it is obvious that nature's capital should be further developed and managed with the help of economic procedures. However, from other perspectives, there are justified doubts as to whether economics has the necessary instruments to undertake comprehensive balancing processes with areas not directly accessible to monetary valuation. After all, it is part of the core of an economic approach that a monetary valuation is only compared with other monetary valuations, regardless of whether these are based on valid assessments. Nor does economics answer questions about the relationship between private property and the ownership of common goods such as natural capital, and who may dispose of them to what extent and at whose expense.

The largely unresolved issues in dealing with the 'unmanaged commons,' which include natural assets and climate and environmental protection, among others, are of great importance to the topic discussed in the book. However, the relevance is not limited to off-farm commons. The processes within a farm system must also be included in the consideration and assessment since they are jointly responsible for the impairments of common goods to a considerable extent.

3.5 Agriculture and the Dilemma of the Commons

The dilemma of the commons was not only a problem of agriculture in prehistoric times but has risen over the centuries to become a central area of conflict in the agricultural economy. Before going into the necessary details in the following chapters, the subject area should be circumscribed, and the fundamental problems worked out. After all, agriculture's overuse of common property is not limited to community members' use of community land. Agriculture makes significant use of nature's eco-systemic services or impairs these eco-systemic services for other users. No less important is the use of the environment in its broadest sense, in that agriculture extracts resources from it, such as water, for its use. On the other hand, the environment is used as a landfill for the residues and pollutants generated by agricultural production processes. This applies, for example, to nutrient discharges, drug residues, and germs resistant to antibiotics from animal husbandry. In this way, agriculture causes considerable damage to the community, both at the local and regional level, as well as at the global level.

Agriculture claims to have "cultivated" the goods of nature during its development. As a result, it was and is possible to make food available to a large and ever-increasing number of people as "agri-cultural" services. In a world in which many people, especially in urban metropolises, have lost touch with food production, farmers feel compelled to recall that the production of food by farmers makes possible the survival of those who can purchase the **"means of living"**. From another perspective, agriculture uses community goods to produce private goods in the form of food and fodder, which it offers for sale to the community to secure its economic existence through the income from the sale of the products. Because of the scarcity of common goods, there is conflict in the nature of things. It is the task of politics to shape the framework conditions so that overuse and conflicts of use are contained. As long as **conflicts** occur locally, legal regulations can limit them comparatively well. However, the consequences of production processes have long since ceased to be merely local areas of conflict but expanded into regional, national, and global problem areas. The areas of conflict affect different interests of different groups of people and are accompanied by gradually varying impairments of these interests. There is much disagreement about how best to reconcile the various particular interests. It ranges from sweeping accusations of environmental destruction by agriculture, as sometimes put forward by nature conservationists and environmentalists, to the rejection of any criticism of agriculture and its harmful effects, as still advocated to the public by agricultural lobbyists. Meanwhile, the representatives of rural policy, who are responsible for the containment of areas of conflict between different social interests, are exercising noble restraint. This also applies to representatives of agricultural science, who are supposed to provide a basis for assessing the extent of harmful effects. The possible reasons for this restraint are reflected elsewhere in this book. Meanwhile, the representatives of agricultural policy, who are responsible for the containment of areas of conflict between different social interests, are exercising noble restraint. This also applies to representatives of agricultural science, who are supposed to provide a basis for assessing the extent of harmful effects. The possible reasons for this restraint are reflected elsewhere in this book.

Derivations

The areas of conflict outlined above between the interests of different social groups give rise to a wide range of questions. This concern, among other things, is the methodological approach to the necessary differentiation between particular and overriding public interests. The question of whether and to what extent there are different advantages or disadvantages for various interested parties also requires profound consideration and corresponding instruments that provide reliable information in this regard. No less important is the question of which reference systems and yardsticks are used to assess the extent of impairments and benefits. At the beginning of such consideration processes, an agreement should be reached on defining central terms and reference systems. It is also necessary to clarify who is legitimized to act as a representative of interests and to make social demands on others.

It is also essential to clarify ownership and the responsibilities that derive from it. Where does private ownership of farms begin, and where does it end? A thing is not per se private property, common property, or common good. It is often made so by social rules and norms. For example, water in state-managed public waters is a public good, and water in a supermarket bottle is private. Thus, it is not always the good but also the technical and financial possibilities as well as the political will and power relations that determine whether something is a public good or is declared a private good. Furthermore, with the need to reduce impairments, the political question arises regarding how the polluter-pays principle enshrined in legislation can be applied in agricultural practice.

3.6 Common Good

It is possibly due to the in-depth study of the subject that the author has been left with the impression that in public discourse, the term **"common good"** is used much more frequently than in earlier periods. This may also be due to a general increase in concern that the common good, even in high-income countries, is undergoing a process of erosion rather than prosperous expansion. The term often appears in connection with criticism of a one-sided pursuit of particular interests and insufficient attention to public welfare interests. However, when the term is used, it is rather seldom preceded by the specific meaning addressed when it is used in the respective context. As an antithesis to particular interests not defined in more detail, the term does not provide sufficient clues to distill insights or instructions for decisions and practical action. This book is precisely about the possibilities and limits of **operationalizing** the concepts of the objective intended by the term **"Common good"** in the context of the production of agricultural products.

3.6.1 The Common Good in Historical Retrospect

In retrospect, the concept of the common good appears at central points in many political philosophies and various contexts of justification. Ancient Greek philosophers, including Aristotle and Plato, set out early conceptions of the common good. The common good is what is shared and beneficial for all or most members of a given community or what is achieved by citizenship, collective action, and active participation in politics and public service. The content associated with the term in each case is closely related to the respective conception of political justice. It is far beyond the author's competence and the intention of the book to present the origin of the term and the diverse conceptual contents, which are intended from different perspectives, comprehensively. However, since the term **"common good"** is of central importance both concerning the discussion of the consequences of agricultural production processes and for questions relating to a reorientation of agriculture, some aspects of the use of the term are considered relevant from the author's perspective are outlined and precede the further explanations.

The thesis that society benefits when individuals primarily pursue their interests is still the starting premise of many economists today. This is attributed to Adam Smith, who we have already met in Sect. 2.1 as the founder of the market economy. Professionally, he was primarily active as a moral philosopher. In this field of experience, he realized that many people, especially wealthy people, succumbed to the delusion that material prosperity would make them happy. In his 1776 work, "The Wealth of Nations," he explained how this delusion kept the economy going as everyone tried to get maximum returns from their land or craft. Products are produced and consumed in excess through greed and the desire for maximum revenue. As if by an "Invisible Hand"—an often-quoted metaphor closely associated with the name of Adam Smith—the "wealthy" would distribute the generated goods among the people through the consequential effects of selfish action in such a way that others would also benefit from the increase in goods. If all actors were oriented towards their goods, this would lead to a self-regulation of economic life, an optimal quantity and quality of production, and a just distribution (Rothschild 1994). For the distribution processes to function, there is a need for a system of natural freedom, i.e., a market that is not over-regulated by the state. The state should largely stay out of it because it is accused of being overburdened with regulation and distorting market mechanisms by favoring certain groups. The danger of one-sided preferential treatment of certain actors (corruption) is as relevant today as it was then. According to Smith's ideas, which are still shared by many economists today, the market and the mechanisms for achieving distributive justice and prosperity work better through the driving force of human egoism than through the regulatory attempts of politicians. However, this should not be misunderstood to mean that selfishness, as such, was endorsed by Adam Smith. Instead, he recognized it as an anthropological peculiarity that should be turned into an advantage for the prosperity of nations.

The reasons for the success of the metaphor of the "invisible hand," intended differently by its moral philosophical mastermind than it was subsequently interpreted by many economists, are apparent. The metaphor names a regulatory principle and offers an interpretation without explaining anything regarding the content. This does not detract from its efficacy but may be the precondition for it. The metaphor as an as-if-explanation has a resounding success because it names cause-effect relationships that largely elude comprehensibility and proof. In this way, it relieves those who use the metaphor of the need for further justification. It is suggested that something would fall away for the good of all others if the economic actors have their interests in mind first and foremost. The metaphor corresponds to the symbol of economic liberalism, which advocates giving the economy a "free hand" since it claims to organize distribution better than the state can. In retrospect, it would be more appropriate to focus less on the (distributing) hand and more on the attribute "invisible" as a reference to an obscuring constellation that eludes comprehensibility.

In a historical outline, Piketty (2020) traces the developments of capitalist society. From his perspective, it is particularly significant how the social significance of private property has changed over the centuries. For example, the properties of the nobility were also associated with sovereign duties of protection towards the "subjects" and special powers such as jurisdictions, rights to levy duties, compulsory servitude, etc. The nobility's property corresponded to a mixed private-sovereign relationship. The property was not consistently privatized until the bourgeois revolutions. Freed from sovereign functions, the property was then legally available to everyone. Private property developed extremely unequally, contrary to the promises of equality made by the French Revolution. One of the primary purposes of the now-established central state was to protect private property and, as far as possible, to spare it from taxes.

In England, Jeremy Bentham (1748–1832) and John Stuart Mill (1806–1873) developed the concept of **utilitarianism**, which is still used today in connection with the common good. In the first chapter of his "Introduction to the Principles of Morals and Legislation," Bentham explains the central concept of utility as follows:

> *"By the principle of utility is meant that principle which approves or disapproves of any action according to its tendency to increase or diminish the happiness of that group whose interests are at stake [...]. By 'utility' is meant that quality in an object by which it tends to produce well-being, advantage, pleasure, good or happiness."* (quoted from Bensch and Trutwin 1984)

Utilitarianism is a form of purpose-oriented ethics that appears in different variants. According to Nida-Rümelin (2020), this school of thought credited itself with having freed itself from all intuitions and laid a new, rational foundation of ethical judgment with a universal utility-optimization principle.

> *"This principle is regarded as immediately obvious and indisputable, for it is, after all, obvious that human action pursues only two ends: to increase (one's) pleasure and to lessen (one's) suffering."*

According to this, actions that increase the totality of **well-being** and reduce the aggregate of pain and discomfort are morally right. Conversely, actions that facilitate well-being and increase discomfort are ethically wrong. Utility refers first to the individual's situation in balancing well-being and discomfort. But it also extends to maximizing aggregate total utility, the sum of the well-being of all concerned. Accordingly, utilitarianism propagates an increase in the common good. In doing so, it politically advocates a vision of a paternalistic welfare state whose laws ensure "the greatest possible happiness for the greatest possible number." Utilitarianism is based on some core principles that set it apart from other normative theories. The standard for judging consequences is their objective value in utility. Thus, utilitarianism can also be regarded as the ethical basis of economics (Common and Stagl 2012).

Utilitarianism judges the moral value of an action based on its consequences. Consequentialist ethics thus stands in contrast to deontological ethics, which judges the actions themselves and not their consequences as commanded, permitted, or forbidden. It also contrasts with virtue ethics, in which the character and motivation of the actor are the crucial components. According to utilitarianism, ethically correct actions depend on balancing the good and the pain or harm these actions cause. Accordingly, the consequences of an action should maximize the total utility of all ethical subjects. Actions that increase total well-being and reduce harmful effects are preferable to those that reduce well-being and increase toxic effects. Pleasure is that which increases utility for an individual; harm reduces it. Utilitarianism demands that action be chosen that optimizes the sum of utility. Accordingly, there is no leeway for shaping one's own life or personal responsibility for one's projects or special interests, thus provoking an essential objection to utilitarianism (Nida-Rümelin 2020).

A crucial conceptual contribution on how economic processes can be designed and accompanied in such a way that they serve both the individual and society was presented by Gustav Friedrich von Schmoller (1838–1917). The economy should not be left to the "laissez faire" principle but contained by a social framework. He thus joined the circle of those who paved the way for the concept of the "**social market economy**." This concept goes back to Alfred Müller-Armack (1976) and contains a social and economic policy model aiming to "combine *free initiative with social progress secured precisely by economic performance based on a competitive economy*." In the first instance, the notion of a social market economy merely affirms that market forces are not to be granted unlimited freedom but are constrained by social concerns. This concedes that there are forces at work here that are at odds with each other to a particular (i.e., unspecified) degree and form an arc of tension. An excessively one-sided orientation in one direction or the other suggests that this would be at the expense of the other objective. At the same time, it is expressed that both market forces and the social concerns of society are considered significant. Accordingly, the primary challenge is to balance both concerns.

In modern political philosophy, the common good of the state comes to the fore. A close intellectual association has developed between the common good and prosperity. The prosperity of societies or individuals is primarily measured by material or monetary indicators. The main reference point for determining societal well-being is the economic growth of the respective economy, assessed by changes in **gross domestic product** (GDP). GDP refers to the total value of all goods and services produced for consumption within an economy during one year. GDP is thus a crucial criterion for political success, primarily since economic growth rates are also related to GDP. Accordingly, the political side suggests in a generalized way that when GDP increases, the prosperity of society and its members also increases.

However, for some time now, representatives of various schools of thought have criticized GDP because it also includes services that represent damage and should, therefore, be deducted from GDP. It is argued, for example, that the current calculation practice encourages the destruction of natural capital twice over:

> *"If, for example, it has to be decided whether forest should be cleared for the construction of a new motorway, the advantages of the "alternative" motorway outweigh the disadvantages of the "alternative" forest simply because the ecosystem goods and services of the alternative "forest" are not fully and adequately included in the assessment. So, the highway is built, and the forest is cut down. This increases GDP, but at the same time, forest services such as carbon storage, noise abatement, air pollution control, and recreational space that contribute to our well-being are lost without this loss of quality of life, which is negatively reflected in GDP. On the contrary, engineered structures may compensate for some of the lost ecosystem services: A noise barrier replaces the sound-reducing effect of the forest and a swimming pool functions as a local recreation area. Both measures lead to increased GDP—this time because natural capital has been destroyed."* (Naturkapital Deutschland—TEEB DE 2012).

The few indications should make it understandable that GDP is an inadequate measure of the common good. If used in most countries, the reasons lie primarily in the irrefutable statistical advantage of aggregating domestically produced but heterogeneous goods in a single measure. Since GDP is considered a measure of economic growth, it is the most important statistical variable in national accounts. In addition, GDP is used to compare a country's economic performance internationally. Thus, the value and changes in GDP guide economic actors and, not least, policymakers. Critics object that GDP points in the wrong direction. However, what is lacking in relevant information about the common good can only be revealed by separate and elaborate considerations of individual economic sectors (e.g., agriculture) and only in processes of weighing up advantages and disadvantages. However, there is no data basis for weighing the pros and cons.

From today's perspective, we know that not everyone benefits from market-based activities primarily based on self-interest. Thus, an increase in GDP does not benefit all community members. GDP tells us nothing about the real relations in the distribution of gain. Adam Smith's initial hope that a more equitable distribution of goods would occur if each self-interest was pursued rather than if the state undertook this task has not been fulfilled. Instead, the property relations between different population strata have diverged widely over the past periods (Piketty 2014). Nevertheless, many economic actors continue to assign a quasi-religious meaning and purpose to the market economy (Hörisch 2013). This may also be due to the fact that state intervention in economic and distributional processes, as practiced in the socialist countries of the former Eastern bloc, has discredited itself and, as an alternative economic concept, has an even more deterrent effect on many people than the current injustices in the distribution of goods.

With his theoretical concept of opportunities for fulfillment and the focus on the well-being of people and its various dimensions and influencing factors, the economist Amartya Sen, who was awarded the Nobel Prize in Economics in 1998 for his work on welfare economics, paved the way for the development of holistic welfare standards as early as the beginning of the 1980s. With the "**Capability Approach,**" he provided the theoretical framework that represented an alternative to the common, mainly economically influenced models of thinking about poverty, social inequality, and human development (Sen 1980). Sen's critique was directed against the traditional understanding of welfare economics, according to which well-being was either conflated with prosperity or wealth ("opulence"; e.g., income) or with utility ("fulfillment of desires") and was almost equated with it. A basic idea of his concept is the goal of social development and progress, the enlargement of the realization of chances, and the people's liberties (Sen 2000). Sen was convinced that human well-being (and thus also inequalities, deprivation, and poverty) cannot be adequately captured and mapped by the classical models of thought and measurement methods since this is influenced by far more than just financial or material resources. Community members must be provided with essential resources for their development. The contribution to promoting the common good is also measured in terms of protection against disruptive factors that could counteract the development of communities. Accordingly, adequate nutrition, freedom from preventable diseases, and opportunities to participate in public social life are essential aspects of the common good (Volkert 2005).

Whereas reflections and disputes between social groups had hitherto revolved primarily around questions of fair distribution and equitable access to resources, as well as participation in general prosperity, with the book "Silent Spring" by Rachel Carson (1963), the first dark clouds appeared on the horizon. Initially published in 1962, the non-fiction book is often described as the starting point of a nascent environmental movement. At the latest, with the publication of the report of the Club of Rome (Meadows et al. 1972) on the "Limits to Growth," it became clear, first to a slight but then to an increasing number of world citizens, that the prospects of a further increase in the "prosperity" that could be distributed among a simultaneously growing world population were dwindling. At the same time, there was growing concern about the damage caused by various phenomena due to industrialization, at least for those who did not disregard it. Fifty years later, a relevant part of the world's population still does not want to acknowledge the negative consequences, which, in the meantime, have accumulated into a veritable climate crisis. As discussed below, many people likely do not want to acknowledge the extent of the damage because they fear its recognition could have consequences that run counter to their interests.

Concerns about the dangers of declining prosperity received further impetus from the "Brundtland Report." The report, entitled Our Common Future, was published in 1987 by the United Nations World Commission on Environment and Development, chaired by former Norwegian Prime Minister Gro Harlem Brundtland (World Commission on Environment and Development 1987). The report is best known for defining the term "sustainable development." Since the report's publication, **"sustainability"** has become an essential topic of social communication processes. Hardly any speech directed at the public and addressing the community's interests nowadays does without this term. However, the frequent use of the term has not led to this being reflected in corresponding decision-making and action structures. Various actors have long since developed ideas about what content is best associated with the term from their perspective and how it can be occupied for their interests (Arts 1994). Kaufer (2015) uses the example of the agricultural and forestry sectors to show how private capitalist actors perceive state measures of greening as a deterioration of competitiveness and the conditions for the appropriation of surplus value. He shows how the issue of sustainability is used as a legitimizing cover for underlying conflicts and production-related interests and how influential actors block political change to secure the benefits from the current state or to push change in the direction of their interests. From this, he derives the demand that the development of a critical science and social theory must consider social power relations to be credible and appropriate to reality.

The Western world has elevated individualism and the priority orientation to one's interests to the highest maxim (Sandel 2020). With the focus on one's interests and in the effort to secure one's advantages, many people lose sight of the negative consequences of the selfish view or have to deny them with additional effort. Often, no mental connections are drawn between one's actions and the implications for the community, as if one lived self-sufficiently and was not part of a community to which one had to pay a particular tribute. An eloquent example is the large number of citizens and business enterprises that go to great lengths and trickery to reduce their tax burden to an absolute minimum. It is also unhelpful when powerful interest groups try to gain interpretive sovereignty over socially relevant processes by occupying central terms and concepts and gaining advantages in public discourse. In this way, the areas of conflict smoldering beneath the surface are concealed rather than revealed.

Derivations

Neither Adam Smith's reflections nor the theoretical concepts developed subsequently have been able to provide comprehensive explanations for a prosperous coexistence of social, economic, and individual interests. The situation has proved far too complex, i.e., permeated by too many different interests. It should be noted that it has not yet been possible to reconcile the overarching perspective of the state with the special, because mighty, perspectives of the economy and with the very heterogeneous views of the citizens of the state and to integrate them into a coherent overall concept. The simultaneity of positive and negative effects of economic and social developments produces an ambivalence that can only be accessed through comprehensive weighing processes within defined system boundaries. The different and sometimes opposing interests and the conflicts that develop from them are highly context-dependent. General rules cannot clarify them; they can only be explained by well-founded analyses of the conflicts' context. The continued search for general laws and regulations about procedural processes is a "business-as-usual" strategy (see also Sect. 7.4.2). Not only does it not lead to conflict solutions, but it also prevents the specific conflicts from being examined in depth. Furthermore, there is a lack of methodological procedures for assessing the areas in which there are overlaps and a congruent mix between the particular interests and the interests of the common good.

In the initial conceptual considerations concerning the interests of the common good, the focus was still on achieving a more equitable participation of all in the existing prosperity, which is increasing through economic growth. Given the limited resources available to distribute and the extent and increase in global harmful effects, today's leading issue is the protection of interests and increasingly offensively represented claims to appropriate participation. The existing conflicts of interest are the starting point of covert or openly fought distribution struggles. However, the "pressure in the cauldron" inevitably increases without adequate countermeasures. This is metaphorically and factually accurate if one realizes we are all exposed to global warming and its associated consequences. The abovementioned interrelationships should have clarified that we must abandon the idea that the "free" market economy will solve the problems. The problems do not solve themselves by invisible forces but require context-dependent regulation, as Ostrom and her colleagues have pointed out. So far, however, the complexity of the processes and the difficulties of penetrating them intellectually stand in the way of regulation, not to mention the willingness to allow oneself to be restricted in one's scope for action.

3.6.2 Agriculture and Common Good

From the perspective of many farmers, agriculture contributes to the common good simply because it provides people with the **aliments** they need to survive. "***We'll feed you!***" is the slogan that farmers have been carrying on banners at demonstrations for some years now. It is, as it were, the counter-slogan to what the critics of the intensification processes in agriculture chant at their now traditional annual demonstrations in Berlin: "***We've had enough!***" This addresses the oversaturation with adverse external effects of the agricultural production processes and outlines the field of tension that will be illuminated and reflected upon from various perspectives in the following section. On the one hand, there is the food supply, on which all people are indispensably dependent. Consumers need food, and farmers have it in abundance. In return, farmers need adequate product remuneration to cover production costs and ensure economic viability. This has not been adequately provided for some time now. Many farms are draining their assets by borrowing money from the banks, with the prospect of ever being able to pay it back increasingly dwindling. As a result, many farmers still in business are threatened with the same fate as many of their former competitors. The obvious question arises: Why are food users denying farmers what they desperately need: full-cost market prices? Since we are all food users, the question is addressed to everyone. One may be inclined to give simple answers to it. As always, however, the reality is much more complex than first assumed. Instead of drawing hasty conclusions, we must break the issue down further.

From the opponents' perspective of an agro-industrial form of agricultural production methods, farmers are responsible for the extent of the damage and destruction caused by these methods. The focus of their attention is the interest in preserving protected goods, which are perceived as common goods, from impairment by agricultural production processes. The opponents are concerned about the adverse effects on the climate, the environment, nature, livestock, and consumers. Prioritizing the goods to be protected can differ significantly depending on individual access and personal concern. What all opponents have in common, however, is a general doubt about the harmlessness of agricultural production processes for concerns that the critics articulate and demand with a claim to represent the interests of the common good.

Both the producers of food and the self-proclaimed protectors of the commons are driven by needs, albeit very different ones. Both sides see themselves as part of a community of like-minded people and are simultaneously members of overarching communities as they exist within communal, regional, and national boundaries. Both sides are engaged in a conflict of interests about limitations on future options *(capabilities),* as they were thematized and reflected by Amartya Sen for the common good (see Sect. 3.6.1). **Some see themselves as deprived of their future options by the demands of others.** The demands for maintaining the existing agrarian economic system are incompatible with the demands for drastic changes. Agriculture appears to have been trapped in this dilemma for decades, without agricultural policy having yet been able to resolve it. The conclusion is evident that the conflict constellations that have manifested themselves over long periods cannot be mitigated or even determined by a one-sided patronage policy. What is needed is an overarching perspective that sufficiently acknowledges the interests of all those involved but readjusts and weighs them up with overriding public interests.

When talking about the common good in the context of agriculture, it is essential to clarify which interests and communities of interest are involved here and to what extent. Furthermore, the question arises as to what benefits or would benefit the individuals and the members of a community and what stands in the way of this. A community of interest consists of a variety of individuals with different interests. In addition to the individual interests, there are more or less significant overlaps. These may be explicitly formulated or interpreted rather implicitly as individuals' sense of belonging to a group or community of interest. Accordingly, disputes and competitive behavior also occur continuously within communities.

Agriculture is very diverse and is associated with various tasks and functions. This makes it difficult to gain an overview. Despite all the complexity of the processes and effects that emanate from agriculture, farmers see themselves first and foremost as entrepreneurs who manage an agricultural enterprise. Given the heterogeneity of the conditions under which they operate their enterprises, it is, therefore, difficult to reduce to a common denominator the interests that shape the decisions and respective actions of individual entrepreneurs. Despite all their differences, the agricultural enterprises nevertheless have one thing in common: as entrepreneurs, they are all players in a free market economy, and—whether they like it or not—they have to adapt to the requirements of the market economy to ensure their economic viability. This applies also to organic agriculture, which initially might evoke a different impression. Here, too, the laws of the free market economy dominate. Organic farming is therefore predestined for clarification, taken up later, on how far one can get with ecological ambitions within the free market economy or how small the scope is under the prevailing economic framework conditions.

Above all, these conditions enclose competition. Farmers are competing not only on a global but also on a national and regional level. They compete for access to resources, including access to leased land. The more limited resources are in demand, the higher the price rises. Similarly, the more products are produced and offered for sale in an already saturated market, the more dramatically market prices fall. Although farmers are economic competitors, they also form a community that pursues common interests. These consist, not least, of ensuring that the profession's interests do not take a back seat to other interest groups. Various agricultural association structures have developed in the past that influence opinion-forming and political decision-making processes. However, they are not alone in trying to exert influence. The numerous enterprises and their employees that operate in the upstream and downstream areas of agricultural production processes and profit significantly from agriculture have no interest in anything changing to their disadvantage. In addition, consumers also have a say in the matter. The list of stakeholders involved is thus far from complete. Their respective roles and interests are reflected in more detail in Chap. 6.

The concepts of the common good outlined above are characterized by considerations of how the available goods or resources can be distributed more "justly." One of the core questions revolves around what is understood by "just" in the respective context and how a just distribution can be operationally translated into concrete actions. Is it just if all members of a community to be defined in advance have equal access to a resource, or is it more just if the individual members have access to the resources according to their respective needs, or if they are allocated them accordingly so that they can better realize themselves and thereby achieve well-being? Or is it only the free market system that creates just conditions because those who perform more also earn more in a double sense? It is obvious that at least two different perspectives are involved here: the perspective of the givers, i.e., those who have resources or the power to distribute them according to their standards or to withhold them from others, and the perspective of the recipients of resources, who make their own needs the yardstick for their wishes or demands. The phrase "I give so that you may give." describes reciprocity as a primary strategy of social behavior, just as "one hand washes the other." However, the principle of reciprocity makes transfers quite confusing. In addition to the perspective of givers and takers, another perspective is added, which was essential for subsidence farming but has been lost from view by many actors in subsequent developments. This refers to the overarching perspective, which looks at the functioning of the community. It looks from a certain distance and a higher perspective (top-down) at how **allocating** resources should be better organized to consider particular and public interests. In the various distribution processes, the question arises as to the frame of reference and the standards and criteria used to measure what is available for distribution, the degree of need of individuals, and the primary objectives of the allocation. In short, how can what is available be appropriately used to achieve overarching goals?

However, it is not only the allocation of resources that matters. Of no less importance are the adverse external effects on public interests caused by agricultural production processes and the relationship between positive and negative external effects, respectively. These are addressed comprehensively in Chap. 6. In this regard, it should first be noted that farmers have not yet been charged for farming at the expense of others. **Externalization** even functions as a competitive advantage when the yield-enhancing effects of intensification processes are of private benefit, and the external adverse impact is charged to the community (e.g., increased emissions of pollutants). Nevertheless, very many farmers are in a precarious economic situation. As long as the costs of the burden on the commons are not charged to farmers and thus internalized in production costs, it cannot be expected, given the overall constellation, that they will be adequately incorporated into farmers' decisions and actions. An elementary prerequisite for offsetting is the recording, on an individual farm basis, of the damage and harmful effects that farmers cause to the community. Only then can these be set against the benefits that accrue to the community, primarily through food production. There is no shortage of proposals for measuring the extent of the harmful effects so that they can be charged to those who cause them. Even if these are not yet methodologically fully developed—as is often argued by interested parties in the agricultural industry—methodological issues do not constitute the core problem. There is simply a lack of willingness to face the challenge of internalizing adverse external effects. Neither the executive nor the legislative nor the judiciary has shown any preparedness to act in this matter, which fundamentally impacts the common good. As long as the executive, in the form of the agricultural ministries of the federal states and the federal government, acts as a protective power of the agricultural industry and does not act as a protector of the common good, there is no containment of the further increasing impairment of the common good by agricultural enterprises in sight.

Derivations

As has been briefly outlined, socially relevant questions and problems linked to a fair distribution of resources and the harmful effects of agricultural enterprises are not a phenomenon of individual farms but a systemic problem of the rural economy. All the more reason to ask how the corset of the market economy framework in which agricultural enterprises are constrained can be broken. This seems necessary to ensure a just distribution of recourses and to contain the negative consequences through regulation so that social—and no less critical—ecological concerns can be better met. At the heart of the impairments to the public interest are, on the one hand, severe conflicts of interest and, on the other hand, the ambivalence of both positive and negative consequences, which emerge between the agricultural enterprises in a very heterogeneous and thus not generalizable form. The existing confusion, which plays into the hands of the current beneficiaries of the status quo, can only be countered by quantifying the individual farms' positive and negative external effects.

3.6.3 Agronomic Perspective on the Common Good

From an economic perspective, monetary value creation is at the heart of a financial enterprise. It is realized through the activities of the company and its employees and is measured as the difference between the market value of the goods and services produced by the company and the costs of these goods, the materials procured from other producers, and the intermediate inputs provided (Chernatony et al. 2000). This calculation measures the contribution of different producers to the total value of the enterprise's output only through the cost of purchased inputs. In macroeconomics, the added value is calculated by looking at the output produced domestically through the use of domestic and foreign factors of production by adding up the contributions of individual enterprises to the gross domestic product (GDP). In this way, the added value data provides information on the contribution of an individual industry or enterprise to the overall performance of the economy. This approach is based on the assumption that the higher the added value, the more effectively and efficiently a company can serve the market and respond to the needs of its customers (Porter 2004). Given the focus on monetary value creation in and for the company, issues such as "common goods" and "public welfare interests" must inevitably take a back seat. This is no different in agricultural enterprises. Nevertheless, some fundamental differences in industrial enterprises must be discussed in more detail.

In an expert opinion on an agricultural policy oriented towards the common good, the Scientific Advisory Council on Agricultural Policy of the Federal Ministry of Food and Agriculture, which is primarily composed of agricultural economists, took up the issue of the common good with a view to the future Common Agricultural Policy in the EU (Scientific advisory board of BMEL 2018). The Advisory Council emphasizes the multifunctionality of agriculture, which produces food and biomass for energy and material use. In addition to the services rewarded via the market, it shapes cultural landscapes and fulfills social functions in environmental, climate, and animal protection. However, the services in the latter areas are not specified in more detail. In the view of the Council, a dilemma arises if "private goods," such as food, are reimbursed via the price mechanisms of the market. In contrast, the market does not reimburse other goods or socially relevant services. A regulatory order about the fulfillment of social functions would increase production costs due to the associated expenses. Since many EU Member States are active in global markets and domestic market prices are determined mainly by the price level of the world market, primary producers in the EU would be placed at a competitive disadvantage by requirements to promote socially relevant goods.

> *"In the EU, the declared aim is, on the one hand, to shape agriculture within the framework of internationally integrated product markets, but, on the other hand, not to participate unreservedly in a global competition in which those who demand the least costly social services from agriculture are the most successful. In other words, we want to be integrated into world markets regarding our product markets, but we want to decouple ourselves from global animal and environmental protection levels."*

Whether this is an additional dilemma or one associated with the one described above can be left open. It is indisputable that a central dilemma of the agricultural economy is described here. On the one hand, one wants to participate in global competition, if only to be able to sell the national surplus quantities; on the other hand, one is thereby at the mercy of undercutting the competition, which is at the expense of animal welfare, environmental protection, and other public welfare interests, which one would also like to avoid if possible. In such a dilemma, where two incompatible objectives are pursued, one is reminded of the classic example of Buridan's donkey. The latter stands between two precisely equal haystacks. Since he cannot choose either for lack of a plausible reason to eat from the left or the right pile, he eventually starves to death. Like all parables, this one has limited applicability to the issue. Nevertheless, it provides clues for reflecting on the position of agricultural economics based on the analogies and the differences between the parable and the current dilemma.

As a rule, dilemmas are characterized by the fact that both options available for a decision in a situation lead to an undesirable result. Buridan's donkey cannot decide because it lacks a distinguishing characteristic. However, the donkey can choose between two positively evaluable options to secure self-preservation, provided it can favor one option. In contrast, the situation described by the agronomists does not involve two equal and positively valued options. Participation in world markets has a positive connotation because it is associated with economic value creation. The goal of "*decoupling from global animal and environmental protection levels*" implies the avoidance of undesirable and, thus, adverse external effects of the production processes, which should be limited as far as possible. However, this isn't easy to achieve because it entails additional costs and thus comes at the expense of participation in global markets. In the trade-off between a positive goal (value creation) and an opposing goal (expenditure of costs to avoid the impairment of common goods), costly avoidance goals are naturally complex. Contrary to the situation in the parable, however, in which the weighing of two options produced a stalemate, the Scientific Advisory Board spares itself the time-consuming but crucial weighting process. As a result, it does not even get into the embarrassment of a possible outcome that has left Buridan's donkey starving because—fixated on the internal perspective—it has been unable to make up its mind. Had the donkey been able to take an independent position and view the dilemma from the outside, the resolution would have been simple.

When deciding between two options, it is relevant to how different their dimensions are and who benefits most from each option. In the parable of Buridan's donkey, the donkey benefits in both cases if it can only choose between the two haystacks of equal size and distance. In the agricultural dilemma reflected here, agricultural economics is called upon to decide whether to prioritize promoting market activities or the interests of the common good or seek standard solutions to the different problems that agriculture faces. At stake are not only different dimensions but, above all, other interests. The question is, therefore, which of the respective options will primarily benefit which interests, and who will be disadvantaged by them? Unless a deliberation process is undertaken, there are no reasons for changing economic structures. In a preliminary assessment, participation in the global marketplace mainly benefits those who profit from exports, i.e., the processing and trading companies. Although tied to fierce competition, retailers have various options to purchase raw goods as cheaply as possible at world market prices and sell food products with the highest possible margin.

Consumers are also among the supposed winners. However, they are also citizens of the state. As such, they are among the losers because the undesirable side effects of the production processes are at the expense of common goods and thus run counter to public interests. Seen in this light, people also find themselves in a dilemma and thus in a "donkey role" because they have to choose between the role of consumers and the benefits of allegedly low food prices and the role of citizens, which expects them to work to avoid harming the common good. Most people will likely lack the basis for decision-making and expertise to reach a balanced judgment. Therefore, it is understandable that many people opt for the advantageous consumer role.

Those whose interests are barely mentioned, let alone considered in the Scientific Committee's opinion, are the primary producers. Given the market situation, primary producers can only profit if world market prices are high enough to more than cover their production costs. Because for many farms, a profit situation has not existed for long periods and cannot be expected in the foreseeable future, most primary producers are among the losers. They are not helped by export orientation and participation in global undercutting competition, nor do primary producers benefit from subsidies paid out of public funds for the implementation of animal welfare and environmental protection measures if, at the same time, ruinous competitive conditions are maintained. Suppose agricultural economists start from a dilemma between integration into world markets and the goal of decoupling from global animal and environmental protection levels. In that case, this dilemma primarily affects primary producers. These are in a quandary, as both options lead to hopelessness for most of the primary producers still in business. Under the current framework conditions, neither option yields sufficient profit to secure their economic existence in the long term.

Unfortunately, the Scientific Advisory Board has failed to take a closer look at the role of primary producers or has not considered it sufficiently necessary. They are not only caught in a dilemma; they also have to deal with the **conflict of objectives** between different interests (low-cost food production and avoiding undesirable side effects). Who else is to take on this task? A solution or at least a containment of the conflict between different interests can only be achieved where conflicts are caused within the operating system. The Council is silent on market events' effects on internal production processes and the causation or avoidance of undesirable side effects. This is understandable insofar as the participants would then have had to reflect on the role of the "free" market economy and the theoretical edifice of agricultural economics and possibly question the prevailing **theories**.

As a rule, a dilemma can only be resolved if both goals, considered imperative but incompatible, are viewed from a superior and distanced perspective. Above all, one cannot avoid carrying out comprehensive analyses and presenting well-founded arguments that make the pros and cons for one or the other option visible. Only this preparatory work makes a well-founded weighting possible, i.e., one supported by plausible reasons. Such a weighing-up would also offer the opportunity to identify the conflict areas and the common intersection between options. If such a procedure is avoided, there will be no discussion of the potential of modifying both options to increase the common intersection. The Council recommends that agricultural policy, to promote the social functions of agriculture, should set the framework so that the entrepreneurial activities of farmers simultaneously serve the common good. However, the question of how this could be operationalized and set in motion without a well-founded analysis of the conflicting objectives is not even raised. Instead, the Council's explanations remain trapped in agricultural economic explanatory approaches.

This applies in the same way to the ideas expressed about the common good from the perspective of agricultural economics. The Scientific Advisory Council 'Agricultural Policy' defined **services of general interest** as services,

> *"which are provided voluntarily by self-responsible entrepreneurs, such as farmers, or by the rural population and which benefit the general public in such a way that individuals cannot be excluded from using them. The voluntary nature of services implies that it is only possible to speak of services for the common good if they go beyond the scope and/or quality prescribed by regulatory law. Thus, mere compliance with regulatory standards does not constitute a public service but rather a duty to avoid a burden on the public good. Non-excludability from use implies that public welfare services have the character of a public good. Thus, for example, producing marketable agricultural products is not a public good." It says, "Despite the lack of remuneration via the market, not all public good services are worthy of financial reward. Government support for public good services can be justified only when they are in short supply."*

It is easy to see that the definition of the common good is embedded in an economic theoretical framework. Even if the theoretical edifice shapes the explanatory approaches, this does not mean these are fully explained. Economic theory building offers several perspectives, albeit one that is very powerful. Regulatory law, non-excludability of use, and scarcity form central pillars of the agrarian economic definition of "common goods." If one follows this ductus, the food produced by farmers, as long as it is not scarce, is explicitly not a common good. Food production in Europe is far from scarce. It is precisely the surpluses of food produced by which primary producers spoil the market prices. However, the exclusion of food production from the services of public interest is unlikely to please the farmers who are protesting in the streets for more recognition and respect from the population because they create the existential basis of life for the population by producing food. From the point of view of agricultural economics, this is merely a private matter.

Agricultural economists distinguish between "non-divisible" and "divisible" goods. This definitional classification leads to food being assigned to the "non-divisible" private goods since several people cannot simultaneously possess or consume it. This approach implies that the state should withdraw from the private sphere of food provision since food is not a common good. In contrast, **public goods** such as the environment are characterized as divisible and thus a common good. However, it can be seen that the use of the environment as a landfill by many businesses enriches the common air and water with pollutants. Consequently, everyday use harms others and limits clean water and air access. These indications conclude that the division into divisible and non-divisible goods, originating from economic thinking, does not contain any utility value for dealing with public interests.

This probably has to do with the fact that the utility theorem of economics is primarily reduced to the utility and the utilization of resources for a specific interest group as the governing maxim. With this approach, the complexity of economic processes and interactions with other social concerns can hardly be comprehensively grasped. Neither the antagonistic and synergistic interactions nor an overarching whole, commonly called the "polity," find their proper place. Instead, agricultural economists use regulatory law as a reference for assessing services to the common good. Services must go beyond what is prescribed by regulatory law and be available in a limited way to claim recognition and, if necessary, remuneration. This implies that all areas not regulated by regulatory law cannot be counted as public services. On the other hand, if services of general interest are defined as exceeding what is specified by regulatory law, this means, by implication, that all services of general interest that are provided voluntarily by farmers over and above compliance with regulatory law grant farmers the right to claim public funding for these services. This would justify farmers receiving further payments from the public purse and the already substantial subsidy payments for the so-called "services" that farmers provide for the common good. What rational person can object to someone also receiving a reward of public money for a service that appears to benefit everyone? If, however, farms, on the one hand, contribute to the common good but at the same time significantly impair the common good in other places, it becomes difficult, and not only for agricultural economics, to take this into account in an appropriate way and to set it off against each other.

Derivations

The sketched position of agricultural economics on the common good exposes this discipline as a prisoner of its explanatory approaches. Suppose one follows the explanations of Maréchal et al. (2008) regarding the role and significance of agricultural economics; in this case, there is no contradiction between economic theory and the pursuit of individual benefit. Accordingly, one should not expect a discipline that explicitly advocates the maximization of particular interests to come up with any ground-breaking concepts on how the interests of the common good can be adequately considered in food production. The approach of defining the additional services provided by farms beyond a pre-existing norm as a contribution to the community's interests is captivating in its simplicity. The statutory minimum requirements act as a reference system; the degree of deviation from the norm provides the standard of assessment for this purpose. According to this, a significantly higher level of minimum requirements is of higher value than a slightly higher requirement profile. At the same time, the respective requirement profiles define **"quality standards."** Agricultural policy is then tasked with agreeing with as many stakeholders as possible on the standards that should be followed. If a considerable number of groups support these, or if possible by all of them, we will gain a certain legitimacy for setting standards and a justification for public funding if these are complied with. At the same time, the question of whether increased standards can solve various problems and conflicting goals in connection with food production is pushed into the background. To adopt a multi-perspective approach, these and other questions will be addressed in detail below.

References

Arts B (1994) Dauerhafte Entwicklung: eine begriffliche Abgrenzung. Peripherie 54:6–27

Bensch R, Trutwin W (eds) (1984) Philosophisches Kolleg 3: Ethik. Patmos-Verlag, Düsseldorf

Chernatony L de, Harris F, Dall'Olmo Riley F (2000) Added value: its nature, roles and sustainability. Eur J Mark 34:39–56.https://doi.org/10.1108/03090560010306197

Common M, Stagl S (2012) An introduction to ecological economics. In: Common M, Stagl S (eds) Ecological economics: An introduction. Cambridge University Press, Cambridge, pp 1–18

Gordon SH (1954) The economic theory of a common-property research: the fishery. J Political Econ 62:124–142

Gronemeyer M (2009) Die Macht der Bedürfnisse: Überfluss und Knappheit, 2nd edn. Wiss. Buchges, Darmstadt

Hardin G (1968) The tragedy of the commons. Science 162:1243–1248

Hörisch J (2013) Man muss dran glauben: Die Theologie der Märkte. Fink, Paderborn

Illich I (1982) Vom Recht auf Gemeinheit. Rororo, vol 4829. Rowohlt, Reinbek bei Hamburg

Kaufer R (2015) Umsetzung von EU-Umweltschutz in der deutschen Land-und Forstwirtschaft. Dissertation, Georg-August-Universität Göttingen

Meadows DL, Meadows DH, Zahn E, Milling P (1972) Die Grenzen des Wachstums: Bericht des Club of Rome zur Lage der Menschheit. DVA, Stuttgart

Müller-Armack A (1976) Wirtschaftsordnung und Wirtschaftspolitik: Studien und Konzepte zur Sozialen Marktwirtschaft und zur Europäischen Integration, 2nd edn. Beiträge zur Wirtschaftspolitik, vol 4. Haupt, Bern

Nida-Rümelin J (2020) Eine Theorie praktischer Vernunft, 1st edn. de Gruyter, Berlin/Boston

Nowak MA, Highfield R (2011) SuperCooperators, Altruism, Evolution, and why we need each other to succed. Simon & Schuster Inc., New York

Ockenfels A (2020) Größtes Kooperationsproblem der Menschheitsgeschichte. Forschung & Lehre 27:906–907

Ostrom E (1999) Die Verfassung der Allmende: Jenseits von Staat und Markt. Die Einheit der Gesellschaftswissenschaften, vol 104. Mohr Siebeck, Tübingen

Ostrom E (2009) Beyond Markets and States: Polycentric Governance of Complex Economic Systems: Prize Lecture, December 8, 2009
Piketty T (2014) Das Kapital im 21. Jahrhundert. Beck, München
Piketty T (2020) Kapital und Ideologie. Beck, München
Porter ME (2004) Competitive advantage: creating and sustaining superior performance, 1st edn. Free Press, New York, Business
Carson R (1963) Der stumme Frühling. Biederstein Verl, München
Rothschild E (1994) Adam Smith and the Invisible Hand. Am Econ Rev 84:319–322
Sandel MJ (2020) The tyranny of merit: what's become of the common good? Picador, New York
Scientific Advisory Board of BMEL (2018) Für eine gemeinwohlorientierte Gemeinsame Agrarpolitik der EU nach 2020; Grundsatzfragen und Empfehlungen. https://www.bmel.de/SharedDocs/Downloads/DE/_Ministerium/Beiraete/agrarpolitik/GAP-GrundsatzfragenEmpfehlungen.html
Sen A (1980) Equality of What? In: McMurrin SM (ed) The Tanner lectures on human values: Volume 1, 1st edn., pp 195–220
Sen A (2000) Ökonomie für den Menschen: Wege zu Gerechtigkeit und Solidarität in der Marktwirtschaft. Carl Hanser Verlag; Hanser, München, Wien
Service ER (1966) The hunters. Foundations of modern anthropology series. Prentice-Hall, Englewood Cliffs, N.J
Volkert J (2005) Einführung: Armut, Reichtum und Capabilities – Zentrale Inhalte, Begriffe und die Beiträge dieses Bandes. In: Volkert J (ed) Armut und Reichtum an Verwirklichungschancen: Amartya Sens Capability-Konzept als Grundlage der Armuts- und Reichtumsberichterstattung. VS Verlag für Sozialwissenschaften, Wiesbaden, pp 11–19
World Commission on Environment and Development (1987) Our common future: The Brundtland Report. Oxford Univ. Press, Oxford

Utilization of Livestock

4

Abstract

The utilization of livestock for food production and economic value creation has occupied an ever-greater space since the Second World War. Significant productivity increases occurred based on increased performance's biological potential, primarily tapped through targeted breeding and nutritional measures. In addition, technical developments create the precondition for drastically reducing the labor time expenditure per product unit and significantly expanding the transport, processing, and distribution possibilities. Production processes are also undergoing fundamental changes due to changes in economic conditions. The interlocking of intensification processes brings about productivity increases of hitherto unknown proportions and a considerable degree of undesirable side effects. These mainly affect the farm animals. They pay not only with significant restrictions in space allowance but also in their options to exercise species' specific behavior. Above all, they are confronted with a high degree of production diseases, accompanied by pain, suffering, and damage, indicating an overstressed ability to adapt to suboptimal living conditions. To grasp occurrence, relevance, and potential solutions, the driving forces behind the productivity increases and the development of undesirable side-effects need to be understood.

Since time immemorial, people have used animals for their purposes. They serve primarily as food but also as manure suppliers, work animals, and property, representing a considerable exchange value. Who can deny that agriculture and livestock farming are vital to society? Given the quantities of food produced annually by agriculture and the efficiency of its production, the particular success story of agricultural entrepreneurship is obvious. With the advent of industrialization, farm animals were essentially reduced to the function of production factors through which additional income could be generated.

A. Sundrum, *Public Welfare-Oriented Production of Food*,
https://doi.org/10.1007/978-3-662-69040-6_4

The fact that agriculture has a negative side and is ambivalent at its core has already been indicated at various points but needs further illumination. Before going into more detail on some of the adverse effects of agriculture, it is crucial to recognize the recent production performance and productivity gains.

For a long time, the concern for sufficient food availability dominated many people's thoughts and still does today in many world regions. Poor soil conditions, unfavorable weather conditions, and the constant danger that other people could dispute one's harvest made food procurement a priority. Experience taught people that it was important for survival to be prepared and equipped to deal with contingencies. The biblical image of paradise, in which there was no lack of anything and especially not of food, embodies the place of longing and thus the counter-image to people's own experiences of lack, to which they were and still are exposed to varying degrees.

Historically, the lack of food was experienced, especially during and after the Second World War. Vast sums of money had to be paid for food, and many families had to sell their silverware to be able to provide their family members with food. After the war, increasing food production was at the top of the political agenda. The desolate supply situation had to be improved as quickly as possible, and simultaneously, after all the political turmoil, the political loyalty of the population had to be won. Since there was a food shortage and means of production, the available resources had to be used as efficiently as possible. Wherever this seemed likely, farmers were supported by politics in expanding production. It was always a question of farming with more output and less input. Also, in the case of livestock, the aim was to make their life processes as efficient and profitable as possible. In this context, it was not so much individual but rather the interlocking of many process steps that gradually increased productivity.

In the realm of agricultural business, animal-sourced food (ASF) holds a significant position. In numerous countries, livestock production accounts for the largest percentage share of net product. However, ASF is traded globally as a commodity, so livestock production is subject to intense global and price-focused competition. This economic landscape, in turn, leaves little room for farm management actions that are not centered on economic efficiency. The biological, technical, and financial aspects are further explored below to shed light on the factors that primarily contribute to the increase in productivity and the emergence of undesirable side effects.

4.1 Biological Developments

The true locus of change and economic success in livestock farming lies within the animals. The surge in productivity in the latter half of the twentieth century, achieved through systematic and one-sided breeding, tailored nutrient supply, and accompanying veterinary measures, was unprecedented. However, this progress comes at a cost, particularly for farm animals regarding animal health and welfare (AHW). A highly complex issue, AHW is best defined by the Terrestrial Animal Health Code of the World

Organization of Animal Health, which provides an international standard for farm animal welfare (Office International des Epizooties 2008). One hundred seventy-four countries have ratified it and thus can claim to represent the highest level of common sense achieved in terms of AHW. It was agreed that "*animal welfare means how an animal is coping with the conditions in which it lives. An animal is in a good state of welfare if (as indicated by scientific evidence) it is healthy, comfortable, well nourished, safe, able to express innate behavior, and if it is not suffering from unpleasant states such as pain, fear, and distress. Good animal welfare requires disease prevention and veterinary treatment, appropriate shelter, management, nutrition, humane handling, and humane slaughter/killing. Animal welfare refers to the state of the animal.*"

The definition refers to the freedom of severe perturbations and diseases as an essential but insufficient prerequisite of AHW. While severe perturbations and diseases allow a valid statement about the absence of AHW, the absence of apparent perturbations and diseases does not allow a statement about the state of physical and mental health. This is in particular because a varying but, on average, a high proportion of farm animals suffer from inapparent disturbances and disorders, which only become apparent when making efforts to use appropriate diagnostic tools. Although often disregarded in farm practice and revealed only in scientific studies, they can considerably impair AHW. Safeguarding the health and welfare of farm animals and preventing animal diseases protects public health, animal production, and rural economies. As such, AHW can be considered a relevant public good (Degeling and Johnson 2015).

4.1.1 Dairy Cows

In extracting milk from the mammary glands of cows, livestock farmers took advantage of the fact that evolution has produced animals that prioritize the care of their offspring. For many millennia, man has been diverting the cow's milk, which is created to maintain and grow the calf and is tailored to the calf's needs in terms of its ingredients, for his purposes. There was and still is a mutual relationship of dependence between livestock farmers and dairy cows, which offers advantages for both sides. Humans provide food, protection from predators, and the rearing of offspring; in return, the cow provides milk and, at the end of her life, meat and hide, which can be processed into leather. An analogous mode of life has also evolved between an ant species and aphids. One might, therefore, be inclined to regard the relationship between humans and dairy cows as a natural way of life. However, the relationship has gradually changed from a mutual to a parasitic form of life, in which humans live at the expense of farm animals as they demand more than is good for them.

The decisive key for the continuous increase in performance was an increasingly supra-regionally organized evaluation and assessment of the animals' performance and the homogenization of the animal breed (Comberg 1984). Breed-related breeding associations increasingly took over the processes of cattle breeding. Strategic breeding, in the sense of mating a male and female animal to obtain offspring with defined characteristics, was based on the knowledge regime of **animal breeding**. Whereas previously, mainly aesthetic criteria were considered when selecting breeding animals, measurable performance characteristics dominated from now on. Farmers hoped that the data would also provide them with information on the potential for optimization within the farm. The increasing competition among farmers helped to overcome the initial skepticism towards external performance monitoring. From generation to generation, attempts were made to help desired traits become more pronounced while at the same time repressing undesirable characteristics.

Nevertheless, not every breeding development and every technical innovation was immediately approved by the farmers. Initially, this also applied to artificial insemination. The development of freezing technology made it possible to preserve the semen collected from bulls in newly built insemination stations more or less indefinitely and to use it for specific mating using artificial insemination. Many farmers rejected this "anti-natural" technique because they feared, among other things, the uncertain long-term consequences of the intervention for the animals. Established cattle breeders, who in the past achieved considerable sales revenues from the sale of breeding bulls, saw the sales of their bull calves swimming away and therefore warned against the risk of excessive inbreeding. However, these objections could not stop the triumph of artificial insemination. With the economic success due to the savings in keeping and feeding individual bulls and the access to the genetics of high-performance animals, all concerns quickly receded into the background. Skepticism was replaced by enthusiasm for animals with memorable performances, which sometimes took on the characteristics of a religious cult. Artificial insemination broke the traditional space and time limits of breeding work and, at the same time, gave an enormous boost to the internationalization of cattle farming. Artificial insemination rapidly increased the importance of individual males. Pabst Ideal was the name of a Holstein–Friesian bull imported from Wisconsin in 1964 and made famous in Europe, the first sire to produce 100,000 offspring by artificial insemination (Thiede 1988).

In addition to the significant advances in breeding, artificial insemination was also a means of containing and overcoming the sexually transmitted diseases of the animals during natural mating. These could cause considerable financial losses to individual farms if the diseases transmitted during mating caused the cows to become infertile. In communities with a high prevalence of the disease pathogens, artificial insemination was, therefore, even ordered by the official veterinarian. The use of artificial insemination was associated with new management requirements. Farmers must recognize the fertility time of the female bovine to order the insemination. However, as animal performance increased, the expression of estrus symptoms decreased. As a result, time-consuming observations and veterinary interventions were often required to maintain the fertility of dairy cows.

Breeding and nutritional measures to increase milk yield showed remarkable success. Between 1950 and 2020, the average milk yield per cow per year increased more than threefold, with considerable variation between farms and regions. The peak values are at a completely different level. Today, the top farms achieve an average yield of more than 14,000 kg per cow per year. In 2017, the world record was held by a cow named "My Gold," which was kept on a farm in Wisconsin (USA) and produced an unimaginable amount of 35,144 kg of milk in one year (Baumgard et al. 2017). For the continuous increase in the milk yield of individual animals, the size and conformation of the mammary gland are of great importance. The mammary gland in cattle consists of four gland complexes with cavity systems in which many milk-forming cells (lactocytes) are located. These are supplied via the blood with the initial substrates required for milk formation. It was observed early on that a cow with a giant udder also gave more milk, provided that the physical functions were not impaired and the appropriate nutrient supply was ensured. Indeed, a close correlative relationship ($r = 0.8–0.9$) exists between milk yield and the number of lactocytes (Märtlbauer and Becker 2016). The purposive selection of dairy cows with high milk yield resulted in offspring with increasingly larger glandular complexes. The numerical increase in the number of lactocytes created the morphological condition of the rise in the productivity of the individual cow. The productive capacity of the mammary gland is so impressive that Brown suggested as early as (1969) that the cow should be regarded as an appendage of the mammary gland and not, conversely, the mammary gland as an appendage of the cow.

This assessment is also justified because, after the birth of a calf, the cow's organism adapts to the new situation with far-reaching physiological and hormonal changes. In this process, the mammary gland is supplied with nutrients as a priority over other organs. This means that the nutrients absorbed from the feed are used primarily for milk production. Suppose the intake of nutrients is too low in relation to milk production. In that case, the organism switches the **metabolism** to an emergency program: The fat and muscle tissue of the cow is broken down to obtain the necessary substrates for milk production. This process is not as efficient as the direct use of feed nutrients and causes waste products that burden the metabolism and can directly endanger the cows' health. At the same time, the supply of nutrients to other tissues is throttled, so they must resort to substitute metabolism and may lose their full functionality.

Livestock farmers exploit the evolutionary reorientation of metabolic processes in mammals after birth towards the prioritized supply of the offspring to deprive dairy cows of many times the amount of milk required to supply a calf. Breeding for animals with a large udder and a large number of lactocytes determines the milk-forming and exploitation capacities. The whole organism tries to meet this by mobilizing all its reserves to meet the supposed demand of the offspring. This is true even if it endangers the health of the mother (Habel and Sundrum 2020). With increased performance potential, the supply of nutrients is, above all, what determines how much milk can be withdrawn from the cows without endangering their health. Breeding for higher performance also changed feeding strategies and farm management. From then on, more and more importance had to be attached to the individual dairy cows' nutrient supply to exploit the ever-increasing genetic performance potential of the animals.

As impressive as the quantitative success story of milk production was and still is, it was constantly accompanied by undesirable side effects. These include an increase in the prevalence of functional disorders and production diseases of the cows and mortality, which occur to a very different extent from farm to farm. After World War II, infectious pathogens primarily threatened dairy cows' health. These included not only the pathogens of venereal diseases but also the pathogens of tuberculosis. By the mid-1960s, mastitis had replaced bovine tuberculosis and was almost eradicated as the leading cause of declining productivity in dairy cows. With the introduction of milking machines, the teat cups of the milking machines moved from udder to udder, resulting in bacterial pathogens being frequently passed on from cow to cow. The resulting inflammatory processes in the udder caused a drop in milk yield and increased treatment costs. In the worst case, the cow's life was threatened. Despite the technical development of milking machines, improved disinfection and treatment methods, and more knowledgeable people in the milking parlor, udder inflammation has remained a permanent problem in dairy farming, closely linked to the production processes. A fundamental conflict between the goal of further performance increases and the improvement of animal health fuels this production disease. The more milk accumulates in the cavities of the gland complexes between milking times, the more critical the breeding trait "milkability" becomes. The larger quantity of milk should be able to be milked out of the udder as quickly as possible. The larger the opening of the milk duct is, the easier the milk can be sucked out with the milking machine. On the other hand, microorganisms with pathogenic properties (pathogenic germs) can penetrate more quickly through the widened milk duct and settle in the udder tissue. The organism reacts to this with inflammation, which can be accompanied by considerable pain in the acute phase.

Local and systemic reactions of the organism on penetrated foreign germs alter the content of immune defense cells and the content of electrolytes and trace elements. Subject to the degree of severity, the inflammation processes in the udder tissue always go hand in hand with losses in milk quality for human consumption. An increased somatic cell count and salinity, a reduction of the casein content, an increase of the pH value, a reduction of the cheese dairy efficiency, and increased lipolysis and proteolysis activity contribute to this. The increased enzyme activity not only impairs the sensory properties of milk but also negatively affects the shelf-life of pasteurized and refrigerated milk storage and cheese yield and storage (Gonçalves et al. 2020).

The EU Regulation (EC-No 853/2004) demands that raw milk must be from animals free of any signs of an infection that can be transmitted to humans via the milk. They must be in good health and may not suffer from visible inflammation and udder wounds, which could disadvantage the milk. Somatic cell counts (SCC) are generally used to indicate udder health. From the veterinarian's perspective, an udder area (quarter) is then seen as healthy if the content of somatic cells in the milk does not exceed the figure of 100,000 cells/ml milk and with a bacteriologic test of the milk that no pathogens are proven. A comprehensive and representative study including 615,187 quarter foremilk samples indicated that 38% of all samples had SCC > 100,000 cells/ml (Schwarz et al.

2010). At the same time, mastitis pathogens were detected in 83% of all samples with SCC > 100,000 cells/ml. The prevalence of subclinical production disease in dairy cows is widely independent of the production method. In a study on organic dairy farms in different European Countries, the median prevalence of subclinical mastitis was 51% (Krieger et al. 2017).

The EU-Regulation (EC-No 853/2004) for bulk tank milk fixes the marginal value of 400,000 cells/ml in the geometric mean from three months for tradable milk. The dairy can refuse to accept the milk if bulk milk samples repeatedly exceed the set limit values. The legal limit for bulk tank SCC in the USA is even higher and fixed on 750,000 cells/ml for Grade A producers. While the inconsistencies within the EU Regulation are apparent, milk utilization from cows suffering from subclinical mastitis is not explicitly prohibited by legislation. The threshold values concerning milk quality ordinance bear no relation to udder health. Although the concentration of pathogen germs and somatic cells in the milk deriving from dairy farms is generally high, this does not comprehend a problem for food safety issues as the milk is pasteurized. To a large extent, the exclusion of health risks for the consumers might be the primary reason consumers and retailers are mainly inert against the health and welfare problems in dairy herds. The milk of cows in which inflammatory processes occur in the udder tissue is used for further processing, like milk from dairy cows with healthy udders. The milk of very heterogeneous quality from different farms is delivered to the dairy and mixed into a homogeneous raw product with defined raw ingredients.

On the farm level, higher efficiencies for milk synthesis are associated with a dilution of maintenance requirements and are suggested as the main reason for the substantial increase in productivity. On the cow level, the primary source for the gains in efficiency relates to inter-animal variation in nutrient partitioning (Baumgard et al. 2017). This potential was exploited particularly by breeding strategies and accompanied by increased scientific knowledge about what is required to exploit the increased genetic performance capacities by providing adequate nutrient supplies and environmental conditions. While the variation in genotype between animals is the starting point of the tremendous increase in performance, it is simultaneously the cause of many AHW problems. The average performance level continuously increased with the help of breeding and feeding strategies. However, the variation between the individual dairy cows remained widely unchanged. Correspondingly, considerable inter- and intraindividual variation exists in nutrient requirements and intake and the availability of energy and nutrients (Rumphorst et al. 2022). Managing dairy cows as herds or feeding groups to minimize efforts and expenditures widely ignores individual variations in animals' requirements. An indifferent nutrient supply to a feeding group of cows, for example, in the case of a total mixed ration, inevitably leads to a situation where many animals are either undersupplied or overfed. At the same time, only a minority of individuals are fed close to their daily requirements. The degree of imbalance caused by feeding strategies that are insufficiently tailored to the individual energy and nutrient requirements is illustrated in Fig. 4.1. The figure shows partial results from a prevalence study conducted in Germany

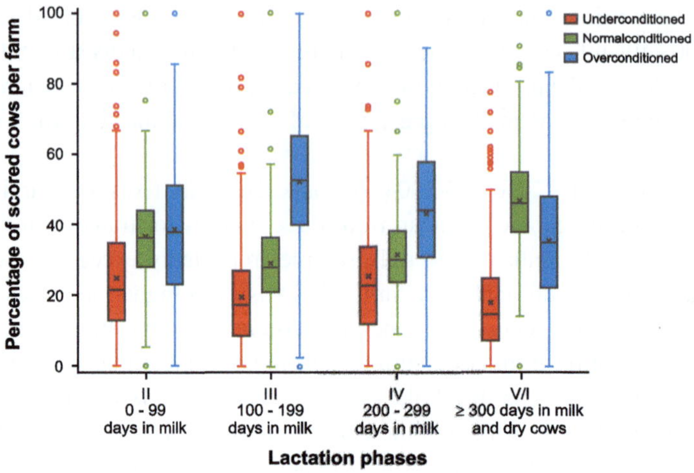

Fig. 4.1 Boxplot diagram of the distribution of cows scored as under, normal, or over-conditioned in relation to the lactation phase on 247 farms in the eastern region of Germany (PraeRi 2020)

in which a representative number of dairy cows per farm were scored on their body condition in relation to reference values of the respective lactation phase (PraeRi 2020). According to the authors, the results of this study can be viewed as representative of a wide variety of loose-housed dairy systems in Europe and North America. The body condition showed a considerable variation between dairy cows and dairy farms, widely independent of the region and the lactation phase, as exemplified by the results of the eastern region. On average, in the dairy farms, ordinary conditioned dairy cows were in the minority, whereas most were either under or over-conditioned. Although various feeding strategies exist to reduce the range of variation in body condition by providing a more needs-oriented energy and nutrient supply, these are, in general, accompanied by increasing efforts in labor time and investments and thus supposedly run contrary to the objective of increased productivity.

Giving productivity the highest priority, it is not only accepted that a high percentage of dairy cows are either under or over-conditioned but also the consequences that might derive from there. The capacity of a cow to cope with nutrient imbalances depends not only on the degree of the gap between supply and demand and the time it lasts but underlies a considerable variation between the farm animals living under the same conditions. The same applies to the individual variation in the capacity to deal with biotic and abiotic stressors to prevent pain, suffering, and damage. AHW problems derive particularly from a mismatch of nutrient allocation between different subsystems when animals are not receiving what they need to prevent dysfunctions (Habel and Sundrum, 2020). Thus, the degree of AHW problems within a livestock farm system is reciprocal to the ability, capacity, skills, and willingness of the farm management to provide demand-driven supply with appropriate resources and protection against biotic and abiotic stressors according to individual needs.

In the face of the generally high but highly variable AHW problems in livestock farms, it is evident that many farm animals do not receive what they need to cope. Clinical and subclinical mastitis is not the only health problem in dairy production. Besides mastitis, fertility problems, metabolic disorders, and lameness represent the central production diseases within dairy farming throughout the world. Production diseases are multi-factorial man-made diseases. They emerge from interactions and synergetic effects between different risk factors and processes that would not necessarily cause clinical signs of disease. Their occurrence indicates an overstrained capacity of the farm animals to cope with the farm-specific living conditions provided by the farm management. The degree and variation between the individual farm animals and farm systems in the manifestation of animal health and welfare problems is illustrated in Fig. 4.2, based on the results of the same prevalence study, quoted in Fig. 4.1.

The causes for the development of lameness are mainly multifactorial. Accordingly, energy and nutrient imbalances are not the only significant lameness factor. On the other hand, dairy cows' health and welfare problems cannot be substantially reduced without considerable improvements in energy and nutrient supply according to the individual requirements in the respective lactation phases.

Impairments of AHW, however, are not restricted to prominent disturbances but also enclose a great variety of inapparent disorders and diseases. These become apparent only after diagnostic efforts during a lifetime or when the carcasses and organs of slaughtered animals are assessed regarding pathological findings. A representative study for cattle health surveillance based on representative meat inspection records of a large cattle slaughterhouse in the Netherlands revealed that most of the culled dairy cows showed pathological findings (Veldhuis et al. 2021). Only 45% of the slaughtered cattle were without pathological *post-mortem findings*. Amongst the most common categories were inflammation of the pulmonary and peritoneal membrane, liver, udder, and lungs, as well as integumentary lesions and liver fluke. Furthermore, dairy industries and individual farmers are concerned about the mortality and culling of dairy animals. This is because the timing and fates of animals that exit dairy farms have essential animal welfare and

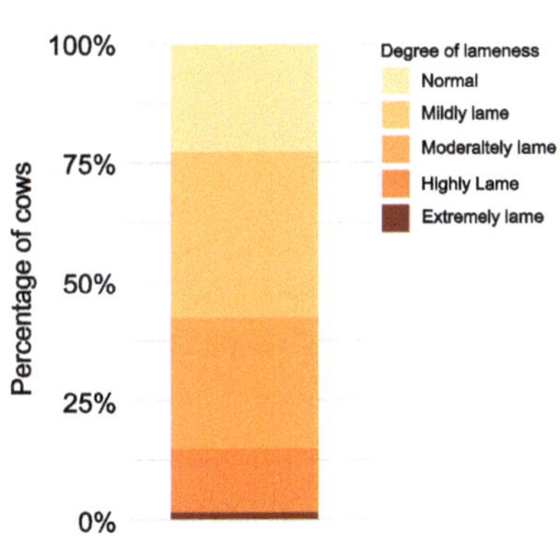

Fig. 4.2 Percentage of dairy cows with a defined degree of lameness scored on dairy farms in the eastern region of Germany (PraeRi 2020)

economic consequences that reflect the conditions under which they are farmed and the efficiency of their production systems. According to a literature review and meta-analysis, increased incidences of mortality and occasional culling of dairy animals have been reported in recent decades in several countries, and these changes have been associated with the intensification of production systems (Compton et al. 2017).

Despite the considerable extent of undesirable side effects, among other things, in the form of functional disorders and production diseases in dairy cows, the animal breeding industry is still pursuing further performance increases. This breeding objective, however, is more and more relativized by the inclusion of "functional traits," with attempts being made to limit the extent of these undesirable side effects. Economic arguments support the adherence to further performance increases. According to the rough calculation, a higher milk yield produced in a shorter lifetime appears more cost-effective than a lower yield produced over a more extended period and with correspondingly higher nutrient inputs and other production costs. However, given the wide variation between farms and farm animals, such a general assessment is invalid. It would need to be examined on a case-by-case basis whether and for which farms the calculation works out in this way. For the dairy cows, the economic calculation does not add up in any case.

4.1.2 Beef Cattle

In addition to milk production, the production of **beef** via young bulls, steers, heifers, culled dairy cows, and veal via calves is of high economic importance. In the post-war years, the initial focus was on providing fattening animals with feed that did not compete with the human diet. In symbiosis with the microorganisms in the rumen, ruminants can digest plant cell wall components and thus gain access to nutrients that have no nutritional value for animals with a single-cavity stomach (humans, pigs, and poultry). In addition, rumen microorganisms can produce protein from simple nitrogen compounds, so ruminants are, to a certain extent, independent of the protein quality in the feed.

After the Second World War, the politically promoted intensification processes caused the animal population to rise considerably. Depending on the farm structure in terms of land, fodder availability, and housing conditions, many farms kept cattle for fattening in addition to dairy cows. This was particularly the case on grassland sites, as this was the best way to use plant growth. Some farms concentrated entirely on fattening young bulls of beef cattle breeds. The drastic increase in production volumes inevitably led to market saturation and, in the following, to a reduction in livestock numbers. At the same time, the competitive pressure further increased the necessity of beef producers to enhance productivity and fattening performance by using energy and protein-rich feedstuffs in the form of cereals and seeds, which otherwise could have been used in human consumption. Furthermore, the competition led to breeding programs favoring specialization in one direction of use and partitioning cattle breeds into those for fattening and those for milk performance. Consequently, among other things, the male calves of dairy cows bred unilaterally for milk production have lost considerably in value, as they were no longer competitive with the offspring of beef breeds in terms of meat yield.

To make the production of veal meat profitable, designated calves are reared on a diet made up of cost-efficient milk replacer, supplemented by moderate amounts of solid feed with a high percentage of concentrate, in general until a slaughter age of approximately six months and a body weight of roughly 200–250 kg. The milk replacer is typically fed in buckets or troughs twice daily, although some farms use automated milk dispensers, allowing more frequent feedings. The diet of mostly iron-poor milk replacer and concentrate ensures low blood hemoglobin levels and the pale color of the veal while simultaneously causing iron deficiency anemia. Anemia is not the only AHW problem that veal calves are confronted with. A striking pathological finding is the high prevalence of esophagogastric ulcers or visible pre-ulcerative changes. In a review, abomasal damage assessed at slaughter ranged from 70 to 93% of the investigated animals (Bus et al. 2019). Ulcers are lesions of the abomasal mucosa that penetrate the submucosa and range from a few millimeters to several centimeters in size. Ulcers can cause perforation of the abomasal wall, leading to inflammation and peritoneum infection. The etiology is multifactorial, with diet being a primary contributor. Calves that grow fastest develop most abomasal lesions. The most mentioned theories about the effect of milk replacers on abomasal damage are overloading and low pH values. The specific milk composition might also affect abomasal damage. The link between abomasal damage and **animal health and welfare**, or more specifically on pain, is poorly understood and has so far received no research attention.

In diets for fast-growing beef cattle, fermentable carbohydrate supply is often provided from a mixture of fiber and highly digestible starch from cereals (barley, maize). Starch ferments rapidly in the rumen and can, therefore, significantly increase the energy supply rate. This is consistent with increased feed conversion efficiency. However, excessively rapid fermentation predisposes rumen acidosis, accompanied by the destruction of many of the normal rumen bacteria and protozoa, with potentially extreme consequences for welfare, including abdominal pain and metabolic acidosis. There is overwhelming evidence that subacute rumen acidosis is the major risk factor for parakeratosis and that parakeratosis is a significant risk factor for liver abscess. Other risk factors include primary foot infections (digital dermatitis) due to endotoxin release. The prevalence of liver abscesses observed at slaughter in batches of feedlot cattle in the USA has ranged from <5 to >90% (Corbière et al. 2008).

While beef production in the United States, Australia, Canada, and Japan is mainly based on the fattening of heifers and steers (castrated male animals), only a low percentage of male animals are castrated in Europe. Fattening bulls are generally kept under confined housing conditions on fully slatted floors. Intensive fattening methods focus mainly on daily gain and high slaughter weights. To create a recognized basis for the comparability of carcasses and their slaughter values among different countries, producers, and processors, different classification and grading systems were established, focusing on different carcass traits of beef. European carcasses are graded using the EUROP class scheme (EC-Regulation No. 1208/81). This includes the subjective assessment of trained graders focusing on five main carcass conformation classes (E to P, finer gradations up (+) or down (−) are possible) and four subcutaneous fat classes of the carcasses in different commercial courses, specified by gender. The higher payment for carcasses with a high lean meat portion promotes animals with a high muscularity and slaughter yield (Monteils et al. 2017). Consequently, European beef breeders primarily produce highly muscled beef carcasses with moderate fat. What is in favor of productive efficiency is, to a relevant degree, contradicting sensory quality attributes. Carcasses with a high proportion of lean meat often show disadvantages in the characteristics of sensory meat quality due to characteristic antagonisms (Hocquette et al. 2010). However, there are significant individual differences in terms of antagonistic expressions. This also applies to the relationships between growth performance, slaughter yield, and meat quality, requiring a qualitative assessment of the individual carcasses.

The EUROP classification system has been regarded for a long as insufficient to meet the demands of the market for qualitative differentiation based on sensory meat quality characteristics to become fair. In contrast, the United States, Australia, Canada, Japan, and South Korea use grading systems for beef that focus on or at least include marbling as one of the main parameters of beef carcass value (Polkinghorne and Thompson 2010). Marbling is one of the most essential quality traits characterizing the quality of beef due to its high influence on sensory traits like juiciness, tenderness, and flavor of the meat (Emerson et al. 2013). Compared to beef produced in the United States, European cattle breeds are very poorly graded for marbling (Schulz and Sundrum 2019). Even though it has long been known in top gastronomy and among many meat lovers that the degree of marbling represents the characteristic with the highest informative value about sensory meat quality, most European consumers are deprived of higher-quality meat products. Instead, the economic interests of slaughterhouse operators and food retailers set the context for how the heterogeneity of carcasses is dealt with.

4.1.3 Pigs

For a long time in agriculture, pigs were mainly kept to utilize waste products and add energy- and protein-rich components to the human diet. Until the 1950s, pig farming emphasized high-fat pigs; after all, it was a matter of "getting enough to eat" and securing food for the population. As markets became increasingly saturated and consumers over-saturated, the market demanded lean **pork**. In 1972, the European Trade Classification Ordinance for pork sides came into force, which has had a lasting influence on pig farming through payment according to lean meat content. Rationalization advantages led to the fact that in the second half of the twentieth century, the breeding of pigs was separated from the rearing and fattening of the animals and concentrated in specialized enterprises. At the same time, hybrid breeding and cross-breeding increased by leaps and bounds. Whereas in the 1950s an average of 380 kg of feed was used to produce 100 kg of live weight, the feed requirement fell to less than 250 kg in 2000. At the same time, the daily increases in the live weight of fattening pigs rose continuously. Meanwhile, fattening pigs reach gains of more than one kg per animal per day. With the rise in daily gains, the fattening period until carcass maturity decreases so that the costly barn space can be used even more efficiently.

Due to targeted breeding measures, the number of piglets born by the sows per litter and the number of weaned piglets sold for fattening per sow also increased continuously. In 2020, an average number of 13 piglets per litter was reached; the number of piglets raised per sow yearly increased to over 28. However, the higher number of piglets per litter is paid for with a reduction in piglet survivability since as the number of piglets per litter increases, the birth weights, and therefore the survivability of individual piglets, decreases. On average, about 12% of piglets born die; however, there are significant differences between farms regarding piglet losses.

Increasing the lean meat content and reducing the fat content of carcasses significantly increased the productivity of meat production. However, this development also came with various drawbacks and undesirable side effects that persist to this day. The increase in muscle mass is mainly due to an enlargement (hyperplasia) of the individual muscle fibers. As a result of the desired enlargement of the cross-sections of the muscle fibers, however, disturbances in the supply of oxygen and nutrients to the muscle fibers can occur. Among other things, increased susceptibility to stress, painful back muscle necrosis, and circulatory problems are consequences of selection for muscle fullness that are relevant to animal welfare (Wendt 2004). In addition, a significant deterioration of the sensory meat quality was associated with greater meat fullness. Based on the trait antagonism between meat and fat content, the fat content in the total carcass and the muscle's fat content decreased with breeding for muscle meat (Hocquette et al. 2010). The intramuscular fat content is crucial not only as a flavor carrier for the species-specific flavor of the meat but also for its juiciness and tenderness. As the lean meat content of carcasses increased, the **meat** became increasingly tasteless. A loss in the palatability of pork pays for the one-sided focus on producing muscle mass. Various options exist to improve the intramuscular fat content and thus the palatability of pork again (Sundrum et al. 2011). Efforts to improve the eating value of pork have no prospect of success as long as the value of carcasses is only assessed and paid for based on the lean meat content.

The pigs mainly pay for the undesirable side effects of the intensification processes in pig production. They are exposed to spatial and hygienic living conditions, which causes them a lot of trouble. There is not enough space here to describe the spectrum of adverse effects relevant to **animal welfare.** Given the database, however, there is no doubt about the extent of the adverse effects, which cannot be reconciled with a welfare state. In addition to the visible changes, it is above all the subclinical pathological processes, i.e., those that can only be identified after specific diagnostic measures. In addition to inflammatory processes in the lungs, those affecting the gastric mucosa are also among the production-related health disorders in pigs. These occur above all in connection with finely ground feed rations low in crude fiber, both in fattening pigs and in sows (Millet et al. 2012; Moesseler et al. 2010). The supposed benefit for the livestock farmer in the form of increased feed conversion is higher the fewer fiber components and structure the feed ration contains. On the other hand, the risk of stomach ulcers increases significantly with decreasing feed structure. In a comprehensive Danish study, gastric ulcers were found to occur in slaughtered pigs, with an average incidence of 29% (Canibe et al. 2016). In sows, lesions in the gastric mucosa were found in more than 50% of the animals on many farms. There were considerable differences between the farms in the frequency of painful changes, which point to the need for an individual farm analysis of weaknesses and a farm-specific solution strategy. While a high fiber diet is suited to increase satiety in pigs, pigs with access to additional roughage such as grass, clover, chicory, maize, or whole crop silage had more significant opportunity to perform species-specific behaviors, such as foraging and rooting behaviors, the supply with roughage is very labor intensive and considerably increases the production costs. The extent of the adverse effects of increased productivity on animal health and welfare is an expression of a **conflict of objectives** inherent in the system, which is tense between the interests of animal use and animal protection and which has not yet been resolved. (see also Sect. 4.2.2).

4.1.4 Poultry

Female hands have long been used to keep and care for chickens. The farmers' wives received direct income through selling **eggs** and **poultry meat**, which they could spend on personal needs. Eggs, in particular, could be sold reliably due to regular customer demand and a longer shelf life. However, there was often little willingness to cull laying hens consistently; after all, every egg was welcome. Since free-range laying hens had to forage for their food and the cost of the additional feed usually did not have to be paid for with egg money, the question of cost versus benefit rarely arose. After World War II, it was a declared political goal to expand the production of poultry products significantly. The first step was to cull all individuals not at a high-performance level and did not earn their own feed money. The USA and the European countries Denmark and The Netherlands were the pioneers oriented towards the industrial ideal of increased productivity.

To reach this goal, breeding was reorganized. With the introduction of the hybrid breeding procedure, the various national regulations became obsolete and were abolished. In a short time, significant domestic and foreign farms put the procurement of young stock on a different basis. The multiple uses of chickens—as egg layers and meat producers—quickly ended and were replaced by breeding for a single direction of use. Breeding for longevity was abandoned and gave way to selecting precocious chickens for only one laying period. The criticism of farmers' concerns at the beginning about breeding oriented solely to profitability did not last long. It fell silent, given the economic successes of the "economies of scale" (lower unit costs with larger production units) made possible by standardization in breeding and husbandry.

The short generation intervals of chickens compared to domestic mammals allow for much faster breeding success than in other farm animals. With the rediscovery of Mendel's laws, a revolution in chicken breeding began. Due to their biological characteristics, they became preferred objects for heredity research. This took place mainly in the United States, and here, it was primarily in the hands of private investors. These investors competed with each other unprecedentedly, so breeding was gradually concentrated on fewer and fewer suppliers. Since 2002, the breeding of laying hens has been in the hands of only three large breeding companies, which share the global market. In addition, there are only a few smaller breeding companies. Broiler breeding is also in the hands of a few companies worldwide. Here, at least a niche market has been preserved, served by the French company Sasso, which offers quality broilers according to the "Label Rouge" regulations. For this quality meat program, brooders are optimized for slower weight gain (minimum of 81-day fattening period versus under 35 days for conventional broilers) and higher intramuscular fat content. They are generally kept on the floor or in free-range conditions.

Starting from the annual laying performance of a wild hen of 8–12 eggs, a significant performance level was achieved through targeted breeding for egg production. In 1960, the average hen laid 148 eggs per year. The egg weight doubled from approx. Thirty-five grams to 60–70 g. In 2000, the laying performance averaged above 300 eggs per year (Zentrale Markt- und Preisinformationen GmbH 2002). The selection for the laying performance of the female animals resulted in the male representatives of these laying hybrids producing only a little meat and requiring comparatively much feed. In Germany alone, over 40 million male day-old chicks of the laying lines are killed after hatching every year, as their rearing is uneconomical. The **Animal Protection Act** requires a "reasonable cause" when animals are to be killed. This includes, for example, slaughter for consumption but not killing due to inefficiency. That is why politicians are urging that this state of affairs be ended.

A central step in fitting the animals into the templates of economic cost optimization was to eliminate the seasonality of egg laying by keeping them indoors all year round with a light and temperature program. The shorter light day in the winter typically results in chickens laying almost no eggs, at least from December to February. Since feed costs are the most significant cost factor, the primary production goal is to produce as many eggs per hen as possible while using as little feed per egg as possible. The goal could be realized in cage management rather than floor management. Because the more significant opportunities for movement on the floor result in a higher nutrient requirement, a hen kept on the floor eats eight to ten grams more feed per animal and day than in a cage (Wegener 1977). The economic successes are based, above all, on a genetic standardization of the animals with simultaneous adaptation to standardized housing and feeding conditions under completely controlled living conditions, with which uncontrollable life processes and weather patterns can be largely excluded. The more constant the living conditions are, the less energy the animals need to adapt to changing conditions, which would be reflected in lower performance.

Despite all the standardization, livestock farmers still do not succeed in eliminating all disturbance variables. These emanate not least from the animals themselves. Being the source of value creation, the farm animals simultaneously limit it. All efforts to normalize and standardize the living conditions also weaken the self-regulating powers of farm animals, which depend on adaptation to change for their ability to function. The cramped housing conditions and high animal stocking densities increase the germ pressure and the susceptibility to pathological infectious agents. Feather pecking and cannibalism are further stress reactions of the animals to the cramped conditions. In return, the management responded with a comprehensive hygiene regime, darkening of housing, beak clipping, and medications, especially antibiotics, to counteract disease susceptibility.

On most farms, the compensation measures outlined have the effect that the performance potential of the animals is not excessively impaired. However, the measures do not protect many animals from an excessive demand on their adaptability and from impairments associated with pain, suffering, and damage. This is exemplified by laying hens. With breeding for high laying yields and the accompanying measures of feeding and hygiene, this could be increased to a very high laying performance of more than 320 eggs per hen per year. However, laying hens pay a high price for this. Approximately 3 g of calcium are needed daily to produce each eggshell. Since calcium absorption from feed is limited, a relevant proportion must be provided from the bones. The high output does not allow the animals to remineralize the bones and replenish the calcium stores sufficiently. As a result, the bones are less compact and, therefore, more prone to fractures, particularly in the sternum of laying hens. In a *post-mortem* investigation, fracture prevalence in the range of 53–100% was observed in flocks from non-caged systems, whereas the prevalence in flocks from enriched cages ranged between 50 and 98% (Thøfner et al. 2021). A longitudinal study based on a radiographic evaluation of keel bone damage in laying hens revealed that 99% of the animals showed at least one keel bone lesion during the study, and 97% had at least one keel bone fracture (Baur et al. 2020). The sternum fractures occurred independently of the housing systems and the production method. Since sternum fractures cause chronic pain and can limit the animals' ability to move, they are highly relevant to animal welfare. Despite the extent of the damage, sternum fractures in laying hens have not yet been taken as an opportunity to address the associated challenges and work towards a significant reduction.

The increase in productivity is even more striking in the case of the production of poultry meat by broiler. From 1957 to 2005, broiler growth increased by over 400%, achieving market weight in 60% less time than broilers 50 years ago and with a concurrent 50% reduction in feed conversion ratio (Zuidhof et al. 2014). Achieving the best food conversion rate with the premium cut of breast meat was and still is the economic driver for the broiler industry, leading to an increase in the amount of breast meat on an individual bird by two-thirds. Broilers are typically reared in barns with a stocking density of around 39–42 kg/m^2 (19.5–21 birds/m^2 when slaughtered at 2 kg). Such high stocking densities negatively affect their walking ability. The rapid growth and increased muscle size have come at a direct cost to the birds as they suffer an increased incidence of musculoskeletal and cardiovascular disorders such as leg weakness, ascites, and breast muscle myopathies (Bessei 2006). According to a study from the UK, around 27.6% of commercial intensively-reared broilers presented leg abnormalities, and 3.3% were almost unable to walk (Knowles et al. 2008). Given the health problems in broiler production, Danish legislation prescribes footpad dermatitis (FPD) surveillance at slaughter to indicate on-farm broiler welfare. The average official FPD flock score was 27 (7.5–53.0) for organic flocks and 48 (4.0–106.5) for conventional flocks. Footpad lesions were primarily chronic, representing a wide range of severity (Lund et al. 2017).

The vast majority of the EU chicken meat sector uses fast-growing strains. Evidence shows that selection for rapid growth reduces immune competence and increases susceptibility to diseases (Swaggerty et al. 2009). On the other hand, breeds exist that can alleviate many of the negative predispositions. By utilizing these higher-welfare breeds, giving birds more space, and enriching the environment, a clear improvement in the level of welfare can be expected (Göransson et al. 2020). Higher welfare broilers have lower mortality rates than typical commercial fast-growing broilers and are less susceptible to leg disorders and heart problems. Furthermore, higher welfare birds are more active and perch, walk, and scratch more than conventional breeds. In 2017, higher welfare breeds accounted for less than 5% of broilers slaughtered in the EU. However, higher welfare breeds represented 90% of the Netherlands' fresh retail market for chicken. In the UK, higher welfare breeds represented almost 10% of broiler production, while in France, the Label Rouge higher welfare breeds make up 15% of the market share (Thornton 2016).

Concerning foot pad dermatitis (FPD), an even worse situation is seen in the case of fattening turkeys. Turkeys assessed at slaughterhouses revealed the undesired results that 97.7% of the examined turkeys were diagnosed with different degrees of FPD. Most frequent were necrotic lesions measuring up to 2 cm in diameter (64% of all turkeys). Extensive necrotic lesions of the foot pads and necrosis of superficial scales (toms: 41%; hens: 20%) were less frequent (Freihold et al. 2019). Other AHW problems are breast buttons and breast blisters, which are pathological skin variations with a multifactorial etiology. Turkey toms exhibited a prevalence of 42% and were affected significantly more frequently by these alterations than the turkey hens (9%) (Mitterer-Istyagin et al. 2011). Significant differences in prevalence were partially recognizable between individual farms and the fattening cycles on a farm.

Apart from severe AHW problems, meat quality defects are increasing. Unfavorable correlations between production traits and meat quality were found, where production rises (e.g., body weight, growth rate) resulted in lighter and redder color and, in some cases, increased moisture loss (Hiscock et al. 2022). The relationship between growth and meat color is essential since cut-up poultry products are typically packaged on trays covered with transparent film, displaying the meat color as the most prominent characteristic to the consumer, which can be very sensitive to color variation in packaged poultry products. In addition, poultry meat is associated with many public health risks caused by slaughtering poultry flocks. A prevalence study of public health hazards in Finnish slaughter turkeys revealed that 68% of the turkey flocks carried microbial public health hazards, the most common of which were *Shiga toxin-producing E. coli* (43% of flock testing positive) and *Listeria monocytogenes* (32% of flocks testing positive) (Blomvall et al. 2023).

4.2 Technical Developments

Changes in livestock farming in recent decades are closely linked to biological and technical developments, shifting the relationship between livestock farmers and farm animals. Animal husbandry became animal production, and the farmer became an animal producer. Over time, farmers' relationship with technology also changed. From an initial rejection and skepticism, the attitude changed from tolerance to the defense of technology and finally to a pronounced belief in technical progress. The following discusses the changes and their implications using examples from cattle, pig, and poultry farming.

4.2.1 Cattle Farming

The development of dairy farming after the Second World War is a story of the milking machine. The argument for using a milking machine was economic, measured by the number of cows milked per hour. The reference figure was six to eight cows that skilled milkers could milk by hand in an hour, depending on the daily milk yield of the cows. As labor was already very costly then, techniques that could replace it were particularly welcome. Compared to human milking, today, milking technology is used to milk more than ten times as many animals and several times as much milk per hour. Nowadays, the so-called "milking robot" is used on many dairy farms. It eliminates the need for a human being to be present during milking. Unlike the hands of the milker, machines do not show any signs of fatigue and can milk all four teats of an udder simultaneously. Despite all the technical developments, milking has remained a sensitive matter for dairy cows, as damaged teats and inflamed udders are a constant reminder.

Interactions between machines and cows are less prone to disruption if the animals match the machine or can adapt to its demands. Milking technology is based on a sequence of constant processes, leaving little room for biological variation in the shape of the udder and teats. Dairy cows incompatible with the machine milking process were already sorted out in advance or due to health disturbances, diseases, and performance loss. Cattle breeding also made its business to influence the shape of the udder gland and the teats as well as the milkability of the dairy cows by breeding, to adapt the dairy cows as best as possible to the standardized specifications of the milking machines. However, the cows and the technology proved to be susceptible to faults. Among other things, vacuum fluctuations in the piping system and contamination in the teat cups can be disruptive for dairy cows. Time and again, it can be observed that farmers save on the costly maintenance of milking technology, only to be confronted with delayed, sometimes considerable disturbances to the udder health and negative economic consequences (see Sect. 4.3.1).

Milking technology has proven to be a very effective means of drastically reducing the daily labor time required for the milking process. In this way, it is possible to save working time, significantly increase the number of dairy cows looked after by one person, and realize economies of scale. However, the economic, biological, and animal welfare implications of the triangular relationship between humans, animals, and technology are not a question of the technology itself but rather the result of the objectives for which the technical options are used. "Carry nothing that can flow!" was the maxim used to make day-to-day barn work easier. This applied to the milk transported in pipe milking systems from the cow's location to a central collection container and to the animal's excrement. The slatted floor was the heart of the technical revolution in animal housing for fattening animals. In pen floors with slats, excrements are trodden into a pit below the tread and collected with the urine in a slurry tank. To flow, the slurry could no longer be mixed with straw. Consequently, strawless housing on slatted floors was an option to avoid time-consuming mucking out of pens involving manual labor. Other options existed but struggled to compete for the best cost-benefit ratio of slatted floors. How the farm animals could cope with the slatted floors was—if at all—of subordinate relevance.

Beef cattle kept on slatted floors have a considerably higher incidence of injuries than animals on straw or sloped, partially straw-bedded areas. Comprehensive hazard analysis for beef cattle identified three significant categories of welfare problems attributable to risks associated with housing and management: Respiratory diseases, which are linked to overstocking, inadequate ventilation, and mixing of animals, and failure of early diagnosis and treatment. Digestive disorders are related to intensive feeding of concentrate and a lack of physically effective fiber in the diet. Behavioral disorders are linked to inadequate floor space, co-mingling in the feedlot, and intensive concentrates (EFSA 2012).

4.2.2 Pig and Poultry Farming

The technical innovations in pig and poultry farming mainly concern husbandry technology. Here, the focus is also on how technology can reduce the labor time involved in keeping, feeding, and manure removal to care for ever-larger animal populations. On the other hand, the use of technology initially requires investment. The investment costs must be recouped through productivity increases. Consequently, with every technical innovation brought onto the market by the manufacturers, the question arises for the livestock farmers as to whether the use of technology can also be expected to lead to an increase in economic profit with which the investments can be amortized.

At the beginning of the "technical age in livestock farming," more or less all new stable techniques had in common that they restricted the range of movement of the livestock. After all, stall space and stall technology are expensive and can be refinanced most easily if as many animals as possible can be kept. In sow management, many farmers at the beginning of this development did not want to impose on the mother sows to be fixed in a crate before and after farrowing. This device is used in cramped housing conditions to contain the crushing of the piglets when the sows are laid down. Corresponding proposals were accompanied by critical comments from the professional group and were rejected as too disadvantageous for the animals. However, the skepticism towards fixing the sow disappeared from the agricultural discussion as the working time in the pigsty became more precious (Settele 2020). Agricultural science and official advice made the new forms of animal husbandry appear particularly progressive, innovative, and sustainable. Farmers who had reservations about fixing sows ran the risk of being seen as backward and hostile to progress. Accordingly, as time passed, it was considered less and less objectionable to restrain sows in crates. Although compensation was demanded for the restraint that the sows should be freed from their narrow individual pens several times a day so that they could move around, this was seldom implemented. As herds grew and working hours became increasingly scarce, this time-consuming measure was abandoned. Under the given economic conditions, no options were available to stay up with the competition other than to participate in technological progress.

Other changes associated with the introduction of housing techniques concerned the need to specialize. The higher the investment volume, the more the livestock farmers felt compelled to limit themselves to a specific production section, such as raising juveniles or fattening to exploit the potential of structural-technical innovations better. Specialization and division of labor were central to improving work organization and realizing economic economies of scale. The more specialized the area of use of the barn, the more precisely the barn equipment could be adapted to the respective production processes. Continuous occupancy had to be ensured as far as possible to avoid costly vacancies. The utilization of the structural-technical investments consequently made it necessary to coordinate the different stages of the animals' lives with the existing barn facilities.

While livestock producers were interested in using technical innovations to increase productivity, the technology suppliers were primarily concerned with selling them. From the 1970s onwards, housing techniques were prominent at agricultural exhibitions. Technical innovations such as cage housing for laying hens also inspired technical developments in pig farming. Sows penned in crates were advertised as part of "modern pig production." The same applied to "battery cages for weaners." The advertising claims were that these forms of housing offer "maximum labor savings with a high level of comfort," a "clear and space-saving system," and "simple and time-saving animal observation" (Anonym 1975). In addition, the number of animals kept per house area could be significantly increased in this way.

The farm animals have to pay the price for the considerable simplification of work with very cramped spatial conditions, a concrete-hard subsoil as a walking and lying surface, as well as stable air enriched with ammonia, dust, and germs, which forms above the slurry pits. Pressure loads from hard floors cause auxiliary slime pouches and claw injuries in fattening pigs. In a study at several abattoirs, 92% of animals had auxiliary bursae, which occur as a pathologic reaction to trauma, especially when exposed parts of the extremities are mechanically stressed (Gareis et al. 2016). Of these, 44% were moderately to severely altered. Furthermore, the cramped housing conditions are associated with poor air quality, particularly enriched with ammonia, generated from the slurry the pigs are exposed to throughout their lifetime. Ammonia vapor has a sharp, irritating, pungent odor and is toxic in higher concentrations. Ammonia reacts with other gases to form particles that are harmful to health (secondary formed fine dust), to which microorganisms and toxins can also adhere. The fine dust particles reach the lung tissue via the airways and can cause inflammatory processes here. If we look at the lungs of pigs at slaughterhouses, a considerable proportion of the lungs of slaughtered pigs show inflammatory processes. Cranioventral pulmonary consolidation was the most frequent lung lesion (31%) detected in fattening pigs at the abattoir, followed by dorsocaudal infarcts with pleurisy (12.5%) and pleurisy alone (6.3%) (Pallarés et al. 2021).

Poor barn climate conditions in pig and poultry production are countered with the help of forced ventilation. While increased ventilation is suited to alleviate the situation, it does not solve all the problems. In addition to the inconsiderable additional costs, there is still the problem of an extensive air exchange to improve the air quality and remove heat from the stalls in winter. Heat loss can be reduced with less air exchange. However, this is at the expense of air quality and thus increases the risk for respiratory diseases. The introduction of slatted floors and slurry technology is a prominent example that exemplifies how technology can solve problems while creating new ones. The inherent ambivalence of technology raises the question of which of the various objectives should be given greater priority.

A lack of materials for the animals to manipulate and a permanent stress situation due to social stress caused by a high stocking density is accompanied by further AHW problems. As a result of general inactivity, some pigs repeatedly bite other pigs in the intact tail or the tail stump, remaining after amputation. The consequences are more or less pronounced injuries, accompanied by painful processes of inflammation and necrosis of the tail. In the case of poultry, feather pecking, where birds peck at the feathers of other birds, sometimes pulling them out and eating them, is a significant welfare issue in poultry keeping, resulting in poor plumage, patches of feather loss, skin damage, and even death. The health and welfare of the affected birds are significantly compromised, and the physical and financial losses can be significant. Once feather pecking becomes prevalent in a flock, it can be challenging to stop, and if the behavior becomes severe and leads to cannibalism, it may result in the need to cull large numbers of birds.

To contain the negative consequences of intensive livestock production, farmers usually know no other way to help themselves than to reach for the scalpel. With the introduction of the slatted floors in pig production, they routinely amputated the tails of suckling piglets. The justification for this painful procedure is apparent; this measure aims to protect or at least reduce the number of animals from having their tails bitten by other pigs. This reasoning makes the aggressive pen mates responsible for the damage they cause, not those who have crammed them into confined spaces that cause the pigs to behave aggressively. In the case of poultry production, it has become a standard practice to blunt beaks to prevent the development of the sharp hook on the beak of adult birds. This is referred to as 'beak trimming' in legislation but is more widely referred to as 'beak treatment' in industry and practice. This reduces the injuries and mortality rate caused by feather pecking but does not prevent abnormal behavior.

Keeping animals in confined spaces not only encourages aggression between them; it also puts them at risk of infection to each other. Infectious diseases can destroy economic success in the effort to increase productivity further. Accordingly, veterinary *know-how* is needed and called in to counteract these financial risks. In the past, consistent hygienic measures, such as the clean-out procedure in pig and poultry farming, were essential in reducing the germ load in the stables. After a production phase, all the animals in a separate pen are removed, and the pens are only reoccupied after extensive cleaning and disinfection. In addition, larger and larger quantities of antibiotics are used, administered to all animals of a pen compartment via feed or drinking water at the first signs of illness in some animals to counteract the spread of the infection. In this way, antibiotics became an indispensable means of production to limit the undesirable consequences of keeping many animals confined and the resulting economic losses. At the same time, veterinary medicine has increasingly become an indispensable repair service for the negative impacts of the concentration processes that result from technical developments.

In agricultural circles, the increase in the concentration of animals in confined spaces was first and foremost referred to as "factory farming" (Gaschler 1974). This happened long before protest groups mutated the term into a fighting term. The ambivalence of economic advantages and animal health and welfare disadvantages was well known in expert circles. However, this knowledge did not significantly influence further developments. The benefits of economic growth promised by economists and the confidence in coping with the problems arising through scientific advances in learning and technical developments prevailed. After all, farms that produce livestock are economic enterprises.

Industrial animal farming entails management procedures and housing conditions that considerably compromise animal health and welfare (Nordquist et al. 2017). The relevant disturbance factors include, among others, insufficient availability and inadequate composition of feed, high occupancy density, prevention of the exercise of species-specific behavior, high concentrations of microbiological pathogens (bacteria, viruses, fungi, yeasts), parasites or toxins (endotoxins, mycotoxins, etc.) and high concentrations of harmful gases (including ammonia). Lastly, adverse effects are also caused by conflicts with other animals (e.g., tail biting, feather pecking). All these factors do not affect the animals in isolation but as a whole. The unlimited variety of combinations of factors results in a considerable variation in the respective living conditions, which are also subject to dynamic changes. Accordingly, standardized husbandry conditions can cause significant differences in their effect on the animals.

The complications accompanying the intensification of livestock production and the introduction of slatted floors and cages did not prevent their breakthrough. Regardless of the adverse side effects, the technology in the barn was rarely discredited. The advantages were too significant, among other things, in terms of making work more accessible in connection with cleaning and disinfection and terms of productivity increases, to dispense with applying them to protect the farm animals from excessive adverse effects. Regardless of the negative side effects of the technical developments for the farm animals and the environment, the decision was made solely based on economic criteria while largely unimpressed by the high degree of AHW problems. There is consensus among farmers, agronomists (Glebe et al. 2018), and agricultural politicians (Bundesministerium für Ernährung und Landwirtschaft 2020) in the problem-solving potential of technical innovations and scientific progress. Advances in knowledge in the agricultural sciences give rise to the hope that hitherto unsolved problems in livestock farming can be solved in the future, provided sufficient funds are available for research. After all, the technical development first brought about thc industrialized production processes in large-scale pig and poultry farming and, thus, the supposed competitiveness in international markets.

Given the very long history of technical developments in livestock production and the myriad of studies available in the literature, which indicate a very high prevalence of AHW problems in livestock production, it is evident that technical developments and animal science have failed so far to improve farm animal health and welfare (Sundrum 2024). AHW problems are highly context-sensitive and multifactorial. They develop within the farm-specific interconnectedness of manifold and highly varying factors with such complexity that technical developments or general recommendations cannot solve them alone. While the various disciplines of animal science predominantly strive for context-invariant and thus generalizable dispositional knowledge, strategies to address the health and welfare of livestock above all need to be effective and cost-efficient within the farm-specific context and therefore, closely align with the farm-specific causes of the animal health and welfare problems (for further explanations see Sect. 7.4).

To prevent misunderstandings and a supposed accusation of hostility towards technology, it is emphasized that technical innovations are not a problem but a **means to an end**. Correspondingly, it needs to be validated whether they contribute to achieving the intended target. At the same time, the associated side effects have to be assessed. Means of production can become a problem if they are used primarily to reduce production costs. At the same time, harmful effects are largely ignored or only perceived as such when they endanger economic success. Above all, the development and continued existence of AHW problems and the attempts to reduce them occur in an agricultural market system characterized by a global price undercutting competition. This leaves little room for the actors to invest in measures to improve AHW problems, which might seem not profitable enough and, thus, not sufficiently suited to sustain the economic viability of the farm system. As all measures contributing to improved housing and living conditions for livestock are subject to costs, they have only a limited chance of realization under the current market conditions. In this way, husbandry technology aimed at minimizing production costs has maneuvered itself into a dead end. Now that consumer demands are beginning to change, for economic reasons, many farms can no longer adapt to changing consumer demands.

4.3 Economic Developments

Livestock farmers depend on livestock farming to provide for their families livelihoods and keep their farms viable and fit for the future. As entrepreneurs, they operate under economic conditions that they, as individuals, can hardly influence and, therefore, have to adapt to. Only farmers with their marketing channels are in a position to influence sales prices proactively. Since integrating the national agricultural sector into the jurisdiction of the EEC at the end of the 1950s, the course for the rural economy has no longer been set at the national but the European level. The communitarisation of agricultural policy aimed to create a common market for agricultural products, thereby increasing production productivity. To this end, the Common Agricultural Policy used various price and market mechanisms. Productive farms were promoted, and less productive ones were pushed to the sidelines. From the outset, European agricultural policy pursued two contradictory **objectives**. On the one hand, supported prices were intended to provide farmers with an income that kept pace with the rest of the economy. On the other hand, prices were designed to provide an incentive to optimize productivity. Subsidies stimulated production; the increase in output depressed prices; price guarantees absorbed the income threatened by this; these led to further subsidies, which again stimulated production (Wehler 2008). Products that could not be sold domestically were sold on world markets, with solid support from export subsidies.

While the goal of aligning farm incomes with developments in other sectors of the economy was primarily missed, the strategy was entirely successful in increasing productivity. At the same time, the number of farms decreased drastically. The decline in actively operating farms is generally called "structural change." Structural changes occur in all sectors of the economy where productivity gains bring about the exit of those farms that can no longer keep up with competitors in terms of production costs. "Up or out" is a frequently used shorthand formula that sums up the compulsion to **grow**, primarily based on the theory of economies of scale. Rising labor, land, and technology costs have pushed many farms away from diverse mixed farms towards ever greater specialization with the aim of cost degression. According to the logic of agricultural economics, highly mechanized farms with large livestock numbers should prevail. However, this was and is not the case everywhere. The narrative of structural change as a "macroeconomic construct" is mainly silent on how the consequences of structural change, specialization, and industrialization are perceived, driven, and processed at the micro and meso levels (Ballmann and Schaft 2008).

From an economic perspective, competition characterizes the relationship between livestock farmers more than ever. This does not have to contradict cooperative relations in the neighborhood. It is not so much the farms in the immediate vicinity perceived as competitors, but rather the farms in other European countries, especially in third countries. On the other hand, the global market allows retailers to arrange the purchase of raw goods according to their specifications and to dictate the price conditions to the producers according to the supply situation. The global market is oriented almost exclusively to the purchase price of raw goods. Accordingly, the international competition is fought out for **cost leadership**. Those who provide price quotations with the lowest operating production costs gain at least temporary advantages over their competitors. In this way, the global market situation perpetuates a so-called "race to the bottom," however, with no bottom in sight. Somewhere in the world, some producers can always produce raw goods more cheaply than others. This never gives most entrepreneurs peace; they must constantly look for ways to reduce production costs. Opportunities for cost minimization are manifold. The following explains some of the most essential strategies for dairy, pig, and poultry farming.

4.3.1 Cattle Farming

To extract increasing quantities of milk from dairy cows and a higher meat yield from beef cattle, a significantly improved feed basis was required in addition to an increased genetic performance potential. Soon, this was no longer limited to the farm's coarse fodder, such as grass and the preserved products, hay, and silage produced during the vegetation phase. Crop rotations were changed on arable land, and energy-rich maize plants and feed grains were grown where possible. At the same time, the purchase and import of highly concentrated feedstuffs increased. Depending on the price, these were brought in from near or far abroad by a rapidly flourishing feed industry and processed in feed mills. Soybeans, in particular, played an increasingly important role in improving the supply of high-quality protein components. Given the limited feed intake capacity of the animals, the nutrient concentration per kg of feed became the decisive parameter. Improved nutrient supply not only led to an increase in milk yield and daily gain. At the same time, efficiency in the use of resources also increased.

The economic incentives to intensify production processes did not fail to be effective. Significantly more milk was produced than could be sold at home and abroad. Surplus product quantities had to be stored temporarily, forming the proverbial "butter mountains and milk lakes." The milk quota was introduced in Europe in 1984 to limit the excessive volume of milk production. Each farm was only allowed to deliver a fixed quantity of milk to the dairy based on the previous delivery quantities. As a result of the milk quota system, rising costs for the means of production could no longer be compensated by increasing the production volume but had to be absorbed by rationalization. As a result, most dairy farmers saw their economic goal in exploiting the milk quota with as few high-yielding cows as possible to remain competitive. At the insistence of the agricultural lobby, the milk quota regulation in Europe was lifted again in 2015. To prepare for the new competitive situation following the abolition of the milk volume cap, many dairy farmers expanded their production capacities and, in this way, contributed to a further increase in the volume of milk available on the markets and the associated fall in milk prices. Given the **conflict of objectives** between increasing productivity of milk production by increasing milk performance and by minimizing expenditures and efforts in labor time, it is not surprising that production diseases remained the main problem in dairy farming.

4.3.2 Pig Farming

In pig farming, the stages of life are divided into production phases. The specialization in piglet production and pig fattening has contributed significantly to a considerable expansion of production capacities and productivity increases. At the same time, however, specialization has increased dependence on developments in the upstream and downstream sectors. In the event of a price collapse, pig farms are exposed to a high economic risk. Despite the dangers, specialization has become a system-wide trend in pig farming.

The legal minimum requirements decisively influence the design of the respective husbandry conditions. These have become the norm as competition forces livestock farmers to implement the requirements at the lowest possible cost. Any movement area exceeding the legal minimum requirements is associated with additional expenses and, consequently, a competitive disadvantage, thus avoided as far as possible. The expansion of pig numbers did not occur evenly across the country. It was concentrated in regions with particularly advantageous conditions, such as feed availability, which was the first limiting factor in animal production. Infrastructures for importing high-quality fodder components opened the door to expanding livestock farming, even in previously disadvantaged regions due to unfavorable soil conditions and farm structures.

Agricultural technology innovations and the close cooperation of livestock farms with upstream and downstream companies also fueled the development. Drastically increasing demands for means of production and *know-how* promoted the foundation of companies specializing in producing corresponding techniques and service offers, which settled in the region. The fixation of the legal minimum requirements for the keeping conditions provided the livestock farmers with the necessary planning security for the required investments. Without fearing legal disputes with animal rights activists or changes in the law, it was possible to build large-scale housing facilities. The livestock farmers could concentrate on the most crucial goal in the production processes: increasing productivity by increasing animal performance and reducing production costs.

Structural change has taken hold in pig farming even more than in cattle farming. The number of animals per farm increased drastically. The difference between domestically produced and imported meat volumes and domestic consumption must be exported. Accordingly, the sector became increasingly dependent on the conditions of demand and supply in the global market and the associated development of world market prices. Consequently, the pork sector became a huge marshaling yard, which, despite the associated transport costs, is driven by cost advantages for some players involved. Competitiveness is also highly dependent on the costs. Pork producers must pay for feed and piglets, which account for more than 90% of the production costs of fattening pigs (Rohlmann et al. 2020). According to the authors' estimates, only farms that belonged to the top 25% have operated profitably on average in recent years, whereas most farms were not profitable. However, farms that generate losses do not immediately drop out of production. They usually 'live off their assets' for a few years or can compensate for the losses through income from other business branches or through subsidies paid. Often, a decision on continuing or discontinuing a holding branch is only made when new investments are pending, or the farm succession has to be arranged.

The market is, however, sensitive to the development of prices for feed and piglets and to epidemic diseases occurring in various regions of the world. In the past, this has been particularly the case for African swine fever (ASF). The main distribution area of ASF is African countries south of the Sahara. It was probably introduced from Africa to Georgia, where the first outbreaks of ASF were reported in June 2007. Since then, the spread in a north-westerly direction took its course. Wild boars and domestic pigs kept in confined stables became infected in various East European Countries. In 2019, ASF infections occurred in Asia, particularly in China, which is by far the world's largest pig producer. Consequently, many animal populations had to be culled to contain the disease. Imports from the US and Europe compensated for the lost product volumes. In this way, the Asia–Pacific epidemic benefited all producers in countries unaffected by the disease.

Once the virus crossed the border, the consequences for pig farmers were fatal, particularly for those in export-oriented countries. In response to the disease outbreak, many countries stopped importing pork from countries with ASF cases. This resulted in a drastic fall in market prices, making it impossible to cover the costs of producing pork for long periods. Individual pig farmers could only have protected themselves against the foreseeable economic consequences by shutting down the farm in good time. To protect pig farmers from the consequences of the price collapse, agricultural policy would have required concerted intervention in the market and a reduction in pig stocks to buffer the market's dependence on exports. Since this was not desired by agricultural policy or deemed unenforceable, the role of pig farmers changed overnight from temporary beneficiaries to the losers of an economic system thrown into severe turmoil by the border crossing of a virus.

4.3.3 Poultry Farming

In the case of poultry, the methods of production developed from an economically insignificant branch of business reduced to secondary income into an independent branch of industry. Poultry farming broke away from its former embedding in an agricultural business and became a largely independent enterprise primarily dependent on outside capital. Changes in tax depreciation practices gave investors the necessary outside capital to enter poultry production in the expectation of corresponding returns. The more outside capital flowed into the sector, the more the existing agricultural enterprises came under competitive pressure. The production of eggs and poultry meat became cheaper, the more extensive and denser poultry operations became. Another requirement was that poultry production became wholly emancipated from the cyclical rhythms of nature. Producers said goodbye to free-range farming, which was still standard then, and shifted production processes indoors. *Economies of scale* changed farming in three ways. The better the barn space was utilized, the lower the cost of keeping the animals. In the cages stacked on top of each other to form a battery, around twenty laying hens could be kept per square meter, which was more than twice as many animals as with floor husbandry. The more animals cared for by one worker, the lower the labor costs. The more effectively the feed was turned into eggs and meat by the animals, the cheaper the feed costs.

Because of its importance to the economics of chicken production, much research has also been done into poultry nutrition. The progress in knowledge about the nutritional-physiological relationships in the gastrointestinal tract of chickens exceeded the knowledge that existed in this respect in humans (Boyd and Watts 1997). Additionally, there was no lack of effort to use various aids. Among other things, antibiotics were added to many industrially produced rearing and fattening feed mixtures as feed additives to favor the state of health and, thus, on the laying performance.

While the adjusting mentioned above screws reduced production costs and thus unit costs, market-induced cost pressure forced companies to optimize continuously. While the initial aim was to refinance investments, increasing their competitiveness by reducing production costs became necessary for survival because the overall demand for eggs and poultry meat was not growing indefinitely. Marx already knew that the "rational calculation of capital," forcing competing enterprises to exploit potentials for cost reduction at the "penalty of their demise"; he thus anticipated the systemic cause of "product, process, and procedure innovations," the production of eggs and poultry meat based on outside capital. (Welskopp 2014). The compulsion to reduce costs leads to fierce competition in which only "*he who seizes every advantage, however small, for the recovery of costs, can survive*" (Mehner 1966). Given the economic conditions outlined above, many farms cannot compete in producing poultry meat and eggs.

Analogous to pig farming, viral animal pathogens have also become a sword of Damocles for many poultry farms. According to experts, avian influenza outbreaks are mainly linked to the migration of wild birds, an unknown number of which are infected with the virus. The farms directly affected are usually insured, so the financial losses incurred because of the necessary killing of all animals susceptible to infection are compensated. However, the insurance companies do not reimburse the loss of income caused, among other things, by trade restrictions in the affected regions and vacant stables during a quarantine period.

4.4 Winners and Losers of the Increase in Productivity

The productivity gains achieved through the interlocking of biological, technical, and economic developments in cattle, pig, and poultry farming are overwhelming in several respects. From the perspective of the postwar years, no one could have credibly predicted that such large quantities of raw animal goods could be produced so efficiently and cheaply in such a confined space. However, the negative consequences of the intensification of "animal production" for the living beings involved, the changes in the immediate living environment, and the environment on a regional, national, and global level are also overwhelming. The livestock keepers sell the raw materials of food of animal origin in their respective enterprises; however, they are produced by the animals through biological services based on metabolic processes. It is the evolutionary biological processes of self-preservation, growth, and reproduction that humans have harnessed for their purposes for thousands of years. In the context of the industrialization of agriculture, however, this process has taken place utilizing quasi-revolutionary changes (Settele 2020).

Farm animals are part of an evolved food chain. They can exist because they can feed on plants that, through the ability to photosynthesize, can produce organic biomass. The herbivores (primary consumers), in turn, are the food source of carnivores (secondary consumers), which in turn can become together with the herbivores, the prey of consumers as tertiary predators. In this natural food chain, humans, as omnivores, have better options to adapt to the available food sources than species that have specialized in specific food sources. In the meantime, there are only a few spots in this world that humans have not yet populated. In the biblical sense, man has subjugated the earth, taking advantage of other living creatures. In doing so, he has not shied away from subjugating other humans. Already the "old Romans" knew: Man is a wolf to man ("*homo homini lupus*"). The Latin aphorism comes from the comedy Asinaria by the Roman comedy poet (Plautus and Danese 2004). The complete sentence reads in translation:

> *"For man is a wolf to man, not a man; that is true at least as long as you do not know each other."*

The quotation refers to the decisive difference people make in their dealings with others and living beings insofar as familiar or unfamiliar to them. This applies particularly to companion animals, often treated like family members, and to actual family members in a strict and broader sense.

In the course of the industrialization of livestock farming, the treatment of farm animals has changed fundamentally. Farm animals have disappeared mainly into anonymity in the ever-increasing animal populations. In general, neither the livestock keepers nor the other beneficiaries of the established mode of production see them as individuals or fellow creatures. The banishment of the animals to the stables, necessary solely for productivity reasons, removed them from the gaze of citizens and consumers. According to the motto: "Out of sight, out of mind," it seldom occurs to most citizens and consumers to think about the treatment of farm animals. In the circles of livestock farmers, who initially felt unease when cages or crates were introduced, the rational argument of economics and the supposed argument of no alternative are now dominating. According to this argument, only those livestock keepers would have a chance to prevail over the competition willing and able to follow the economic mandate of further productivity increases and exhaust all possibilities of cost minimization for this purpose. However, this is valid only to the point where they become victims in the race to the bottom.

Neither in the past nor present has there been a lack of arguments with which individuals have attempted to justify the handling of farm animals to themselves and others. Before going into more detail on various contexts of justification elsewhere, the perspective will first be broadened here, and the use of animals by humans is placed in the overarching context of evolutionary biological processes. It is indisputable that how humans deal with farm animals is linked to various advantages and far-reaching negative consequences. If the latter are to be curbed, the processes and the forces driving them must be understood and comprehended in the context of this embeddedness. Livestock farmers exploit the biological processes of farm animals. Initially, this is primarily to the detriment of farm animals. However, the negative consequences extend far beyond this. They affect the fauna and flora of the entire globe, the environment and climate, and people themselves. The extent to which this is happening is discussed in more detail in Sect. 5.1.

As outlined before, farm animals have been confronted for decades with considerable impairments accompanied by pain, suffering, and damage. These can be classified as undesirable side effects of intensification processes in livestock production (Rauw et al. 1998). Despite the high degree of disturbances and diseases, this has not been changed and has not found its way into the economic evaluations to date, mainly ignoring the self-interests of farm animals. What matters to farm animals in welfare terms is their subjective experience. Animals perceive and process the incoming information from the internal physical/functional state via their nervous system and the external factors of their living conditions generated by sensory inputs individually and thus differently from each other. They weigh the importance of these factors according to their judgment concerning their well-being (Boissy 2019). Accordingly, well-being is an animal-individual computation result of the brain, which is subject to dynamic changes even in the course of the day. Each disturbed or disrupted feature generates sensory inputs, which the brain processes to form specific negative affective states. These are associated with physiological and behavioral reactions that act to restore the body's internal stability. Negative and positive affective states provide orientation concerning the actual situations and the estimation of whether objects are attractive or should be avoided to improve the chance of survival (Gygax 2017). Thus, farm animals are self-referential concerning their perception and reactions to stimuli. The individual operates as an integrated whole entity, organized to avoid, solve, or compensate for dysfunctions and disturbances.

What also has to be considered is that individual animals react very differently to the same or different living conditions. They differ not only in age, sex, or live weight but also in their immune competence, physiological condition, and rank position. Thus, they have different capacities to adapt to their respective living conditions. In a herd exposed to the same suboptimal living conditions, this is reflected by the proportion of animals that can harmonize their needs with the given conditions without recognizable impairments. Other animals in the herd may show behavioral disturbances. Still, other animals show signs of clinical illnesses; some herd animals exhibit subclinical health disorders, which only become apparent when diagnostic measures are conducted but can nevertheless result in a considerable impairment of welfare. From this perspective, AHW expresses the capability of the individual to cope with the living conditions. In contrast, dysfunctions indicate that animals are overstressed in their ability to cope, caused by discrepancies between what the individual animal needs and what the living conditions provide (Sundrum 2015). As requirements and needs vary considerably between farm animals, and each animal has its specific limitations and capacity to adapt, the individual animal represents the reference system for realizing AHW and assessing the success of farm management in striving for this goal.

What makes it particularly difficult to reduce the prevalence of production diseases is their multi-factorial genesis in the interactions between individual reactions of farm animals confronted with highly heterogeneous and always-changing living conditions. Thus, it is required in the first place to identify and remove the leading causes and risk factors of production diseases, therefore considering the whole farm context. The degree of AHW problems within a livestock farm system is reciprocal to the farm management's ability, capacity, and skills to provide appropriate resources and protection services against biotic and abiotic stressors according to individual needs. On many farms, the bottleneck for any improvements might be due to limitations in the ability to think through the complexity of the production processes on different scales. Many farmers are trapped in their perspective and are often left alone when assessing, diagnosing, and treating farm animals without a valid reference point and any control by a veterinarian authority or certification bodies to determine whether they successfully improve the animal's health status. Considerations of single or very few aspects include the risk of oversimplifying the matter and jumping to premature conclusions. Because relevant resources such as labor time and investment capacities are limited, excessive demands in securing AHW provoke conflicts between other objectives. The need for additional labor efforts and increasing costs for health improvements often are constraints that prevent farmers from seeking advice and making use of recommended measures. Furthermore, uncertainties concerning the effectiveness of preventive measures explain why farmers are often reluctant to increase prevention efforts to reduce disease prevalence rates and the failure costs deriving from there.

On the other hand, carcasses and organs, as well as products from previously diseased animals, are sent for further processing in the same way as those from healthy animals. If necessary, as with a rotten apple, the defective parts, which may result from inflammatory processes, are cut out, or the whole organ, but more rarely, the entire carcass, is discarded. Finally, it would mean a considerable financial loss for primary producers if carcasses or products from previously diseased animals could not be sold or would gain a considerably lower payment. However, as long as products from previously diseased animals are gaining the same price as those from healthy animals, there are not enough incentives for the farmers to considerably reduce the prevalence of diseased animals.

The fact that a large proportion of livestock farmers themselves are affected by the negative consequences of competition fueled by the continued efforts to increase productivity further is rarely addressed. Instead, the focus is primarily on the competition's winners, whereas the losers are widely disregarded. The development is ruining those farms that have lost out to their competitors in the competition to increase productivity unless they have taken themselves out of the running in time. Many farmers—convinced of their abilities, with a great deal of faith in God or for lack of alternatives—have accepted the competition and invested in the future with the help of outside capital. However, most of those who have embraced competition are among the losers. The dramatic decline in the number of businesses speaks a clear language in this regard. As in other sectors of the economy, the reasons for giving up a farm or a branch of a farm are manifold. In addition to the initial structural conditions (unfavorable locations, insufficient land, outdated stables, lack of liquidity, etc.), various social and human factors influence decisions, particularly in the case of an impending generation change. Furthermore, animal diseases in the form of epidemic or production diseases can undermine the economic basis of existence. Last but not least, unexpected market developments that affect the costs of the means of production or the market prices of the products sold can hit the farms at an unfavorable time and result in the withdrawal of a farm from production.

All these factors have in common that they do not mark the underlying cause of farm closures. The primary reason is the continuing need for further increases in productivity in a globally competitive environment, irrespective of the progress already made. This continues to exist as long as the market promotes only those companies that can produce more cheaply, and the negative consequences and side effects that the production processes cause are not considered. The latter would require an institution that monitors the undesirable side effects, prepares the bill for negative externalities, addresses it to the polluters, and sanctions its disregard. In addition, it would require the offer of food produced with significantly fewer undesirable side effects. In this way, the companies that have so far profited from the fact that the undesired side effects have not been considered so far would face fair competition.

So far, the proponents of the intensification processes seldom mention the associated undesirable side effects, e.g., in the form of AHW problems, distortions of competition, or even the need for reorientations necessary for the agricultural economy. At most, there is talk about challenges. References to undesirable side effects caused by the production processes are usually sought in vain in public announcements. Thus, a problem description and, above all, a comprehensive analysis is often missing. Instead, agriculture's production value and employment share in the entire agribusiness and the upstream and downstream sectors of the economy are highlighted. While the published figures underline the importance of the agricultural industry for society as a whole, they contrast with the income situation in agriculture. Gross value added per person employed lags significantly behind incomes in other economic sectors. The current gross value added per person engaged in agriculture and forestry is very low. The value added in trade, the hotel and restaurant industry, and transport, which earn their living, among other things, by using raw materials from agriculture, is, in general, significantly higher than in agriculture. At the same time, the share of expenditure on food and beverages in total consumer spending has fallen considerably; for instance, in Germany, it fell from 24.5% in 1970 to less than 14% in 2019. A far below-average rise in food prices offsets considerable income increases on consumers' part. The higher standard of living manifests itself in increased spending on housing, transport, leisure activities, and health care; lower spending on food makes the standard of living possible.

Farmers have received an ever-smaller share of what consumers have paid for food over the past decades. (Bundesinformationszentrum Landwirtschaft 2021). The share of consumer spending on **meat**, **milk**, and **eggs** in 1970 was 44%, 57%, and 85%, respectively. The shares fell dramatically to 22, 39, and 43% in 2019. The prices farmers receive for their raw produce have less and less to do with the prices consumers pay at the supermarket checkout. This is evident in pork, for example. At the end of 2014, producer prices were € 1.33 per kg carcass weight, and the average consumer price was € 6.07. At the end of 2020, due to export restrictions due to African swine fever, producer prices fell to 1.22 euros. (Dynowski 2021). Notwithstanding the slump in quotations at the slaughterhouse, retail prices did not decrease but increased to 7.19 €. This increased the difference from 4.74 to 5.97 € in favor of the downstream areas of primary production. Thus, there is no doubt who benefits most from establishing the political framework conditions of a supposed free agricultural market and globalization.

Anyone who is surprised by these price developments or protests against them, as farmers do time and again, has not yet sufficiently taken note of how pricing works in a free market economy. According to this, especially in the case of anonymous mass-produced goods, it is part of the self-image of the buyers of the food retail trade that they only pay for the raw goods that they have to pay to obtain them. Any oversupply of raw goods plays into the buyers' hands. Since **food retailers** compete with each other, and consumers continue to be very price-conscious in their purchases, they have no reason to deviate from a least-cost purchasing practice. At the same time, as the pricing of pork shows, the grocery industry is not compelled to pass on lower purchase prices to consumers. The small share that reaches farms from consumer spending does not reflect productivity gains and the extent to which production costs have fallen. It documents the importance of oversupply in the markets, which retailers can exploit to pursue their interests. So far, neither the potential beneficiaries nor the large number of losers nor the agricultural economists have addressed the issue of who will have to pay the price for this misguided agroeconomic development.

So far, the general public has not yet noted the extent of the undesired side effects and animal health and welfare problems in livestock production. Among other things, this might be because the issue is very complex. Thus, laypersons and even farmers are widely incapable of assigning significant variations of impairments. Being overstrained to assess the actual quality of animal-sourced food, consumers see themselves forced to either condemn livestock production and the food products deriving from there more or less in general or to disregard all the possible undesired side effects. On the other hand, consumers are confronted with an increased risk of coming into contact with zoonotic pathogens deriving from animal-sourced food, causing foodborne infections and intoxications. Many food infections go unrecorded because patients fail to report the illness. Due to the multiplication of micro-organisms in overcrowded housing systems, enormous amounts of antimicrobial effective substances are used to contain the spread of germs within the flock, promoting an increased development of antimicrobial-resistant germs (Ritter et al. 2019). These reduce the effectiveness of antimicrobial substances in farm animals and humans, who come in contact with resistant germs in several ways, notably through animal-sourced food.

Although agricultural and animal science has provided a comprehensive database, authorities are reluctant to communicate the extent of adverse side effects to the public and draw corresponding conclusions. Extensive monitoring systems are lacking in many of the relevant animal-source food-producing countries. Where data exist, they are not made official. On the other hand, consumers are increasingly interested in the attributes of the food they consume. A recent survey in many European countries revealed that over nine in ten Europeans consider it essential to protect the welfare of farmed animals, and even an absolute majority think that it is crucial (European Union 2017). Around nine in ten Europeans think that providing farmed animals enough space, ensuring that they have enough food and an adapted environment, and ensuring that people who handle them have sufficient skills and training is essential to ensure that farming and breeding practices meet our ethical responsibilities to animals. More than eight in ten Europeans also believe that the welfare of farmed animals in their country should be better protected than now, suggesting an increasing concern regarding farmed animals. Moreover, six in ten Europeans are willing to pay more for products sourced from animal welfare-friendly farming systems. Finally, six in ten Europeans look for labels identifying products sourced from animal welfare-friendly farming systems when buying food products. However, the current plethora of labels on the market does not adequately address the issue of AHW problems; rather than providing actionable information, most labels add to consumer confusion (see Sect. 6.5). In addition, there is a tendency toward "absence labels" that can contribute to a negative consumer perception of conventional products that may or may not include the attribute in question.

Whereas the demand for animal-sourced food (ASF) is still increasing in developing countries, many developed countries record decreases in the consumption of ASF, particularly pork and beef (Knap 2022). This might be attributed to an increasing awareness of consumers and citizens about the adverse side effects of intensified production processes. Reports in the media about suboptimal conditions in livestock farming concerning animal health and welfare have increasingly shaken consumers' confidence in food safety and quality (Alonso et al. 2020). They are further confused about environmental pollution and climate change links with livestock production (Chai et al. 2019). Within a short period, negative media attention has grown to unprecedented levels. Consequently, we see a dramatic deterioration in the image of livestock production and ASF in developed countries. The number of flexitarians, vegetarians, or strictly vegan consumers has risen substantially, particularly in the young generation. By changing consumption behavior, consumers wish to distance themselves from the adverse effects of livestock production by increasingly favoring plant-based over animal-sourced food.

Derivations

The particular extent of food shortages during and after the Second World War prompted politicians in various European countries to reach a supranational agreement and to set the course for increasing productivity in agriculture and livestock farming. The strategies provided the blueprint for increasing productivity in the industrial sectors. These had shown how production processes could be intensified, and their efficiency increased dramatically through mechanization and specialization. The price guarantees granted at the beginning created an enormous incentive for farmers to increase their income by increasing product quantities. In further development, competition forced the farms to align themselves closely to the production goal of increasing productivity. All agricultural sectors had to subordinate themselves to this goal to a greater or lesser extent. At the same time, the overriding **objective** created the preconditions for a concerted interlocking of the different production areas. Outside the farms, an infrastructure was developed that ensured the smooth supply of means of production, mainly animal feed, and the collection and processing of raw materials of plant and animal origin. In animal houses, the main tasks, feeding the animals, disposing of excrement, and milking, like dairy cows, were at the center of mechanization and automation efforts. Increasing specialization promoted the implementation of technical simplifications of routine work. Accordingly, fewer and fewer people were needed for the work processes. Year-round housing was inevitable in pig and poultry farming to take full advantage of mechanization. Farm animals were relegated to the interior of ever-larger barns built outside villages. This made the farm animals mainly invisible to most people.

The policy-driven expansion of agricultural research institutions created the conditions for a detailed illumination of biological processes and, thus, for the possibilities of harnessing them to increase productivity. At the same time, structures for performance measurement and advisory services were created, by which advances in knowledge could be communicated to practitioners. However, policymakers failed to counterbalance the wide range of support measures with regulatory measures that could have curbed excessive growth. People were too susceptible to trust the advice of agricultural economists that the market would regulate the supply and demand situation. Today, we know that the market is failing to regulate itself in a way that benefits public interests in the long run.

At the beginning of the post-war years, the conflicting interests of producers (increased incomes) and consumers (low product prices) were reconciled by the ongoing increase in productivity. The hitherto unknown availability of butter and meat at low prices became a metaphor for rising living standards. However, by the late 1970s, demand for staple foods was already largely saturated. Given farmers' enormous surpluses, support from public funds and the development of new markets were needed. Consequently, prices are no longer set in a regional context but based on world market levels. However, the prices that can be achieved on the global market often do not cover the production costs of many producers. Nevertheless, many farms feel compelled to hold on to this line of business because of the lack of alternatives and cling to the hope that world market prices will rise again. Farmers have no direct influence on market prices but have to deliver their milk to the dairies and slaughter animals to the slaughter companies. The latter are in fierce competition for contracts with food retailers and are therefore under price pressure. Livestock farmers are at the beginning of the supply chain of raw goods; as far as rewards are concerned, they are just the residual cash recipients. Farmers produce our food but can hardly secure their economic existence from it.

The tremendous increase in productive efficiency in livestock production is based mainly on selecting high-producing animals within breeding programs and developing technological and scientific solutions to deal with the extensive consequential effects of performance increase. Due to the coevolution of biological, technological, and economic progress, the capacity of farm animals to transform plant biomass into valuable food for humans has been largely exploited. To call it an enormous success of the human spirit illuminates only one side of the coin. This perspective disregards the law of diminished marginal utility, which refers to the phenomenon of each additional unit of gain leading to an ever-smaller increase in utility and the cost-benefit ratio. This law applies to all processes that try to increase utilization with increased efforts and investments. As shown in Sect. 5.2, the proportion of costs and benefits can become harmful, mainly when the economic expenditures and all adverse side effects are considered on the cost side. Furthermore, questions arise: who belongs to the winners and who to the losers? Furthermore, is their relationship the result of fair or unfair competition? When the majority of those who are responsible for the farm animals in the first place become losers, questions arise about whether this is an unchangeable fate to which those concerned resigned or if there is a chance to escape the potential consequences.

The quote cited above indicates a possible way out. The man is no longer a wolf to man when they know each other or when they belong to the same community, respectively. Members of a community can enclose the negative and ruinous implications of competition. A fundamental prerequisite is that other members are not only perceived as competitors but also as part of a common superordinate community characterized by clear rules to which all members must comply to remain part of the community. In communities, rules can be implemented and enforced when those who violate the rules are confronted with an exclusion from the community. These interrelationships will be addressed in further detail in the following chapters. First, however, the extent of impairments of the public interests caused by the intensification of agriculture will be further discussed.

In a global food market, competition primarily revolves around cost-efficiency in primary production and in the following agri-food supply chains. The economic framework leaves little scope for actions that are not focused on economic efficiency and hampers all those efforts in farm management from which a financial return appears unlikely. A persistently high level of animal health and welfare problems in livestock farms indicates that many farmers fail to provide what the individual farm animals need to cope and what an increasing portion of citizens and consumers demand. With the focus on productive efficiency, adverse side effects are not adequately considered when weighing up conflicts of interest relating to farm animals, food quality, consumers' expectations, or public welfare interests. This can apply even to the negative impacts of AHW problems on the economic viability of the farm system, which substantial financial losses can undermine without being adequately realized. Thus, AHW problems are not only a threat to the survival of the farm animals and the economic viability of the farm system but are also partly responsible for a decrease in the demand for animal-sourced foods in many European countries. In the long run, this poses a severe threat to the business model of livestock farming. The negative image of livestock production decreases the demand for animal-sourced foods and raises questions about societal support and the continuance of generous public subsidies. In the long run, livestock farms with an inferior AHW level might lose their social license to produce animal-sourced food.

References

Alonso ME, González-Montaña JR, Lomillos JM (2020) Consumers' concerns and perceptions of farm animal welfare. Animals 10. https://doi.org/10.3390/ani10030385

Anonym (1975) Tierfreundlich, arbeitssparend, kostengünstig. Leistungsgerechte und arbeitssparende Haltungsverfahren für Sauen. Mitteilungen der DLG 15:857

Ballmann A, Schaft F (2008) Zukünftige ökonomische Herausforderungen der Agrarproduktion: Strukturwandel vor dem Hintergrund sich ändernder Märkte, Politiken und Technologien. Archiv Für Tierzucht 51:13–24

Baumgard LH, Collier RJ, Bauman DE (2017) A 100-year review: regulation of nutrient partitioning to support lactation. J Dairy Sci 100:10353–10366. https://doi.org/10.3168/jds.2017-13242

Baur S, Rufener C, Toscano MJ, Geissbühler U (2020) Radiographic evaluation of keel bone damage in laying hens-morphologic and temporal observations in a longitudinal study. Front Vet Sci 7:129. https://doi.org/10.3389/fvets.2020.00129

Bessei W (2006) Welfare of broilers: a review. WPS 62:455. https://doi.org/10.1017/S0043933906001085

Blomvall L, Kaukonen E, Kurittu P, Heikinheimo A, Fredriksson-Ahomaa M (2023) Food chain information and post-mortem findings in fattening Turkey flocks. Food Control 150:109739. https://doi.org/10.1016/j.foodcont.2023.109739

Boissy A (2019) How to access animal sentience? The close relationship between emotions and cognition. In: Hild S, Schweitzer L (eds) Animal welfare: from science to law. La Fondation Droit Animal, Éthique et Sciences, Paris, pp 21–31

Boyd W, Watts M (1997) Agro-industrial just in time: the chicken industry and Postwar American capitalism. In: Goodman D, Watts M (eds) Globalising food: Agrarian questions and global restructuring. Routledge, London, pp 192–225

Brown RE (1969) The conversion of nutrients into milk. University of Nottingham. In: Swan H, Lewis D (eds) Proceedings of the University of Nottingham, third nutrition conference for feed manufacturers. London, Churchill, 23–42

Bundesinformationszentrum Landwirtschaft (2021) Infografiken. https://www.landwirtschaft.de/landwirtschaft-verstehen/haetten-sies-gewusst/infografiken/. Accessed 12 July 2021

Bundesministerium für Ernährung und Landwirtschaft (2020) Empfehlungen des Kompetenznetzwerks Nutztierhaltung. https://www.bmel.de/SharedDocs/Downloads/DE/_Tiere/Nutztiere/200211-empfehlung-kompetenznetzwerk-nutztierhaltung.html

Bus JD, Stockhofe N, Webb LE (2019) Invited review: Abomasal damage in veal calves. J Dairy Sci 102:943–960. https://doi.org/10.3168/jds.2018-15292

Chai BC, van der Voort JR, Grofelnik K, Eliasdottir HG, Klöss I, Perez-Cueto FJA (2019) Which diet has the least environmental impact on our planet? A systematic review of vegan. Vegetarian Omnivorous Diets. Sustain 11:4110. https://doi.org/10.3390/su11154110

Comberg G (1984) Die deutsche Tierzucht im 19. und 20. Jahrhundert. Tierzuchtbücherei. Ulmer, Stuttgart

Compton CWR, Heuer C, Thomsen PT, Carpenter TE, Phyn CVC, McDougall S (2017) Invited review: A systematic literature review and meta-analysis of mortality and culling in dairy cattle. J Dairy Sci 100:1–16. https://doi.org/10.3168/jds.2016-11302

Corbière F, Cassard H, Foucras G, Meyer G, Schelcher F (2008) Hepatic abscesses: pathogenesis, treatment and prevention in cattle. Le Nouveau Praticien Veterinaire Elevages Et Sante 9:36–41

Degeling C, Johnson J (2015) Citizens, consumers and animals: what role do experts assign to public values in establishing animal welfare standards? J Agric Environ Ethics 28:961–976. https://doi.org/10.1007/s10806-015-9571-x

Dynowski K (2021) Im Laden klingelt die Kasse. Allgemeine Fleischerzeitung 2021

EFSA (2012) Scientific opinion on the welfare of cattle kept for beef production and the welfare in intensive calf farming systems. EFSA J 10:166. https://doi.org/10.2903/j.efsa.2012.2669

Emerson MR, Woerner DR, Belk KE, Tatum JD (2013) Effectiveness of USDA instrument-based marbling measurements for categorizing beef carcasses according to differences in longissimus muscle sensory attributes. J Anim Sci 91:1024–1034. https://doi.org/10.2527/jas.2012-5514

Freihold D, Bartels T, Bergmann S, Berk J, Deerberg F, Dressel A, Erhard MH, Ermakow O, Huchler M, Krautwald-Junghanns M-E, Spindler B, Thieme S, Hafez HM (2019) Investigation of the prevalence and severity of foot pad dermatitis at the slaughterhouse in fattening turkeys reared in organic production systems in Germany. Poult Sci 98:1559–1567. https://doi.org/10.3382/ps/pey473

Gareis M, Oberländer S, Zipplies J, Reese S, Schade B, Böhm B, Schwaiger K (2016) Prevalence of auxiliary bursae and injuries of claws in fattening pigs at time of slaughter-results of a study at four slaughterhouses. Berl Münch Tierärztl Wochensch:428–436. https://doi.org/10.2376/0005-9366-16032

Gaschler A (1974) Problembereiche der Massentierhaltungen auf wirtschaftlichem und gesundheitlichem Gebiet. Mitteilungen der DLG:420–424

Glebe TW, Bitsch V, Kantelhardt J, Oedl-Wieser T, Sauer J (2018) Agrar- und Ernährungswirtschaft zwischen Ressourceneffizienz und gesellschaftlichen Erwartungen. Ber Landwirtsch 96:1–29. https://doi.org/10.12767/buel.v96i2.211

Gonçalves JL, Kamphuis C, Vernooij H, Araújo JP, Grenfell RC, Juliano L, Anderson KL, Hogeveen H, Dos Santos MV (2020) Pathogen effects on milk yield and composition in chronic subclinical mastitis in dairy cows. Vet J 262:105473. https://doi.org/10.1016/j.tvjl.2020.105473

Göransson L, Yngvesson J, Gunnarsson S (2020) Bird health, housing and management routines on Swedish organic broiler chicken farms. Animals 10. https://doi.org/10.3390/ani10112098

Gygax L (2017) Wanting, liking and welfare: The role of affective states in proximate control of behaviour in vertebrates. Ethology 123:689–704. https://doi.org/10.1111/eth.12655

Habel J, Sundrum A (2020) Mismatch of glucose allocation between different life functions in the transition period of dairy cows. Animals 10:1028. https://doi.org/10.3390/ani10061028

Hiscock HM, Leishman EM, Vanderhout RJ, Adams SM, Mohr J, Wood BJ, Baes CF, Barbut S (2022) Describing the relationships among meat quality traits in domestic turkey (Meleagris gallopavo) populations. Poult Sci 101:102055. https://doi.org/10.1016/j.psj.2022.102055

Hocquette JF, Gondret F, Baéza E, Médale F, Jurie C, Pethick DW (2010) Intramuscular fat content in meat-producing animals: development, genetic and nutritional control, and identification of putative markers. Animal 4:303–319. https://doi.org/10.1017/S1751731109991091

Knap PW (2022) Drivers of animal source food consumption: a biophysical approach. Front Sustain Food Syst 5. https://doi.org/10.3389/fsufs.2021.732915

Knowles TG, Kestin SC, Haslam SM, Brown SN, Green LE, Butterworth A, Pope SJ, Pfeiffer D, Nicol CJ (2008) Leg disorders in broiler chickens: prevalence, risk factors and prevention. PLoS ONE 3:e1545. https://doi.org/10.1371/journal.pone.0001545

Krieger M, Sjöström K, Blanco-Penedo I, Madouasse A, Duval JE, Bareille N, Fourichon C, Sundrum A, Emanuelson U (2017) Prevalence of production disease related indicators in organic dairy herds in four European countries. Livest Sci 198:104–108. https://doi.org/10.1016/j.livsci.2017.02.015

Lund VP, Nielsen LR, Oliveira ARS, Christensen JP (2017) Evaluation of the Danish footpad lesion surveillance in conventional and organic broilers: misclassification of scoring. Poult Sci 96:2018–2028. https://doi.org/10.3382/ps/pex024

Märtlbauer E, Becker H (eds) (2016) Milchkunde und Milchhygiene, 1st edn. utb, vol 8664. Ulmer, Stuttgart

Mehner A (1966) Welche Rolle soll die Legehennenhaltung in der Landwirtschaft spielen? Der Tierzüchter:364

Millet S, Kumar S, de Boever J, Meyns T, Aluwé M, de Brabander D, Ducatelle R (2012) Effect of particle size distribution and dietary crude fibre content on growth performance and gastric mucosa integrity of growing-finishing pigs. Vet J 192:316–321. https://doi.org/10.1016/j.tvjl.2011.06.037

Mitterer-Istyagin H, Ludewig M, Bartels T, Krautwald-Junghanns M-E, Ellerich R, Schuster E, Berk J, Petermann S, Fehlhaber K (2011) Examinations on the prevalence of footpad lesions and breast skin lesions in B.U.T. Big 6 fattening turkeys in Germany. Poult Sci 90:775–780. https://doi.org/10.3382/ps.2010-01142

Moesseler A, Koettendorf S, Grosse Liesner V, Kamphues J, Mößeler A, Köttendorf S, Große V, Kamphues J (2010) Impact of diets' physical form (particle size; meal/pelleted) on the stomach content (dry matter content, pH, chloride concentration) of pigs. Livest Sci 134:146–148. https://doi.org/10.1016/j.livsci.2010.06.121

Botreau R, Laurent C (2017) A set of indicators to better characterize beef carcasses at the slaughterhouse level in addition to the EUROP system. Livest Sci 202:44–51.https://doi.org/10.1016/j.livsci.2017.05.017

Nordquist RE, van der Staay FJ, van Eerdenburg FJCM, Velkers FC, Fijn L, Arndt SS (2017) Mutilating procedures, management practices, and housing conditions that may affect the welfare of farm animals: implications for welfare research. Animals 7. https://doi.org/10.3390/ani7020012

Office International des Epizooties (2008)—World Organisation for Animal Health, Terrestrial animal health code. 21. Auflage. Paris: OIE. http://www.oie.int/international-standard-setting/terrestrial-code/access-online/

Pallarés FJ, Añón JA, Rodríguez-Gómez IM, Gómez-Laguna J, Fabré R, Sánchez-Carvajal JM, Ruedas-Torres I, Carrasco L (2021) Prevalence of mycoplasma-like lung lesions in pigs from commercial farms from Spain and Portugal. Porcine Health Manage 7:26. https://doi.org/10.1186/s40813-021-00204-3

Plautus TM, Danese RM (2004) Asinaria. Editio Pautina Sarsinatis, vol 2. QuattroVenti, Sarsinae

Polkinghorne RJ, Thompson JM (2010) Meat standards and grading: a world view. Meat Sci 86:227–235. https://doi.org/10.1016/j.meatsci.2010.05.010

PraeRi (2020) Animal health, hygiene and biosecurity in German dairy cow operations—a prevalence study (PraeRi). Final Report, June 30, 2020. https://ibei.tiho-hannover.de/praeri/pages/69#_AB

Rauw W, Kanis E, Noordhuizen-Stassen E, Grommers F (1998) Undesirable side effects of selection for high production efficiency in farm animals: a review. Livest Prod Sci 56:15–33. https://doi.org/10.1016/S0301-6226(98)00147-X

Ritter GD, Acuff GR, Bergeron G, Bourassa MW, Chapman BJ, Dickson JS, Opengart K, Salois MJ, Singer RS, Storrs C (2019) Antimicrobial-resistant bacterial infections from foods of animal origin: understanding and effectively communicating to consumers. Ann N Y Acad Sci 1441:40–49. https://doi.org/10.1111/nyas.14091

Rohlmann C, Verhaagh M, Efken J (2020) Steckbriefe zur Tierhaltung in Deutschland: Ferkelerzeugung und Schweinemast. Johann Heinrich von Thünen-Institut: Braunschweig, Germany

Rumphorst T, Scheu T, Koch C, Sundrum A (2022) Balancing trade-offs in milk production by making use of animal individual energy balancing. Dairy 3:345–363. https://doi.org/10.3390/dairy3020027

Schulz L, Sundrum A (2019) Assessing marbling scores of beef at the 10th rib vs. 12th rib of longissimus thoracis in the slaughter line using camera grading technology in Germany. Meat Sci 152:116–120. https://doi.org/10.1016/j.meatsci.2019.02.021

Schwarz D, Diesterbeck US, Failing K, König S, Brügemann K, Zschöck M, Wolter W, Czerny C-P (2010) Somatic cell counts and bacteriological status in quarter foremilk samples of cows in Hesse, Germany—a longitudinal study. J Dairy Sci 93:5716–5728. https://doi.org/10.3168/jds.2010-3223

Settele V (2020) Revolution im Stall: Landwirtschaftliche Tierhaltung in Deutschland 1945–1990. Dissertation. Kritische Studien zur Geschichtswissenschaft, Band 239. Vandenhoeck & Ruprecht, Göttingen

Sundrum A (2015) Metabolic disorders in the transition period indicate that the dairy cows' ability to adapt is overstressed. Animals 5:978–1020. https://doi.org/10.3390/ani5040395

Sundrum A, Aragon A, Schulze-Langenhorst C, Bütfering L, Henning M, Stalljohann G (2011) Effects of feeding strategies, genotypes, sex, and birth weight on carcass and meat quality traits under organic pig production conditions. NJAS: Wageningen J Life Sci 58:163–172. https://doi.org/10.1016/j.njas.2011.09.006

Sundrum A (2024) Why has animal science not led to improved farm animal health and welfare? Front Anim Sci 4. https://doi.org/10.3389/fanim.2023.1214889

Swaggerty CL, Pevzner IY, He H, Genovese KJ, Nisbet DJ, Kaiser P, Kogut MH (2009) Selection of broilers with improved innate immune responsiveness to reduce on-farm infection by foodborne pathogens. Foodborne Pathog Dis 6:777–783. https://doi.org/10.1089/fpd.2009.0307

Thiede G (1988) Landwirt im Jahr 2000: So sieht die Zukunft aus, 1st edn. DLG-Verl, Frankfurt am Main

Thøfner ICN, Dahl J, Christensen JP (2021) Keel bone fractures in Danish laying hens: Prevalence and risk factors. PLoS ONE 16:e0256105. https://doi.org/10.1371/journal.pone.0256105

Veldhuis AMB, Smits D, Bouwknegt M, Worm H, van Schaik G (2021) Added Value of Meat Inspection Data for Monitoring of Dairy Cattle Health in the Netherlands. Front Vet Sci 8:661459. https://doi.org/10.3389/fvets.2021.661459

Wegener R-M (1977) Wie es ohne Käfig geht. Landwirtschaftliches Wochenblatt 134:33–35

Wehler H-U (2008) Deutsche Gesellschaftsgeschichte: Band 5: Von der Gründung der beiden deutschen Staaten bis zur Vereinigung 1949–1990. Deutsche Gesellschaftsgeschichte, vol 5. Beck, München

Welskopp T (2014) Unternehmen Praxisgeschichte: Historische Perspektiven auf Kapitalismus, Arbeit und Klassengesellschaft. Mohr Siebeck, Tübingen

Wendt M (2004) Erbdefekte und Unerwünschte Selektionsfolgen. Dtsch Tierärztebl 52:357–360

Zentrale Markt- und Preisinformationen GmbH (2002) Agrarmärkte in Zahlen: Europäische Union; Tier und Pflanzenproduktion. ZMP, Bonn

Zuidhof MJ, Schneider BL, Carney VL, Korver DR, Robinson FE (2014) Growth, efficiency, and yield of commercial broilers from 1957, 1978, and 2005. Poult Sci 93:2970–2982. https://doi.org/10.3382/ps.2014-04291

Beneficial and Harmful Effects of Agricultural Industrialization

5

Abstract

Many interactions exist between the intensification of production processes and the undesirable side and harmful effects that counter the interests of the common good. Economic advantages for primary producers and adverse external effects for the common good promote an ambivalence that varies greatly depending on the specific context of each farm. Accordingly, each farm is the reference system for assessing the discharge of pollutants into the environment in relation to the quantity of desirable products sold. The individual farm management decides which measures are implemented and thus is responsible for the ratio between positive and negative effects that may arise from this. Above all, this has to do with the farm-specific marginal utility processes. The following remarks outline the extent of the harmful effects caused by agriculture, primarily due to a one-sided economic orientation of the production processes. If the damaging effects are to be contained, it is essential to understand their magnitude and diversity and the driving forces behind their causation.

In the post-war years, intensification processes began in agriculture, especially livestock farming, dwarfing all previous developments. The course was set based on the experiences gained in the USA with the industrialization of agriculture from the 1920s onwards, which led to impressive advances in productivity. Despite a certain saturation of the markets, efforts to further intensify agriculture continue unabated. This raises questions about the driving forces and the negative consequences associated with intensification. In this context, the need to feed a growing world population is often emphasized, and this should be prioritized. However, the enormous increase in product quantities has so far been able to do little to change the fact that large sections of the

A. Sundrum, *Public Welfare-Oriented Production of Food*,
https://doi.org/10.1007/978-3-662-69040-6_5

population in the so-called "developing countries" continue to suffer from hunger, and many people die due to a lack of adequate food.

In the past, the increase in the global availability of food and even the comparatively low world market prices have not prevented hunger. Hunger emerges from a variety of causes, including political and social causes such as wars, poverty, and a poor education system, in addition to regional climate and weather influences. The production of feed for animals for export to the industrial nations and the destruction of regional markets through overproduction by the industrial nations are also relevant in connection with food shortages in some regions of the world. On the other hand, an excess of food supply over demand on the world markets ensures that world market prices are comparatively low. This makes food more affordable for many people; on the other hand, world market prices are too low for many primary producers to cover their production costs fully.

The developments outlined indicate that we are dealing with very complex interactions not only at the farm level but also at the global level. The simultaneity of positive and negative consequences and the complexity of the processes at the different levels disqualify any form of generalization. The only way left is the uphill path of gaining knowledge by approaching complex issues from different perspectives. In the previous section, various beneficiaries of productivity increases have been addressed. The main beneficiaries include the upstream and downstream sectors of primary production, which have considerably increased their sales and profits due to the expansion of the agricultural output. Moreover, all consumers have benefited, as they now have to spend significantly less of their increased income on food than in the past due to the immense increases in productivity and the associated cost reductions. They can afford high-priced luxury foods and, to a large extent, other consumer goods. From this perspective, all other sectors of the economy gained advantages from the productivity advances in agriculture because this is what created the purchasing power for different consumer goods in the first place. While the protagonists of the intensification processes never tire of emphasizing their successes, the disadvantages and undesirable side effects are generally negated or relativized. A scientific debate on the extent to which the negative consequences can or must be classified as inherent to the system (i.e., not as somehow avoidable, but as a characteristic inherent in the system of industrial agriculture) is only getting off the ground very slowly. Before outlining the extent of the adverse and damaging effects in more detail, we shall first consider the consequential effects emanating from the five main factors of intensification: land use, increased use of means of production, specialization, regional concentration, and mechanization.

5.1 Intensification of Production Processes

5.1.1 Land Use

The surface area of the Earth that can be used for agriculture is limited. In 2016, this area amounted to around 49 million square kilometers worldwide; this corresponds to approximately 37% of the Earth's surface. While it was still possible to significantly increase the amount of agricultural land worldwide between 1961 and 2000 (from 37.1 to 48.1 million square kilometers), the increase in usable land since then has been relatively marginal (World Bank 2020). Ongoing efforts to expand land use are being made at the expense of largely untouched biotopes, such as rainforests in South America or Asia.

Soils contain varying amounts of nutrients, which are removed by cultivating plants. Where it is worthwhile to maintain or increase yield performance, these are at least partially substituted by fertilization measures. However, this is never done entirely, and it is often so inadequate that soil fertility and yield performance successively decline in many regions. The more the soils are additionally treated with machinery, the more the physical, chemical, and biological soil conditions and their regenerative capacity change. If there is excessive nutrient removal by the cultivated plants and insufficient substitution, the logical consequence is depletion, which is analogous to exploiting mineral resources such as ores, coal, gravel, or rare earths.

Land use potential depends on soil, site, and climate conditions. Around 70% of agricultural land consists of grassland. The remaining land can be tilled and thus used for the cultivation of various arable crops. It is used for the cultivation of food and fodder and, increasingly, for the cultivation of plants for energy production. In the past, fodder production comprised about 40% of the globally available arable land. In addition, raw materials for biofuels and material biomass use were produced on 11% of the land (Raschka and Carus 2012). Roughly speaking, about 36% of the world's **cereals** and 70% of the world's soya harvest, 40% of fish catches, and even 33% of dairy products were fed to animals. About 57% of the harvested cereals in Europe ended up in the feed troughs. The orders of magnitude indicate that the various benefit options compete with each other and that the market prices and the yield potential to be achieved decide which crops are cultivated.

At the turn of the millennium, the number of livestock globally was estimated to be about 4.3 billion, of which 1.65 billion were cattle and 900 million were pigs (Smil 2002). Farm animals represent the largest share of the **biomass** of vertebrates on earth, with about 620 million tons. The increase in the biomass of farm animals by a factor of 3.5 over the past 100 years illustrates the extent of the changes that have occurred. In contrast, the biomass of wild mammals is estimated to be just 40 million tons, with a marked downward trend. Continued deforestation in developing countries for cropland and pasture depletes wildlife habitats and biodiversity.

The immense increase in the number of farm animals not only displaces the number of wild animals and species that can still survive under these conditions. The forage crops fed to livestock also reduce the amount of nutrient resources available for human consumption. Generally speaking, humans and livestock compete for land to produce food or feed. Competition for land concerns primarily arable land that can be used for growing both food and feed. In contrast, only ruminants can use grassland growth as a source of nutrients. Since ruminants can use grassland growth and forage from arable land and by-products resulting from food production, there are no sharp dividing lines regarding land competition. It is roughly estimated that about 86% of the feed produced globally consists of components unsuitable for human consumption (Mottet et al. 2017).

On a global average, producing one kilogram of beef requires about 2.8 kg of feed that humans can use directly. For pigs and poultry, the relative benchmark increases to 3.2 kg. Monogastric animals depend on high-quality feed components, a large proportion of which humans can utilize. Compared to ruminants, however, the nutrients are used much more efficiently.

However, these comparative figures should not disregard the fact that they are average figures, which only allow a first approximation of the proportions. In reality, the averages are based on considerable dispersion and do not allow general statements about the specific **ratio of advantages and disadvantages** of the respective use option; this **always depends on the local context**. While within many farm system boundaries, the benefits of intensified livestock production outweigh the disadvantages, livestock production has considerable disadvantages for the public welfare interests in other farms. A balancing exercise should also consider that the production of food of animal origin does not only contribute to the supply of essential macro- and micronutrients to humans. Products of animal origin are also substitutes for the consumption of plant products. If added to the soil, the excrements contain essential nutrients for plant growth. In many countries, livestock are also used as draught animals and serve as the reserve and property of extended families. Accordingly, the effects of the respective land use should not be reduced solely to the quantities of products produced.

For farmers, land use is part of the effort to generate an income and a livelihood to feed the family and to secure their economic existence. The potential for utilizing the land and the plants grown on it depends not only on the possibilities for utilization within the farm but also on the market situation in the form of the prevailing price level. The supply of raw materials on the local, regional, or global markets determines the price the farmer can obtain for his crop products when selling them or the prices that he has to pay for production inputs such as fertilizers and animal feed. On mixed farms, farmers are constantly faced with the choice of using different proportions of their own or rented land for growing and selling products ("cash crops") or for producing fodder to feed their livestock on the farm. The central question is: "Make it or buy it"? This question has to be answered repeatedly based on the cost development of the means of production, the expected yields, and the market prices for the sales products. It is, therefore, not surprising that agricultural sales products have been traded as speculative objects on the stock exchanges for some time now. The art of speculation is the correct estimation of the opportunity costs, i.e., lost revenues, which result from the fact that existing possibilities (opportunities) are not taken up. It is, therefore, an economic concept for quantifying the benefits of lost alternatives. However, financial considerations do not enclose possible disadvantages for public interests.

To better classify the significance of land use for the interests of the common good, it helps to change the perspective from the supply side to the demand side: how many people can be fed depending on the type of land use? Using some figures, the ratios can be roughly approximated as follows: A predominantly vegetarian diet requires an area of about 800 m^2 of arable land per person to cultivate various plant products, assuming a high yield level (Smil 2000). In contrast, a typical diet of people in the Western world with a meat consumption of about 50 kg per year requires an average land area of about 4000 m^2. The difference makes it clear that the Western world's dietary style is primarily a matter of access to limited arable land. The consumption habits of the Western world are based, to a relevant extent, on the import of products (e.g., maize and soya) produced by exploiting land outside its territory. The import of products into the Western world is consequently accompanied by a reduced availability for other world populations, which cannot afford to use the products produced in their own country.

The Food and Agriculture Organization of the United Nations (FAO) monitors the production and distribution of agricultural products and foodstuffs worldwide. Where possible, the FAO should help to ensure food security for the world's population and improve living standards. However, FAO cannot enforce improvements against market mechanisms of action, nor can it modify the framework conditions of markets so that they are conducive to the overall objectives of providing food for the world's population. As will be discussed in further detail, the market conditions are not designed to feed the world's population in the sense of an orientation towards **public welfare interests** but to ensure the functioning of the markets (see Sect. 8.1). The drastic increase in the use of land for the cultivation of biomass for energy production in recent years, which public funds have massively promoted, has created an additional financially strong competitor for limited usable land. From the perspective of developing countries, the production of energy from biomass or the feeding of maize and soybeans from intensive cultivation to ruminants is, above all, one thing: quite a waste of resources. In developed countries, bioenergy production ensures the economic viability of many farms because the cultivation of food and feed can often no longer cover the production costs. While the use of grassland and by-products of the processing industry represents a strategy of resource use and food production that is not at the expense of human nutrition, it is often not competitive with the use of nutrients that would also be suitable for direct human consumption.

In summary, it can be concluded that land use and the degree of intensification of production processes are not determined by efforts to make the most significant possible contribution to feeding the world's population. They are defined by market activity, which determines the purpose of using available cultivated land and products.

5.1.2 Increased Use of External Resources

In addition to water, **reactive nitrogen** (Nr) is the key element of biosynthesis and an indispensable prerequisite for all biological processes. In all biological growth processes of microbes, plants, and animals, including humans, its availability determines the extent to which the genetic growth potential of organisms can be exploited. The availability of Nr, however, is limited. This may seem surprising at first, given that nitrogen is available in vast quantities around the globe. The total mass of nitrogen (N) in the earth's atmosphere, soil, and water is about 4,000 gigatons. It thus exceeds the combined mass of carbon, phosphorus, oxygen, and sulfur. The crucial point is that more than 99% of the nitrogen is present as molecular nitrogen (N_2). This is inert and, therefore, not usable by more than 99% of living organisms. Consequently, the availability of reactive nitrogen compounds and water availability limits biological growth processes in priority to other nutrients. Consequently, biological growth processes could only be intensified with an expansion of the availability of Nr. Against this background, the most significant human intervention on the globe to date has been to break the triple bond of molecular nitrogen (N_2) to produce reactive nitrogen synthetically and thus make it usable for the intensification of biological processes.

The German chemists Fritz Haber and Carl Bosch achieved this at the beginning of the twentieth century through a combination of chemical expertise and engineering skill. For the first time, they produced ammonia from atmospheric nitrogen and hydrogen, thereby chemically synthesizing a reactive N compound. The Haber–Bosch process, named after the inventors, was explosive in two senses. Using large amounts of energy provided by burning fossil fuels allowed the production of explosives and mineral nitrogen fertilizers (urea, ammonium nitrate, ammonium sulfate, ammonium phosphate). The extent of the release of Nr, which picked up speed after World War 2 and has been increasing ever since, is shown in Fig. 5.1.

Whereas with explosives, the induced effect is explosive, reactive N compounds have a slow but far more profound impact. Without the availability of Nr in the form of mineral fertilizers, much of the world's population alive today would not exist because they could not have been fed (Smil 2012). Figure 5.2 illustrates that the global overpopulation of humans and farm animals could not have occurred without the Haber–Bosch process.

It seems idle to judge whether the Haber–Bosch process is a blessing or a curse of humankind. Humans invented the process; humans must balance negative and positive effects. No other chemical substance symbolizes the ambivalence of the advantages and disadvantages of human intervention in geo-biological processes better than Nr. If this

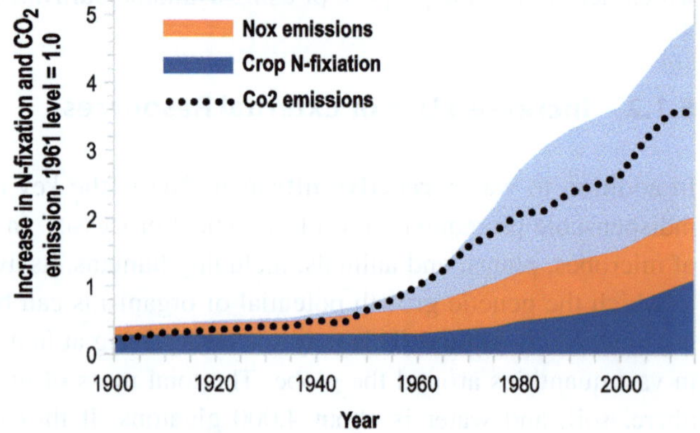

Fig. 5.1 Trends in anthropogenic releases of reactive nitrogen compounds since 1900 about anthropogenic CO_2 emissions (Battye et al. 2017)

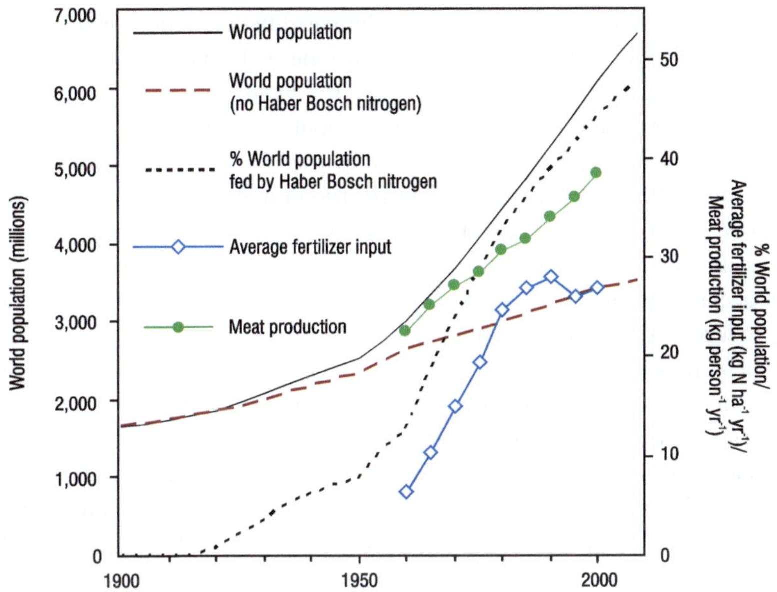

Fig. 5.2 Ammonium feeds the world population (Erisman et al. 2008)

is the key to **growth**, then this is also the key—according to the essential thesis of this book—to a containment of growth processes that is compatible with the common good. Before going into more detail, explaining the effects of Nr and other substances at various agricultural production levels is necessary.

If today there is talk of a new geochronological epoch, the Anthropocene (Ehlers 2008), an age in which humans have become one of the most critical factors influencing Earth's biological, geological, and atmospheric processes; then, it was heralded by the Haber–Bosch process. It established not only the enormous biological growth achievements that have since taken place globally but, in equal measure, the destructive downsides resulting from the negative consequences associated with synthetic Nr. The production of nitrogenous fertilizers, for example, is very energy-intensive. Consequently, production is accompanied by high CO_2 emissions and the release of considerable quantities of the greenhouse gas nitrous oxide (N_2O). This is also a reactive nitrogen compound. Other Nr compounds include Ammonia (NH_3), ammonium (NH_{4+}), oxides of nitrogen (NOx), nitric acid (HNO_3), nitrate (NO_{3-}), urea, amines, amino acids, and nucleic acids. Conversion from one form to another depends on various preconditions and is associated with global consequences for all bio-geo-chemical processes (Galloway et al. 2003). These take place both inside and outside agricultural systems. The reactive nitrogen compounds that cross the intra-farm boundaries of farm systems diffuse into all four environmental spheres: Biosphere, Atmosphere, Hydrosphere, and Lithosphere.

The driving force behind a high fertilization intensity with mineral nitrogen, which the fertilizer industry supplies relatively cheaply, is the farmers' desire to maximize crop yields. The Haber–Bosch process essentially eliminated the limitations that had previously existed in the availability of Nr for plant fertilization and, thus, for plant growth. Most farmers try increasing farm income with higher crop yields to compete with global competitors. Consequently, the use of this production input has increased exponentially. So far, the most relevant limiting factor on the quantities used is the price farmers must pay. In Europe, the Nitrates Directive was enacted in 1991 to limit the input of the reactive N compound released during N conversion into groundwater to 50 mg per liter (see Sect. 5.3.3).

At the same time, large quantities of nitrogen-, potash- and phosphate-containing excrement accumulate in livestock-intensive regions in Europe and Germany. Often, the amounts produced exceed the absorption capacity of the farm's land. If possible, the surplus excrement, mostly in **slurry** (manure and urine mixture), is delivered to livestock-less arable farms in the neighborhood or the nearer and broader surroundings. As the capacities of farms receiving manure are also limited in regions with high livestock numbers and the immediate vicinity, many livestock farms often have no choice but to have the manure transported and disposed of over longer distances by specialized service providers.

Reactive nitrogen compounds and many other bulk elements are not only found in mineral and farm fertilizers. They are also essential ingredients of feedstuffs. Generally, vast quantities of externally produced feedstuffs are imported to livestock farms. In particular, soybeans, rich in essential amino acids, are used as protein supplement feed in animal nutrition. In the meantime, the mineral nitrogen fertilizers for plant nutrition and the essential amino acids for animal nutrition can be produced synthetically. The price-worthiness of the available feed components and their complementarity concerning the requirement profile of the total ration is decisive for the respective ration proportions.

The improved availability of fertilizers and feedstuffs enables the intensification of on-farm production processes. However, only some nutrients imported into the farm are organically bound as plant or animal sales products or the soil. The N-compounds contained in the feed and not utilized by the organism are excreted via the feces (organically bound form) and urine as urea (mineralized form). The excrement can be used as "farm manure" for plant nutrition. For centuries, farm manures were the only Nr-source of plant nutrition outside the soil reserves aside from nitrogen fixation by nodule bacteria when growing legumes. This is still the case in organic farming, where mineral N fertilizers are banned as part of the overall concept. Plant growth and the yields that can be achieved with it are correspondingly limited here.

The synthetic Nr compounds are accurate miracle weapons for increasing plant and animal production productivity, if not for the undesirable side effects. The cultivation and transport of soybeans and other feedstuffs and the synthesis of mineral fertilizers and amino acids are associated with high energy consumption and correspondingly high CO_2 emissions and other greenhouse gases such as nitrous oxide. Soy cultivation also has negative consequences in the countries where it is grown, such as deforestation, land use conflicts, and environmental pollution. These are just a few of the upstream harmful effects of the production processes involved in procuring Nr-containing means of production. Due to the intensification of crop and livestock production based on increasing quantities of fertilizer and feed, the quantities of excrements and the nutrient surpluses on farms and, thus, the substance discharges into the environment have also increased drastically. The effects of these discharges on the environment are explained in Sect. 5.3.

5.1.3 Specialization

The specialization of farms in one or a few branches is the counter-concept to the mixed farms that have characterized the structure of farms in past centuries. Mixed farms offer various possibilities for utilization and income, partly compensating for disturbances and impairments in sub-areas and thus ensuring greater supply security. However, this is only true if the farms are in the black with the total income. Under a general decline in the price of agricultural commodities, a diverse farm structure has a detrimental effect because they can no longer keep up with the productivity advances of specialized farms. Therefore, a certain degree of specialization is inevitable to compete for cost leadership. Because specialization made relevant productivity-increasing effects possible, it was associated with the hope of participating in general economic progress and income development. Limits to specialization arise where production constraints such as limits to crop rotation and the need to look after livestock are encountered and where the further expansion of a farm branch can no longer reduce production costs. While the advantages of specialization for productive efficiency are indisputable and are readily brought to the fore (see Chap. 4), the disadvantages within and outside of the farm systems and the negative consequences for common goods are rarely discussed.

In crop production, specialization is expressed, among other things, in machinery adapted to the production direction and specific expertise, e.g., in plant diseases, fertilizer requirements, and weed control. In most cases, unlike in animal production, a certain amount of leeway is maintained. Agricultural land can generally be cultivated with various fruits, depending on the market situation. Extending the cultivation of one crop at the expense of others narrows the crop rotation, which is contrary to maintaining soil fertility. Also, most crops are not self-compatible (i.e., they cannot be grown on the same land for several years). An exception is maize, which is mainly compatible with itself and also, for this reason, dominates crop rotation in many regions. Other limits to specialization are determined by the availability and remuneration level of skilled labor and, thus, by the situation in the labor market. Ultimately, the situation in the sales markets determines whether specialization can be sustained and, if necessary, expanded within a foreseeable time. In the future, greater emphasis will be put on whether specialization offers solutions for the problems caused by the intensification processes.

In the past, the **recycling community** within farms essentially consisted of several so-called "non-market products." These were utilized within the farm and converted into marketable products to benefit the farm. Non-market products are, on the one hand, the main products of fodder cultivation on arable land and natural grassland and, on the other hand, by-products of cash crop cultivation, such as straw in cereal cultivation, waste potatoes or beet leaf in sugar beet cultivation. The quantities produced are directly related to the harvest quantities of the main crop. On mixed farms, different types of livestock were kept in coordination with the available resources of nutrients, stable buildings, and labor, which allowed the resources to be harnessed for farm income. On specialized farms, the bond of the recycling community has largely dissolved. With specialization, individual process steps are removed from the context of the recycling community to maximize productivity. The focus narrows the view on details, while secondary aspects and the superordinate whole threaten to be lost from view. Consequently, many resources within a farm system remain unexploited. This is particularly true for grassland growth in geographically and climatically disadvantaged regions. Since this biomass is hardly suitable for high-yielding dairy cows or beef cattle, many areas must be kept open as part of landscape management measures at great expense, financed from public funds.

In livestock farming, specialization is usually accompanied by a significant increase in the number of animals, if only to amortize the necessary investments more quickly. Initially, when the herd is increased, the labor requirement per livestock unit decreases very rapidly. Thus, the relative labor-saving from the increase in the number of heads alone is very large. However, above a specific herd size, a further increase in herd size does not significantly reduce the labor requirement per individual animal. On the other hand, when herd sizes are increased, the individual animals and their specific needs are increasingly lost from view. Above all, however, specialization in one type of animal or one specific production stage within the production chain (breeding, reproduction, pre-fattening, or finishing) involves a commitment that, once made, is almost impossible to revise. On the one hand, the unique features of barn construction are too specific to the production processes to be made usable in any other way. On the other hand, the necessary investments are extensive and can only be depreciated over extended periods. Specialization, therefore, entails a **path dependency** from which the farms can hardly extricate themselves. With specialization, farms place themselves in economic dependencies on the costs paid for the means of production and the market prices obtained for the products sold. As part of a vertically organized production chain, they depend on the fact that the processes with the supplying and purchasing market partners function on schedule, just as they are compelled to comply with the agreements entered into.

If the relationship between costs and performance becomes unfavorable due to changes in market conditions, the biological processes—unlike technical production processes—cannot be shut down overnight. The animals still have to be fed, the dairy cows still have to be milked, and the **eggs** still have to be collected, even if the market prices for the products sold do not cover costs. Only farms specializing in the final fattening phase of fattening animals may temporarily refrain from introducing new animals unless this is contrary to contractual agreements. Breeding farms must, however, keep breeding animals even if the sales prices for the young animals do not reward the effort involved. Once established, a stock of breeding animals cannot be frozen and thawed like sperm or eggs. The elimination of a breeding stock usually means the end of that branch of the business. These relationships explain why many farms continue to farm even when it is not profitable; path dependency forces them to do so. They then accumulate further debt, hoping the market situation will change again. On the other hand, rising market prices lead to many farms simultaneously expanding their production capacities to earn money and, if necessary, to reduce debts. As a result of the increase in volume, market prices then quickly come under pressure again. The possibility of increasing added value is limited to implementing further measures to increase productivity, with which they seek to gain a temporary advantage over competitors in an ongoing cut-throat and undercutting competition.

As soon as the size of the livestock is no longer tied to the farm's forage areas but relies on external feed sources, the livestock can be significantly expanded. At the same time, this creates new problems, among others, for using animal excrement. A mismatch between animal stocking and land availability means that excrements have gone from valuable farm fertilizer to a waste product that must be disposed of as cheaply as possible. Again, context determines if nutrients become pollutants and reverse from a beneficial to an undesirable consequential effect. The situation is different in organic farming. Due to the fundamental renunciation of mineral nitrogen fertilizers, farm manure becomes a limited resource for plant nutrition. The higher availability of farm fertilizers would have an increasing yield and thus an income-relevant effect. The renunciation of mineral N-fertilizers creates new framework conditions in which farm manures have a different significance than conventional agriculture. Farmyard manure is a limited resource in organic farming that must be used efficiently.

Where specialization is advantageous from the farmer's perspective, at least in certain phases, it becomes an environmental **problem**, i.e., the temporary economic advantages of primary producers are at the expense of the environment and, thus, of the common good. Farmers can only gain the cost advantages of specialization because society does not charge farmers for dumping excess waste into the environment. While every other citizen has to pay fees for household waste disposal, waste disposal is free for farmers. While farmers are forced to specialize and increase livestock numbers due to economic conditions, they are simultaneously imposing additional burdens on the environment and, thus, on society. In this way, supposed profits are privatized, and the consequences of the costs are socialized.

5.1.4 Technification in Agriculture

The intensification of agriculture is inconceivable without the progress that has been and continues to be made in technical innovations. The mechanization of processes has made it possible to achieve an immense increase in productivity on the farmland or in the stables and, at the same time, to make drastic savings in terms of labor and working time in relation to production output. The core of mechanization is expressed in the imposing tractors equipped with more power and functions and in the automation of work steps that once had to be carried out by hand. In the barn, this mainly concerns feeding, the removal of excrement, and milking dairy cows. While the collection of eggs has been mechanized for many years, the labor-intensive collection of fattening poultry before slaughter is now technically supported by large suction devices. There are hardly any limits to technical ingenuity. This may also explain why many people have an almost religious faith in technical progress, which is expected to master even the trickiest tasks in the future and even solve the problems associated with the use of technology itself.

As explained in Sect. 4.1, it is above all the biology, particularly the biology of farm animals, that proves to be the core problem from the technology perspective. Technology is based on process sequences that are always the same and, therefore, predictable. The recurrence of the same constantly collides with the basic biological principle of variation. The basic evolutionary principle of "survival of the fittest" is based on the fact that living beings differ and thus bring different potentials of **adaptation** to changing living conditions. This applies to microorganisms in the same way as it does to plants and animals and, of course, to humans. Since technology, in its automated processes, cannot adapt to the varied forms and expressions of life of living beings, living beings are forced to adapt to the technical requirements. In essence, farm animal husbandry is about adapting farm animals to the requirements of production processes to increase productivity.

Technified processes require standardized conditions. In livestock farming, these exist above all about husbandry conditions, which in pig and poultry farming have been decisively shaped (in both senses of the word) by the statutory minimum requirements. A high degree of standardization is also practiced in the supply of nutrients to farm animals. The adaptation of livestock to living conditions is achieved by selection. In breeding farm animals, not only are those selected that achieve the highest production performance under the given conditions, but also those best suited to the technical conditions. For example, in dairy cattle farming, the animals best suited to the milking technique are selected based on the morphology and position of the udder teats. The more tolerant the farm animals are of the living conditions, the greater the chance of surviving in the operating system and passing on the genes to the next generation. Breeding efforts to adapt animals to technical conditions have considerably narrowed genetic and phenotypic diversity. Many livestock breeds have already died out because they could not compete for the best adaptation performance; other breeds are on the verge of extinction. Within the breeds or hybrid lines of the various animal species, specific genetic origins have prevailed regionally and often globally because they were best able to adapt to the demands of increasing production performance and to the standardization of the technical framework conditions that also exist globally.

Despite all efforts to standardize the basic genetic makeup of livestock, evolutionary variation between individual animals in a herd persists. Even if farm animals no longer have to engage in foraging, competition between animals for food, exercise space, or other opportunities to gain advantage continues unabated. It manifests itself, among other things, in ranking conflicts and the different rank positions that farm animals occupy in animal groups. Individual animals vary to a high degree in their capacity to adapt to the respective living conditions, which, despite all standardization, continue to differ in various aspects from farm to farm. Adaptation is only possible within certain limits and represents an achievement to be made by the organism as a whole. Whether and to what extent the livestock farmers react to an overstressed adaptation capacity with countermeasures is a further cause of the significant differences between livestock farms regarding AHW problems. While from an ethical point of view, disorders, and diseases in farm animals induce an immediate need for action, many livestock farmers arrive at a different assessment from an economic, operational, or other perspective and try to save themselves the associated additional efforts and expenses. Often, the hope may resonate that the problems as undesirable side effects of the production processes can be controlled through veterinary and breeding advances. After all, the main aim is to make economically successful use of effort-minimizing living conditions. Most livestock farmers cling to their belief in the problem-solving options of technical developments and the problem-solving competence of the sciences. The reasons for this belief are manifold. The belief has the unsurpassed advantage that questions concerning possible system-inherent problems and doubts about unstoppable agricultural progress can be easily disregarded.

On closer inspection, not only the many promises of solutions in the past but also current technical developments are of limited use. This is not only the case because their use entails costs. While it is indisputable that technical innovations create new options, no valid statements can be made without a review of the relationship between expenditure and expected benefits and without knowledge of the cost-performance ratio in the operational context. However, a comparison between biological and economic advantages and disadvantages is rarely made in agricultural practice. Instead, it is assumed that technological developments have already proved effective when they have found their way into practice. Socially relevant side-effects on protected goods of the community (such as **animal health and welfare** or the environment) play a subordinate role, if any, in the introduction into practice. Reservations about using technical innovations are often met with accusations of hostility to technology. In this context, technology is a means to an end, and the cost–benefit ratio could be examined. However, whether the use of technical solutions fulfills its purpose can only be assessed in the farm-specific **context** of the application. Any generalization to other situations in other companies is not reliable.

Derivations

The intensification of production processes in agriculture and livestock farming serves the purpose of increasing productivity in the production of agricultural raw materials. In the post-war years, the priority was to secure food supply for the domestic population. After this goal was achieved, farms felt compelled to continue with the processes of growth and intensification to ensure economic viability in a competitive environment intensified by overproduction. The advantages of the development of farms are based primarily on economies of scale, which help to reduce production costs in various ways. Among other things, the share of fixed costs in the costs per produced unit decreases with more units produced. In addition, there are benefits from the division of labor when complex processes can be broken down into simple, easily repeatable tasks. Machines can then complete these more quickly and cheaply than expensive human labor. In addition, there are various possibilities for saving production costs in the logistical handling of large quantities of goods. Last, the knowledge (*know-how*) about the specific options for influencing processes and realizing savings potentials can be made available and used more efficiently.

The impressive increase in productivity compared to the pre-war years was achieved through the interlocking and mutually reinforcing effects of (a) increased input of means of production (especially fertilizer and feed), (b) specialization in sub-sectors of the production chain, and (c) last but not least, mechanization of processes, which significantly increased labor productivity on the farms. Further inter-farm preconditions were regional infrastructures, coordination between the specialized supplier and buyer farms, and the development of agricultural industries in upstream and downstream areas, including establishing technical and biological repair services. The latter include, in particular, veterinary medicine, whose services have been instrumental in ensuring that livestock functioning has not been unduly impaired and that productivity has been maintained. Another indispensable prerequisite was the availability of outside capital, without which the extensive investments could not have been financed.

While the productivity increases far exceeded the initial targets, undesirable side effects of the intensification processes soon became apparent. These manifested internally primarily in animal health problems among farm animals and externally in releasing large quantities of pollutants into the environment. The specialization of livestock farms in sub-areas of the production chain has largely decoupled livestock farming from the previously existing utilization connections in on-farm nutrient cycles. The concentration of livestock in specific regions expresses the economic advantages that can be realized through processes based on the division of labor. However, the financial benefit is questionable if the negative consequences are excessive. To limit these, regulations that consider the interests of the common good would have to be implemented. Given the resistance to regulation by various powerful interest groups, implementation is still far off. This allows the players to continue to proceed within the framework of the statutory minimum requirements according to the motto: What is not forbidden is permitted.

While impairments to the welfare of livestock and nutrient discharges into the environment cannot be avoided as a matter of principle, it is the individual service of a farm concerning the protection of animal health and welfare and the environment that determines what can still be considered tolerable or where there is an urgent need for action to counteract the **harmful effects**. However, no independent authority has judged the extent of the detrimental effects of the production processes. Given the lack of external insight into internal processes and the self-referentiality of the assessment of impacts by those responsible for them, the extent and the significant heterogeneity of negative impacts between farms cannot surprise. The harmful effects of agricultural production processes are inherent in the system in several respects. They are dependent: (a) on the specific structures of the operating system in which they originate, (b) on the thought structures and patterns of those responsible, and (c), not least, on an economic system. This imposes a competition fought out through cost leadership and, thus, through the need to continuously increase productivity. Those who cannot keep up are eliminated. To what extent and over what periods have farm animals been subjected to pain, suffering, and damage and the farm workers to excessive demands, mental anguish, and over-indebtedness, and to what extent the livelihoods of flora and fauna have been destroyed is irrelevant to the economic system or its protagonists. The fates of animals, humans, and the environment are not categories relevant to the theoretical edifice of agricultural economics.

5.2 Marginal Utility

Many farmers continue to intensify production processes and further increase productivity, although, in the past, most farms that have followed this path have failed to secure their livelihoods. Business closures correspond to the inner logic of an economic system that has made competition an end in itself. Even if the competition drives many companies to ruin and can, therefore, be seen as ruinous competition, the economic system in its current form is not questioned. It forces all the players still in the race to continue their efforts until they, too, feel compelled to give up the competitive struggle out of exhaustion or the realization that the battle cannot be won.

Many farm owners cannot escape the inner logic of economic constraints. The question of who is interested in and benefits from the financial system maintained in this form is all the more pressing. Before we discuss the stakeholders' interests in more detail in Chap. 6, we should first discuss how the benefits can be specified and delimited in more detail. Generally speaking, in economics, utility is a measure of the ability of goods to satisfy needs. Something has utility for a higher purpose or the needs of persons. Unlike a sum of money, which can be used to fulfill very different wants that correspond in magnitude to the sum of money, the utility of agricultural processes is not arbitrarily interchangeable but is closely tied to context. For example, high-quality nutrients have quite different benefits for animals at various stages of life and performance. The benefits of feedstuffs intended to increase production output are not the same on every farm but vary depending on the initial conditions. Farm managers have very different opinions about how high the benefit is to be classified for the purpose they have chosen themselves. Accordingly, the purpose and the assessment of benefits are context- and person-specific.

Rarely does the benefit of means of production relate only to a need. Instead, various additional benefits are often involved. Just as utility is the result of a variety of influencing factors, the effects of utility are also interwoven in a network of interactions. Due to the multifunctionality of the benefit and its wealth of facets, the internal benefit is difficult to classify and quantify for outsiders and insiders. In terms of self-interest, utility always remains subjective; after all, needs and weighting are always subjective. No one "from the outside" can evaluate whether my needs have been met in my perception. This is true for farm animals as well. Given the complexity underlying the processes and the benefits, the classification of the effort-benefit ratio is also often left to subjective assessment. However, this does not mean that benefits are not amenable to intersubjective assessment and that efforts to make comprehensible judgments should be spared. Thus, the **public welfare interest** can be prioritized intersubjectively. Consequently, the benefits of specific measures for the respective requirements can be assessed regarding purpose fulfillment. Despite the various methodological challenges, it is argued that assessing benefits and the effort required to achieve them plays a crucial role in determining internal and external benefits and considering public welfare interests.

A benefit is expected with each of the daily practiced work time expenditures and investments made. Sometimes, the advantage is apparent as soon as one stops working. Frequently, however, the usefulness of spending is not immediately obvious, but only after a considerable time delay or not at all. Whether, when, and to what extent the expected benefit has occurred or failed to materialize and whether the effort expended for this purpose has paid off is fundamentally open to scrutiny. What is daily practice in other sectors of the economy with far less complex processes only takes place in rudimentary form in agriculture. Although evaluations of farm branches are part of the standard accounting repertoire on many farms, they are not always available. However, comparing costs and services incurred or generated in a farm branch only provides a rough indication of the actual relationship between costs and benefits in the various areas of the farm.

Effort and benefit of processes and actions follow a utility function. Such a function is roughly illustrated in Fig. 5.3. The course of the curve is based on biological growth processes. **Growth** in biology refers to the increase in size and mass of a living organism

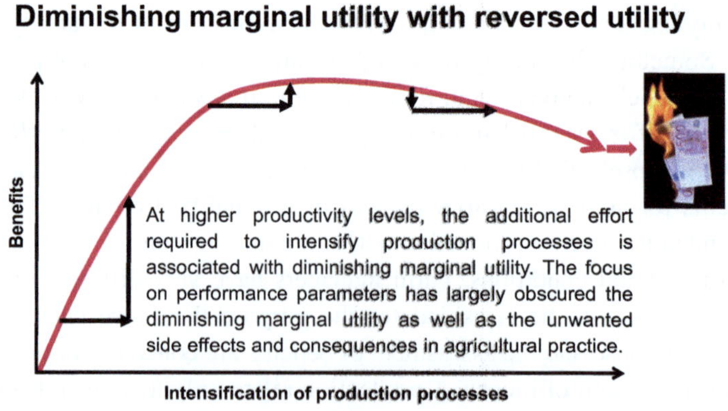

Fig. 5.3 Schematic representation of the relationship between an increase in expenditure and the impact on utility as a function of a marginal utility function (own presentation, photo:© Rainer Sturm/PIXELIO)

or one of its parts. It is based on networked structures such as cells, which interact with each other in a highly complex manner. These systems (interrelated wholes) are kept in function by **metabolism** and a metabolic rate. The latter quantifies the energy required per unit of time to keep living systems (cells, organisms) alive and promote growth processes. When energy is converted, this always has a price because there is nothing for free in life (West 2019). The energy conversion in a system causes consequential effects anchored in natural law. The second law of thermodynamics states that every practical energy conversion produces "useless" energy as a decomposition and waste product. "Unintended consequences" in the form of unusable heat and waste products, called **"Entropy,"** are therefore inevitable. It always occurs due to the interactions of matter through the exchange of energy and raw materials. Continuous supply and energy consumption are required to maintain order in an evolving system. To stay alive and maintain the functions needed, living systems must absorb nutrients and the energy they contain in the form of negative entropy (Schrödinger 1992). The by-product of this is entropy, which manifests itself, among other things, in an increase in heat that must be dissipated. With an increase in metabolic processes, entropy also increases. At the same time, the expenses increase to keep the side effects of the system within limits so as not to impair its functionality. This is, incidentally, one of the decisive reasons that can be used to explain the limits to growth already discovered by Galileo Galilei in the early days of modern science (West 2019).

Before delving into this further, let's first understand the concept of marginal utility and its relationship to the expenditure of various means of production. Not only does biological growth run up against limits, but so does the utility that can be achieved with an increase in the expenditure of various means of production. It is generally assumed that the more effort is expended, the more utility can be increased. However, this only applies below a utility maximum (marginal utility). The cost-benefit ratio depends on the initial level of the targeted parameter. If, for example, the level of production output is low, relatively little effort is required to increase output further. If, on the other hand, the output level is already high, the same amount of additional effort will lead to a significantly lower increase in output. Consequently, a higher effort would have to be made at a higher level to produce the same additional benefit. Suppose utility increases are sought using further expenditure and means of production, although the maximum cost-benefit ratio has already been reached. In that case, these efforts will not only lead to useless spending. An expenditure of time and money that does not produce a benefit is equivalent to destroying value; the expenditure could have made an additional benefit in other system contexts. In addition to the non-benefits, there is a higher level of waste products, which can be associated with increased undesirable or harmful side effects.

The relationships described suggest that it should be of great interest to the management of individual farms to know (a) the current level of the relationship between effort and benefit in the respective production processes, (b) the course of the benefit function as a function of the farm-specific functional relationships, and (c) the level at which maximum benefit growth is likely to be reached. Missing the context-depending point where the positive returns of marginal utility turn into a negative marginal profit undermines the economic viability of the farm system. Thus, this turning point is an essential threshold that should not be exceeded, guiding all management decisions. Whereas knowledge of the interrelationships between effort and benefit is generally readily quantifiable for industrial production operations and constitutes an indispensable basis for management decisions, few farm managers have an overview of the internal interrelationships between efforts and benefits of production processes. The reasons for this are manifold; some of them are discussed below with the help of examples.

5.2.1 Utility Functions in Crop Production

In agricultural crop production, the yield increases achieved in the past decades are based on the breeding influence on the yield potential of plants and the possibility of meeting the growing nutrient requirements with increasing use of mineral fertilizers and the susceptibility to plant diseases with appropriate **pesticides**. In production theory, production function describes the relationship between inputs and the resulting outputs. It indicates the highest quantity of output that a firm can produce using the required combination of inputs. At the same time, the **law** of diminishing returns also applies to biological growth. Above a specific, i.e., context-dependent, level of yield, an increase in input no longer leads to a rise in output. In Fig. 5.4, the influence of N-fertilization intensity

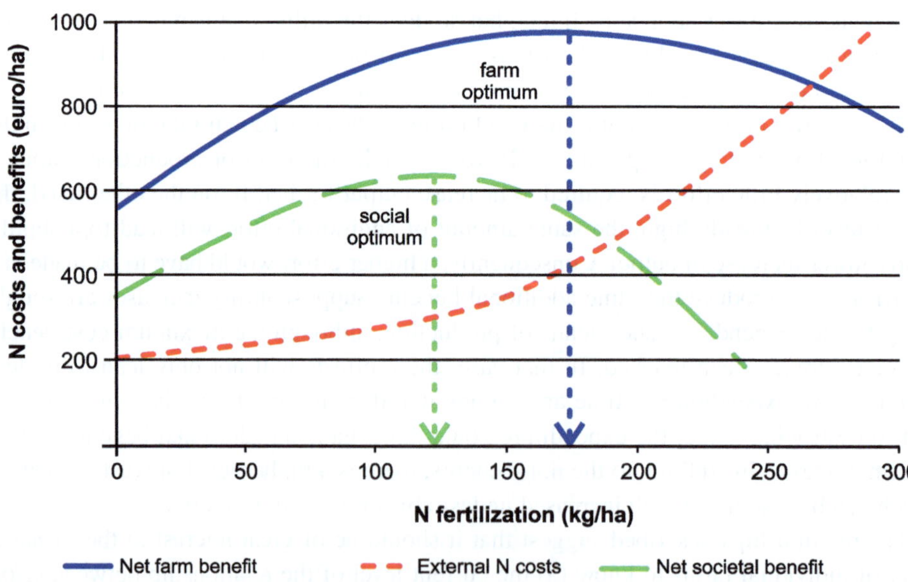

Fig. 5.4 Influence of N fertilization intensity to winter wheat (x-axis) on nitrogen outputs and costs, the optimal N fertilization intensity without (blue) and with consideration (green) of external N costs (red) (Grinsven et al. 2013)

on winter wheat and the resulting benefits for farmers in Euro per hectare are shown as an example (blue curve). According to this, from the farmer's perspective, the highest benefit can be realized with a fertilization intensity of approximately 180 kg nitrogen. As the fertilization level increases, the yield growth decreases, i.e., the benefit decreases. From a business point of view, the optimum fertilization intensity is reached when the revenue generated by the additional fertilization no longer exceeds the costs generated by the additional fertilization (Isermeyer 1992).

However, the nutrient loads introduced into the system with the fertilizers have positive effects and lead to an increase in waste products. In this example, these are reactive nitrogen compounds (including nitrate and nitrous oxide), which the plant cannot use, i.e., cannot be organically bound. These leave the operating system via the air or water pathway (surface and groundwater) and escape into the environment. Outside the operating system, they lead to external effects, including **groundwater** contamination with nitrates and local **eutrophication** of surface waters. Furthermore, they contribute to an increase in the concentration of greenhouse gases (see Sect. 5.3). If the aim were to contain the external effects, the optimal N fertilization intensity for wheat would have to be reduced by about 30% from the "agricultural optimum" of approx. 180 kg N/ha to the "social optimum" of approx. 125 kg N/ha. In this example, the conflict between the interests of farmers in maximizing yield growth and the community's interests in keeping the environmental impact of nitrogen emissions as low as possible is given concrete form. However, although the relationships presented are based on well-established data, they cannot be generalized. Individual farm levels and functional patterns of benefit can deviate considerably from the orders of magnitude presented. For example, not every farm has an optimal fertilizer intensity of 180 kg N due to different soil and weather conditions, and many factors also influence the external effects of fertilizer use.

As significant as the magnitude of the external effects is, so too are the differences between the individual farms, the individual fruits, and the individual fields, as well as between the crop years regarding the production and damage functions. Variation in outputs of products for sale and of pollutants is one of the critical problems preventing accurate quantification of externalities. However, this does not mean there are no valid quantification possibilities. However, such a quantification is only reliable if the actors involved have an interest in a reliable quantification. Under the current conditions, this is not the case. Suppose an attempt would be made to include the external effects in the economic calculations. This would generally result in reduced fertilization intensity and yield expectations and, thus, decreased income from the products sold. This contradicts the farmer's interests because the farmer would voluntarily forego a possible increase in revenue, especially when many farms are already "up to their necks in water."

In this context, farmers and policy-makers should be intensely interested in knowing the maximum benefits when growing fruits on individual fields to limit harmful effects. Finally, efforts to reduce toxic effects are also subject to an effort-benefit relationship. Farmers who practice a fertilization level beyond the farm-specific optimum and are thus responsible for a drastic increase in adverse external effects would have to be prevented from continuing this practice. This is countered by the fact that politicians fear the supposed bureaucratic burden and resistance from farmers. After all, many farmers do not want to be told how to farm by others. The simplest solution would be to make the means of production (including mineral N fertilizers) drastically more expensive by imposing taxes to encourage farmers to handle the means of production more carefully. However, these considerations are not new but were already recommended decades ago by scientific policy advisors several (Scientific Advisory Board of BMELF 1993). The agricultural lobby has generally been able to assert itself when it has been questioned whether the state should regulate interventions in the farm sector. What has been formulated by interested parties as good professional fertilization practice primarily serves to leave farmers as much room for maneuvering as possible. This is then exhausted to very different extents. However, because of the immense dimensions of the damage caused by an excessive intensification of production processes, there is an urgent need for action and regulation. Even with the most conservative estimate, the external costs per hectare of agricultural land are roughly equivalent to what farmers receive from the EU agricultural policy for meeting basic requirements for farm management in terms of good farming practices (Taube 2016).

Harmful effects can, in principle, be caused by all nutrient inputs that are not organically bound in the operating system in the form of sales products, farm fertilizers, or in the soil and leave the operating system again in mineral form. Depending on the respective magnitudes, the discharges harm **biodiversity**, water quality, and atmosphere composition (see Sect. 5.3). If the soil mineral losses associated with crop production are not adequately replaced by fertilization, soil depletion reduces soil fertility. Thus, the farm will damage itself in the long term. Discharging nutrients into the environment creates negative externalities and harms the common good. The damage to the common good does not usually play a role in business calculations. Finally, the agricultural economic benefit theorem focuses exclusively on benefits to users. Harm effects concern farmers only to the extent that they might affect farm benefits. According to the above marginal utility function, it can be assumed that farmers always obtain maximum yield growth at the expense of public welfare interests. From the crop production side, this leads to the conclusion:

> *"Against this background, a "mineral turnaround" in crop production is required, i.e., a transition from the unsustainable use of minerals under short-term economic considerations and extensive externalization of costs to a management of mineral flows in agroecosystems with the goals of maintaining or increasing soil fertility, increasing yield and quality of crop products, and minimizing mineral pollution of crop products and the environment."* (Engels 2015).

5.2.2 Utility Functions in Animal Production

The production processes of animal-origin products are also subject to marginal utility functions, which describe the relationship between the inputs (expenditures) and the resulting outputs (consequences). The relationship is overlaid by various influencing factors that interact with each other at different levels so that the degree of complexity increases again compared to crop production. Like crops, livestock are also location-fixed. Unlike wild animals, which spend much of the day foraging for food, farmers are responsible for providing essential nutrients to confined farm animals. They dictate what feed components go into the troughs, in what quantities and compositions, and from what sources. The utilization of feed nutrients by livestock is, on the one hand, a question of efficiency by which the nutrient resources are made available. On the other hand, there is the question of the efficiency with which the feed is converted into food by the livestock.

Efficiency of making feed available

Feedstuffs are substances that are fed to animals individually or in mixtures. A broad spectrum of individual components can be used for livestock feed. This ranges from green forage plants and their preserves (hay, silage) to straw, roots and tubers, **cereals**, and legumes. Many by-products of industrial processing (e.g., bran, pomace, extraction meal) and feedstuffs of animal origin (e.g., fish meal, meat-and-bone meal, whey) also find their way into the feed mixtures. In addition, minerals, trace elements, vitamins, and synthetically produced amino acids are added. All these components are combined to provide livestock with what they need to maintain themselves and achieve high production levels. Each feed has a characteristic composition of nutrient substrates (proteins, carbohydrates, fats) and substance fractions. At the same time, the nutrient contents vary considerably even in the same feedstuffs, depending, among other things, on the variety, location, fertilization intensity, harvest time, and preparation. At the same time, the feedstuffs can be utilized differently by the various types of livestock in their respective life phases. Consequently, the challenge in using feedstuffs is to reconcile the variation in feed ingredients and ration composition with the individual animal variation in feed intake and the individual animal requirements for nutrients and energy.

At the farm level, site conditions determine what forage can be grown. The grassland growth can essentially only be utilized by ruminants and horses. They have a digestive system in which the fibrous plant structures can be broken down in symbiosis with microorganisms. For **pigs** and **poultry**, this is only possible to a minimal extent due to the single-cavity stomach and the limited capacity of the appendix. However, feed utilization is less and less based on what can be grown and made available on the farm, but rather on what can be grown or bought at a reasonable price and what price can be obtained for the home-grown fruits on the markets. While large-volume coarse fodder (e.g., silage from grass or maize) grown primarily on the farm's land continues to play a significant role in cattle farming, feed production on many pig and poultry farms is largely decoupled from the use of the farm's land for fodder production (see Sect. 5.1).

Components from cereals and legumes dominate the diet of pigs and poultry. Meanwhile, these are also used in large quantities for ruminants. A global feed market has long since developed, shifting huge volumes between producer and consumer farms. Supply and demand determine the price of feed. The transport costs incurred are of secondary importance. Due to the import of feed components that can be produced more cheaply in other countries than domestically or internally, there is a global feed market analogous to foodstuffs, primarily based on price. The focus here is on soybeans, which, as a legume, are a crucial supplier of essential amino acids for feeding pigs and poultry and also occupy a central position in animal production as a protein supplier for feeding cattle. From 2013 to 2015, the EU imported an annual average of approximately 36 million tons of soybeans and soybean meal (European Commission 2018). About 95% of soy imports are used as feed to produce meat, eggs, and dairy products.

EU agricultural policy favors soy imports by not imposing tariffs on feed soy. The imports come mainly from Brazil, Argentina, and Paraguay. The global cultivation of soybeans has increased more than tenfold in the past five years. The expansion of industrial soybean cultivation has severe consequences for the environment and the local population. The cultivation of genetically modified soy also involves the use of large quantities of herbicides, which pollute the soil, contaminate drinking and **groundwater**, and contribute not only to the destruction of the biodiversity of flora and fauna but also negatively affect the microbial composition of the intestines of humans and animals (van Bruggen et al. 2021). Human rights abuses also continue to be commonplace in the dispute over access to arable land (Guereña and Riquelme 2013).

Of the more than 700 million tons of cereals and pulses produced worldwide, about one-third is consumed by livestock. Theoretically, about 3 billion people could be fed from the high-quality components used in animal feed. However, this is only true because they are largely satisfied with a vegetarian diet. However, since the vast majority of arable land that could be converted from feed to food production is located in wealthy countries, this capacity for food production is only available to a limited extent to people mainly dependent on food (Food and Agriculture Organization of the United Nations 2000). Figure 5.5 shows the tonnages of feed suitable for human consumption in European animal production.

Fig. 5.5 Proportions of different feed components in the feed rations of various animal species in the EU, suitable for human consumption (European Union 2017)

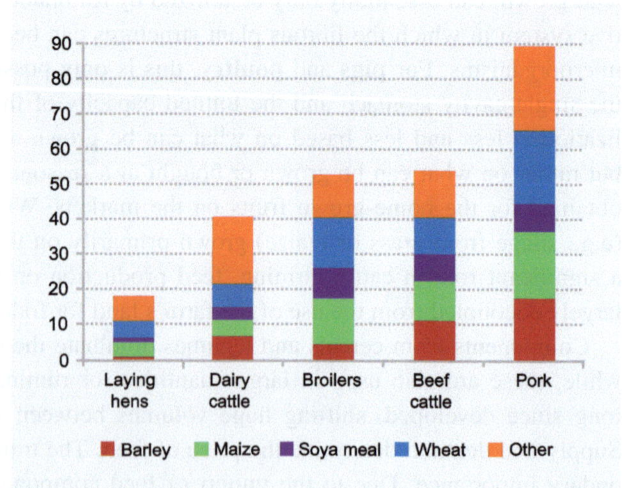

While soybean imports to Europe have risen continuously in the past, the cultivation of domestic legumes in Europe has declined significantly at the same time. Domestic legumes can be used in animal feed as an alternative to soy products. However, due to the lower content of value-added ingredients, especially essential amino acids, they are only competitive to a limited extent compared to soy products. In addition to low-cost soybean imports, yield uncertainties and disease susceptibility make cultivating domestic legumes in Europe unattractive for many farmers. On the other hand, conventional or organic farms that use genetically modified, non-genetically modified, or organically produced soybeans on the market generally make a profit. However, the business calculation does not include the external effects associated with the production of soybeans in the producing countries, the transport costs, and the internal nutrient overloads. At the same time, the additional benefits related to the domestic cultivation of legumes are not adequately considered (Blume et al. 2021). As nitrogen-fixing plants, legumes have the unique ability to enrich the soil with nitrogen, which is then available to subsequently cultivated crops. This natural process can significantly reduce the need for N-containing fertilizers, thereby helping to contain groundwater contamination and reduce greenhouse gas emissions. The N-fixing capacity of legumes offers a far more environmentally compatible way of making Nr available for plant growth than the production and application of mineral nitrogen fertilizer and the spreading of farmyard manure. However, these advantages of domestic legume cultivation can only be fully realized if the external effects caused by the import of feedstuffs are also considered.

The economic implications of feedstuff choices are significant. The efficiency of nutrient conversion into animal products varies considerably between animal species and within the same production directions. For instance, monogastric animals (pigs and chickens) are more efficient in converting feed into animal protein than ruminants. This understanding is crucial for farmers and policymakers in making informed decisions about feedstuff choices (Mottet et al. 2017). However, ruminants are not dependent on high-quality feed components, as is the case with pigs and poultry. In the latter, a large proportion of nutrients are used that could also contribute to meeting the needs of humans. High proportions of high-quality feed components are also used in intensive cattle fattening. This is the case, for example, in feedlots in the USA, where feeding is based to a high degree on concentrate feeds based on grain and soybeans. The opposite concept is extensive cattle fattening in Argentina or Uruguay, where the animals graze on extensive grasslands and are fed hardly any feed that humans could also utilize. Agricultural practice exhibits various feed components, ration proportions, consumption rates, and feed utilization between the extremes of a pure concentrate-based feed ration and a pure grass-based feed ration. It follows that animal species is only one factor among many in the process chain that influences the degree to which forages can be utilized. Again, the context and economic conditions determine the use of external feed components and the associated externalities.

Nutrient use efficiency at the farm level

The dissemination of information through the media has the consequence that the conditions in livestock farming concerning the efficiency of resource use are usually presented in a very abbreviated way. Generalized statements shape opinion formation but are often far removed from the realities of agricultural practice. Due to the significant variation in the initial and boundary conditions on farms and their diverse network of effects, the conditions in agricultural practice are highly heterogeneous. The complex network of effects includes not only the very different efficiency of individual farms in making available their own or externally purchased feed but also in the utilization by the livestock.

Under standardized experimental conditions, in which disturbance variables can be largely eliminated, livestock's growth and performance curves run as a function of genetic growth and performance capacities and the level of nutrient supply. Efforts to increase productivity are reflected in an increase in the efficiency with which the nutrients made available via feed are converted into meat yield or production output (milk per cow, number of eggs per laying hen). In addition to the increases in production output already described in Sect. 4.1, the level of feed efficiency has also improved significantly over the past decades in all directions of use. However, the increase in feed efficiency is also subject to a marginal utility function. The higher the initial level of feed efficiency, the greater the effort required to achieve a further increase.

The utilization of nutrients in feed varies considerably between animal species and, within animal species, between genetic origins and sexes. Even the utilization of the same nutrients by animals of the same genetic origin and sex, which are also kept under the same living conditions, nevertheless varies significantly from animal to animal. Accordingly, it is difficult and costly to record the efficiency of feed conversion. For pragmatic reasons, the information on which farmers' decisions and actions regarding feeding are based is generally limited to average values. This concerns both production performance and feed consumption and, thus, nutrient efficiency. As benchmarks, averages provide some initial guidance. However, the price of this is ignorance and hence failure to take into account the variation that exists between animal individuals in terms of production performance and the individual differences derived from this in nutrient requirements, feed intake, and the discrepancy between nutrient requirements and supply and, ultimately, nutrient efficiency.

As a result of intensive breeding efforts, genetic growth, and performance capacities have now reached a level that can no longer be fully exploited, even under optimized living conditions. The fact that performance now depends less on genetics than on living conditions is expressed, among other things, in that significantly higher performance is always achieved on experimental farms under optimized conditions than achieved in animals of the same genetic origin with the same feed in agricultural practice. Under practical conditions, variation between individual animals is usually much more pronounced than under standardized experimental conditions. As exemplified by the results of studies in pig fattening (see Fig. 5.6), biological stressors in the form of infections or toxin contamination can substantially impair the utilization of nutrients. These manifest themselves, among other things, in a significant reduction in daily gains due to increased

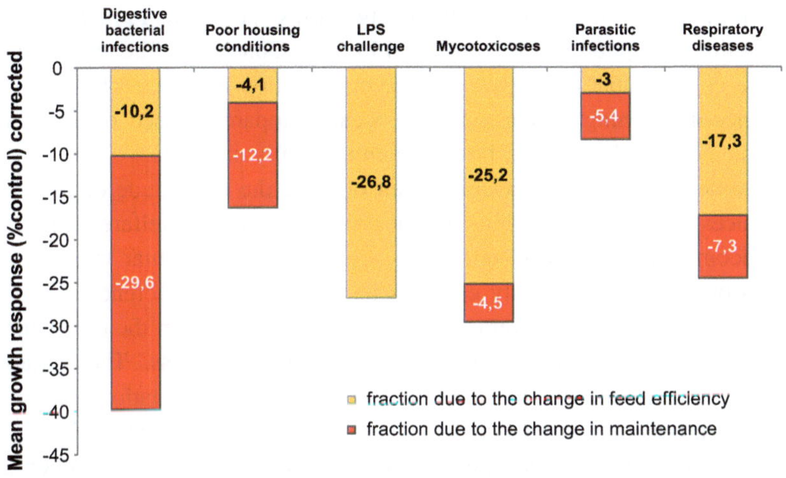

Fig. 5.6 Significant reduction in daily gains of pigs in response to various health stresses due to suboptimal living conditions (Pastorelli et al. 2012)

maintenance requirements (for physiological and immunological processes, among other things) and decreased feed efficiency.

On the one hand, the extent of impairments affects the welfare of livestock and, thus, the public interest. On the other hand, the reduced production performance and economic losses run counter to the particular interests of the livestock farmers. From this, one could deduce that the common good and the specific interests of livestock farmers are aligned. Even if this is the case to a certain extent, two aspects limit the common intersection. Firstly, the suboptimal living conditions do not only result in loss costs. Their elimination is associated with labor-time and monetary expenditures, which livestock farmers often cannot estimate whether they will be profitable. On the other hand, the extent of the negative consequences of suboptimal conditions cannot be assessed precisely. As the processes mainly occur in the non-visible area, livestock farmers often do not notice them. Visibly diseased animals—like the tip of an iceberg—at best hint at the true extent of potential impairments. What remains unclear is the totality of clinical and subclinical effects and the extent to which animal health impairments, production performance, and production effort/benefit ratios vary between individual animals.

The potential implications are illustrated by farm branch evaluations of two pig farms (Hilgers and Heger 2017). Both farms, each with 1500 fattening places, had roughly the same initial situation, including the same genetic origin of the animals and the feedstuffs used. While on farm A everything went according to plan regarding animal health, a respiratory tract infection occurred on farm B. The externally visible disease was a severe infection of the respiratory tract. The externally visible disease situation in the fattening house soon calmed down again, so only a few animals became clinically ill and had to be treated. Fattening on farm B lasted, on average, only four days longer, and daily gains were only 30 g lower. Nevertheless, the individual animals grew very much apart (more significant proportions of heavy and lightweight pigs) and gave farm B, among other things, a significantly poorer feed conversion in total. Concerning the annual period, this resulted in a considerable discrepancy in the contribution margin of 46,000 € per year. This existence-threatening dimension only became apparent based on a detailed data evaluation on an individual animal basis, which is usually omitted.

The example illustrates how averages can obscure the magnitude of individual animal differences and the far-reaching implications for animal welfare, feed efficiency, and farm income. Averages level out differences between particular values; how many are close to or far from a roughly equal mean makes a big difference concerning the consequential effects. It is not average values but the variation in the cost-benefit ratios between the individual animals determining economic success. The variation only becomes apparent when time-consuming data collection, evaluation, and diagnostic measures are carried out. The advantage of regular controls is that deviations and the possibility of counteracting them are detected early before severe consequences occur. Whether the costly expenditures for preventive measures pay off can, at best, be assessed retrospectively. Even retrospective assessments remain speculative because it can never be clarified which negative consequences could be prevented by preventive measures. If adverse effects have occurred, the only conclusion is that the previous precautions were insufficient. It is necessary to establish a company-specific cost-benefit ratio for production performance and to contain undesirable side effects of the production processes.

Negative marginal profits (i.e., losses) that may occur due to suboptimal management are part of private entrepreneurial risk. In contrast, protecting animals from pain, suffering, and damage is in the interest of the common good. While primary producers want to achieve the highest possible income with the least possible effort, the community's interest is in the lowest possible animal welfare-related harm. It is undisputed that primary producers are also interested in a low level of undesirable side effects. However, the optimum between cost and benefit for primary producers is usually at a different level than from the perspective of public welfare interests. Much would already be gained if the farm-specific maximum benefit were not exceeded on the individual farms, if only out of self-interest. The prerequisite, however, would be that the farms have appropriate assessments in this respect. If livestock farmers were forced to carry out on-farm control measures based on the individual animal level and if these were subjected to an inter-farm comparison, the extent of the adverse marginal utility effects could be drastically reduced. Thus, the situation could be improved for the benefit of the livestock as well as the farm income and the public interest.

5.2.3 Efficiency in the Use of Food

Food is used to satisfy needs. The benefits depend both on the food itself and on the needs of the users. This includes the ingredients and the positive and negative side effects of the production processes. Chapter 6 discusses the benefits in more detail about the respective interests. The first point to be made here is that the benefit for producers and consumers is subject to a marginal utility function. The **marginal utility** is the benefit that the consumer derives from consuming a further unit of the good in question.

The **law** of diminishing marginal utility applies to all goods and, consequently, food: If a person consumes another good G2 after the first good G1, the utility of this good G decreases. For example, when consuming milk, the marginal utility decreases after a certain amount. 5 glasses of milk do not have five times the utility of one glass. Similar is the case with financial income. A person can spend about a monthly income of 1000 euros in a given time. However, the utility of further income does not increase proportionally. An income of 100,000 euros a month does not have 100 times the utility for the same person since numerous needs have already been satisfied (Piekenbrock 2018).

The EU population is one of the world's largest consumers and, therefore, users of **meat** and other products of animal origin. In 2017, the per capita consumption of meat was approximately 68.5 kg (31.9 kg of pork, 24 kg of poultry meat, 10.8 kg of beef and veal, 1.8 kg of sheep and goat meat) (European Union 2017). These quantities are not all produced in Europe but are also imported in relevant proportions. In addition, feed is also imported to feed the animals kept in Europe. However, there are considerable differences between European countries, regions, and farms.

The context of the utilization determines the relations of effort and benefit with which the nutrient and energy contents of the raw materials are utilized. In all countries, situations of need and scarcity are juxtaposed with abundance. However, the ratios between the proportion of people in a country undernourished and those overnourished or overweight differ considerably (see Fig. 5.7). While in African countries, a relevant proportion of the population is undernourished due to insufficient income and significantly fewer people are overweight, the ratio is reversed in wealthy nations. The proportion of people who are overweight in the total population has increased dramatically since the 1980s, especially in the USA and Canada, and later also in Mexico and the countries of the EU. Falling food prices are blamed for about 40% of the increase in the US popula-

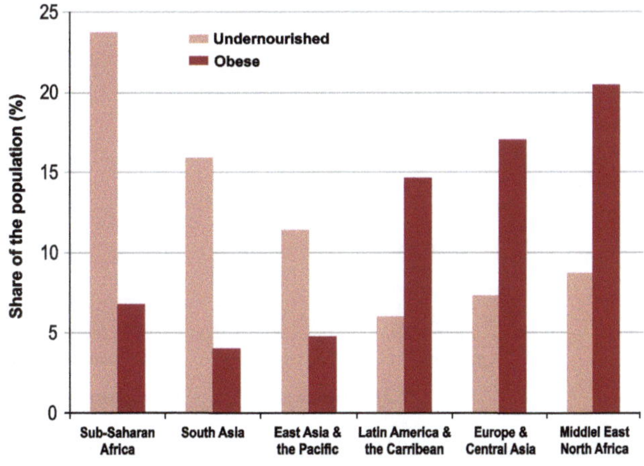

Fig. 5.7 Status of food security in the world (Ng et al. 2014)

tion's average body weight (Lakdawalla and Philipson 2002). Other causes of obesity include decreased physical activity and changes in eating habits. The increase in single households and a reduced willingness and ability to cook for oneself have promoted fast food consumption, the ingredients of which considerably contain fatty meat products. The consequences are manifold. Among other things, epidemiological studies show close links between obesity and a reduced lifespan as well as an increase in various diseases (Cassell and Gleaves 2000). In addition, the costs to the national health care system are increasing (Birmingham et al. 1999). This is accompanied by the state having fewer resources for carrying out other tasks for the common good.

The more efficiently food is produced, and the more it is offered on global markets, the lower the price farmers receive. As prices have fallen, the proportion of food that is not consumed but ends up in waste has also increased. According to FAO figures, around a third of all food produced worldwide is thrown away. In Germany, around 11.86 million tons of food waste was identified in 2015 (Schmidt et al. 2019). Of this, around 6.68 million tons could theoretically be avoided. Food waste in the entire value chain can be divided into five sub-sectors (primary production, processing, trade, out-of-home catering, and private households). The range of food waste in Germany and its avoidable shares according to the value chain sectors is shown in Fig. 5.8. The extrapolation of waste volumes and the identification of preventable shares are primarily based on non-representative samples from the literature. Against this background, the quantities determined and their avoidance potential represent an estimate of their magnitude. According to these figures,

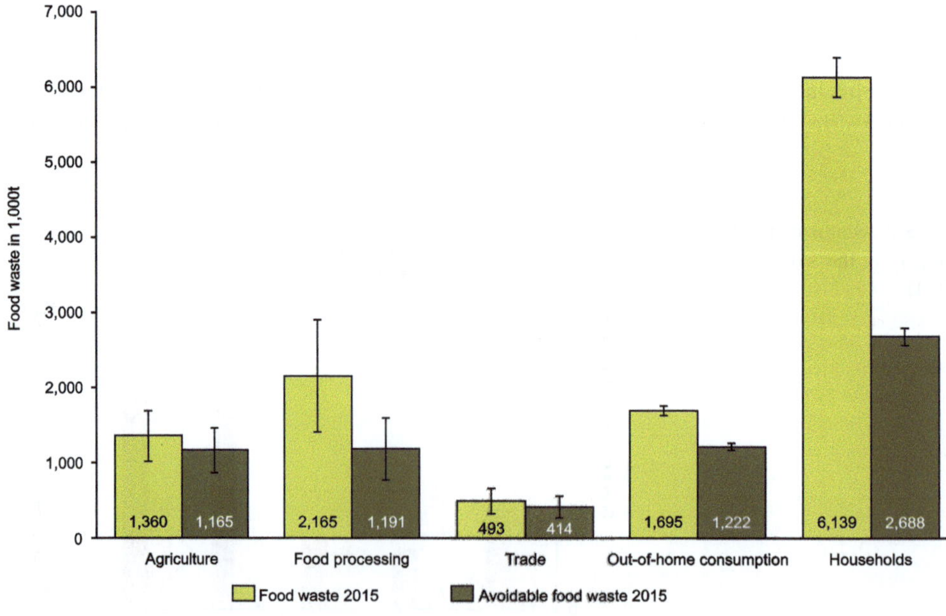

Fig. 5.8 Food waste (2015) and its avoidable share in Germany (1000 t/a) (Schmidt et al. 2019)

primary production accounts for 12% (1.36 million tons), processing for 18% (2.17 million tons), trade for 4% (0.49 million tons), and out-of-home catering for 14% (1.69 million tons). The majority of food waste, 52% (6.14 million t), is generated proportionally in private households, corresponding to about 75 kg per capita in 2015.

According to the available projections, about half of the waste across all sectors would theoretically be avoidable. The term "avoidable food waste" includes foods that are still entirely edible at the time of their disposal or that would have been edible if consumed in time. In distinction, "non-avoidable food waste" is usually removed during food preparation. This includes inedible components (e.g., bones or fruit peelings) and edible ones (e.g., potato peelings).

Derivations

From the perspective of primary producers, the focus of agricultural production processes is primarily on productivity gains. Only if they succeed in keeping production costs per unit of product low can they survive in global undercutting competition. In the past, intensification strategies have led to veritable productivity gains (Sect. 5.1). However, the undesirable side effects and adverse impacts that occur in the process have been and still are largely ignored. These are also closely related to the inefficient use of internal and externally purchased nutrients. Nutrients that are not organically bound in products of plant and animal origin or other forms sooner or later escape into the environment, where they cause considerable environmental and climate damage. Animal individuals who are not supplied according to their needs and thus inefficiently run the risk of developing metabolic disorders and production diseases associated with pain, suffering, and damage and of leaving the production process prematurely with financial losses.

As a rule, the damage caused in the upstream and downstream areas of operational production processes, for example, in producing fertilizers and feedstuffs and disposing of waste materials, is also not considered. Even if on-farm and off-farm nutrients are used efficiently in production processes, this does not mean that animal manure is also recycled and returned to plant nutrition without significant losses. The same applies if efficiently produced foodstuffs are consumed in excess by consumers or if high proportions are disposed of unused as waste. Efficiencies can change drastically both in the internal process chains of a farm and on the way to direct consumption by the consumer ("from farm to fork"). Efficient nutrient conversions in individual parts of the process chain can be canceled out or reversed by inefficient conversions in other sub-steps. Accordingly, efficient conversions of nutrients in crops and livestock do not necessarily benefit public welfare if considerable adverse effects occur in upstream and downstream areas. Individual sub-steps must, therefore, not be considered in isolation, and partial successes of efficient use must not be extrapolated and appropriated or generalized as an advantage for the entire process chain (pars pro toto fallacy). While the scientific, political, and media focus is on the increased efficiency of sub-steps, the overall effects within the operating systems and the entire process chain from the primary producers to the end consumers are decisive for the common good.

Intensification processes that exceed the optimal ratio of effort and benefit in some areas cause negative marginal returns and gains. This raises the question of why diminishing marginal utility and negative marginal returns have not been the focus of more general or economic reflection. This is all the more surprising given that inefficient resource use runs counter to the interests of the common good and farmers' interests. It seems reasonable to assume that ignorance of the undesirable side effects obscures the view of the interrelationships. Negative marginal returns only appear in economic calculations when the undesired side effects are also recorded in detail and included in the overall calculations. It is easier to ignore the extent of undesirable side effects than to face the responsibility of having caused their occurrence. Another complicating factor is that many undesired side effects only appear with delay. Possible causal relationships between one's actions and the occurrence of adverse effects can thus easily be ignored. Above all, however, assessing adverse effects lacks a generally accepted and reliable assessment standard and the possibility of comparison across companies.

Inadequate efficiencies in the internal use of resources only become apparent through comprehensive assessments. From the perspective of public welfare interests, not only are the quantities of sales products relevant, but also the quantity of pollutant discharges into the environment. The operating system as a whole is the reference system that is decisive for assessing resource efficiency and environmental and climate compatibility. The efficiency of the overall operational use of resources is the relevant target variable for management's control and allocation decisions. It forms the intersection between the particular economic interests of the companies and the interests of the community in minimizing the waste of resources and creating harmful effects. Therefore, one of the primary tasks is to identify, considering the specific marginal utility functions, the individual farm utility optima for which, with the degree of expenditure and intensification, an optimal economic return can be realized in combination with minimal harmful effects.

The current economic framework is characterized by ruthless undercutting competition on the global level for the lowest production costs. This competition is at the expense of undesirable internal and external effects; thus, it is incompatible with public welfare interests' realization. As the degree of intensification increases, marginal utility decreases. If the optima of marginal utility functions are exceeded, increases in efficiency in sub-areas come at the expense of efficient resource use in the entirety of the operating system and the entire production chain from production to final consumption. The high proportions of food disposed of as waste counteract all the successes in the process chain concerning increased resource efficiency.

This no longer has anything to do with the frequently invoked "value creation" but illustrates an extent of value destruction that has long since exceeded the tolerable limits. The non-use of produced and usable food is not only associated with the loss of the food itself but, at the same time, with the harmful effects that have arisen during production. The inefficiency in using nutrient resources results from a market process to which all other areas have been subordinated for price leadership and quantity maximization. Consequently, no reorientation will be possible without a change in the framework conditions. One cannot avoid intervening in the market if one wants to correct the undesirable developments of the "free" market economy.

5.3 Extent of Harmful Effects Caused by Agriculture

The primary production processes take place within the boundaries of an agricultural operating system. Unlike in previous centuries, farms today can no longer operate self-sufficiently and self-sustainingly. They are highly dependent on external resources to realize a sufficient level of productivity. Inputs originating from outside the farms include external capital, fertilizers and feedstuffs, machinery, fossil, and electrical energy, as well as means of combating pathogens (above all, plant protection products and medicines). Providing these external inputs is an essential prerequisite for the progress achieved in farm productivity gains. The production and supply of these inputs are energy-consuming and use considerable resources. Insofar as the inputs are used for on-farm production processes, they should be charged to the products for the production of which they are used.

The input variables are matched by output variables that leave the boundaries of the operating system. In addition to the sales products, residual and waste materials escape into the environment if not recycled within the operating system. The environment of the farm is all that surrounds the operating system. The boundaries of the operating system mark the boundary between inside and outside, between internal and external affairs. As owners or tenants of the farmland, farmers can dispose of the internal areas. No one can dictate to farmers what they grow on the farmland or what animals they keep in their barns. What happens in, on, and to the land initially affects only the tenants or landowners. As explained elsewhere, the unrestricted disposal right does not apply to farm animals, even if some owners give a different impression. At least formally, the state grants farm animals a right to integrity.

The substances brought as means of production into the operating system generally leave the system in a transformed form unless they are not accumulated within the system. The desired forms of transformation are the revenue-generating sales products, to which all the attention of primary producers is directed. In contrast, other substance discharges are primarily ignored; after all, they are rarely income-generating. The discharges of residual and waste substances into the environment are only visible to a limited extent and are, therefore, seldom evident, not to mention quantifiable. This makes it easy for primary producers to ignore discharges into the environment. Quantification, in turn, is aggravated by the fact that unlike in the case of industrial plants, substances released into the environment are not concentrated in waste pipes or chimneys and cannot be measured at the 'end of the pipe.' They escape over large areas and diffuse via the air surface and groundwater. The discharges of nitrogen and phosphorus compounds are illustrated in Fig. 5.9.

In addition to nitrogen and phosphorus compounds, various other substances are discharged from farm systems. These include minerals such as zinc and copper, methane, particulate matter, and, last but not least, pesticide and pharmaceutical residues. In addition, some microorganisms multiply very strongly due to the conditions in the animal houses and exhibit varying degrees of virulence and pathogenicity. Pathogens originating from animals can also cause diseases in humans (**zoonosis**). With the use of **antibiotics**, the multiplication of bacteria with resistance genes is massively promoted. This impairs the effectiveness of antibiotics against bacterial infections not only in livestock but also in humans.

Economic success in harnessing economies of scale to reduce production costs per product unit manifests itself primarily in the concentration of livestock farming in regional "livestock production hot spots." These are characterized less by a favorable location for cultivating feedstuffs than by favorable conditions for developing an

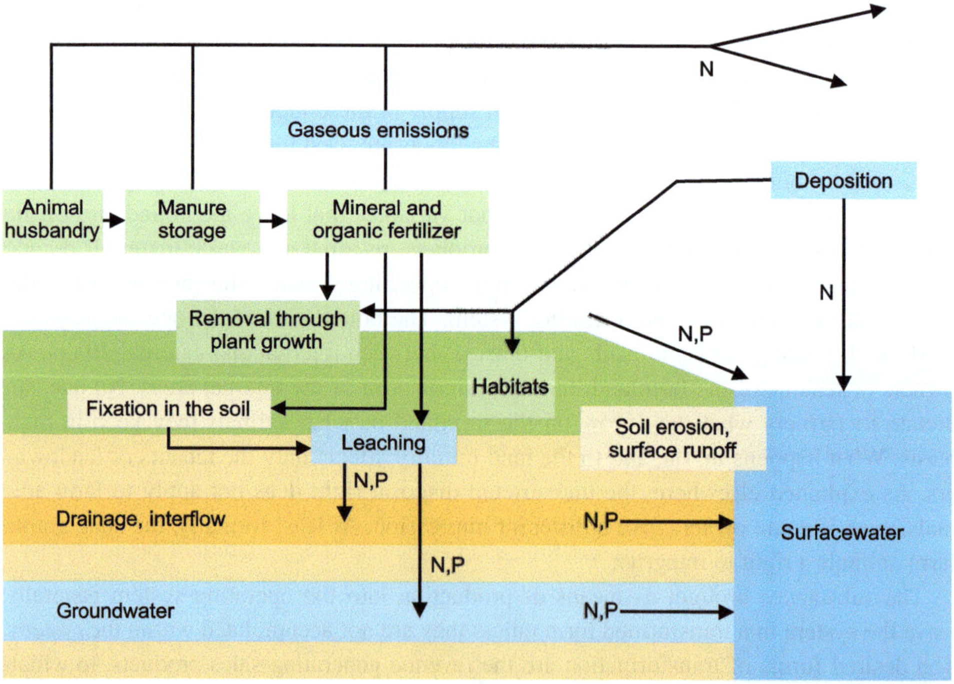

Fig. 5.9 Emissions of nitrogen and phosphorus from agricultural production processes into waters and the air (Copyright Bundesanstalt für Landwirtschaft und Ernährung 2018; Broschüre 1756 Die neue Düngeverordnung)

infrastructure tailored to the production processes. This includes good transport links (seaports, waterways, motorways), which make it possible to transport, temporarily store, or process large quantities of production inputs (including fertilizers and feedstuffs) and raw materials produced at low cost. Consequently, livestock farming was intensified above all, where favorable infrastructural conditions were available or could be expanded. The changes were flanked by generous investment subsidies, land consolidation measures, expansion of the network of field paths, privileges for building in the outskirts, energy use with preferential tax treatment, and much more.

While various growth processes took place over time, this did not apply to the availability of agricultural land. The resulting limited availability of domestic feed could be compensated for by feed that was transported cheaply over long distances by ships, rail, and eventually by trucks from its original production sites to places of high livestock densities (Tamminga 2003). The larger the freight batches, the lower the transportation costs per unit load. Increasing problems, however, are caused by the limited availability of usable land to spread livestock excrement. The capacity of soils, ground and surface waters, and the air to absorb the residues and waste released is also limited. When infrastructure was developed to increase agricultural productivity, many things were considered, but there was no need to dispose of residues and waste properly. The environment was and still is seen as a free dumping ground for the waste of agricultural production processes.

The 'hot spots of livestock production,' which include large parts of North-West Europe, North America, and Asia, are also the hot spots of discharges of pollutants that pollute the environment. They are found on all continents and are now very well-mapped (Steffen et al. 2015). In many European countries, the amount of **reactive nitrogen compounds** brought into the farms via feed exceeds the amount of mineral nitrogen fertilizers (Olsthoorn and Fong 1998). At the same time, the return of animal excrements to the land on which the fodder crops were grown is interrupted, and thus, the recycling of Nr is primarily decoupled. The land loses its original ecological utilization. Agroecosystems have mutated into economic systems, which focus mainly on economic benefits and widely ignore the undesirable side effects.

Focused on economic parameters and essentially free of concerns about undesirable side effects, agricultural businesses and businesses in the supply and processing industries have spurred each other on to intensive growth. They have been driven by a quest for cost leadership over those regions where these conditions do not exist to the same extent. Increasingly, however, the undesirable side effects are proving to be blatant disruptors of competition. Even if not all primary producers and other interest groups want to acknowledge this, the times are over when it was possible to largely ignore them without endangering one's ability to act. Today, the undesirable side effects not only mark the absolute limits of further growth processes. It is becoming increasingly clear that for some time now, they have been responsible for a decreasing **marginal benefit** and, on many farms, for a negative marginal profit. If it were only the self-harm caused by the undesirable side effects, we could leave it at that. However, what escapes into the environment now concerns us all. It is not only the people near the local discharge points who are affected by emissions from agricultural land and during the storage of fertilizers and animal feed. Since the airborne and waterborne groups of substances diffuse in all environmental systems: Biosphere, Atmosphere, Lithosphere, and Hydrosphere (see Fig. 5.10), they have an irreversible, widespread effect, which reduces substance discharges, an urgent task for agriculture.

Due to the different aggregation states, the four environmental systems can be easily distinguished. The same groups of substances react very differently in the respective environmental systems, which are in close and dynamic exchange with each other. Consequently, substance discharges from farms are sooner or later distributed over the globe. Even if the releases are local, the effects are global. This applies, in particular, to the reactive nitrogen compounds (Galloway et al. 2003). The **complexity** of biological, chemical, and physical processes naturally exceeds that of industrial processes based only on physical and chemical laws. Given the complexity, it is impossible to provide a comprehensive overview of the extent of all harmful effects. Using various examples, the primary intention is limited to making the diversity and extent of the negative consequences in the four environmental systems visible and understandable.

5.3.1 Loss of Biodiversity in the Biosphere

Biodiversity refers to the diversity of life on Earth. The results of a large-scale study initiated by the United Nations show that **ecosystems** and their services have been exposed to more significant threats and tremendous pressures in the past 50 years than at any time in human history (Millennium Ecosystem Assessment 2005). This process has accelerated considerably over time as human resource needs increase exponentially. The Earth is in a state of degradation, meaning that 15 of 24 ecosystem services studied were already in a state of advanced and ongoing degradation at the time of the report's publication. Among other things, this damage contradicts the United Nations' plans to end hunger worldwide. Among the most severe destruction of ecosystems is the

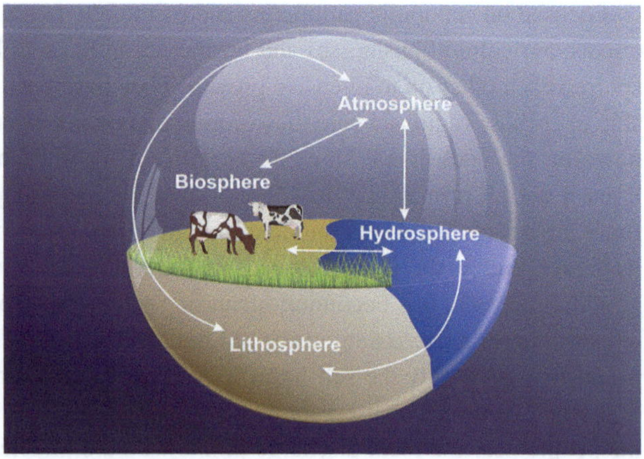

Fig. 5.10 Four interacting environmental systems within planetary boundaries (own illustration)

deforestation of rainforests. These are characterized by an immense wealth of species. For example, the forests of Amazonia are seven times richer in bird species than those of Europe. The diversity there has developed from a general lack of available nutrients. Species diversity is not an expression of luxury but rather the stored ability to cope with the adversities and limited availability of living conditions (Reichholf 2004). Deforestation and subsequent exploitation fundamentally and irretrievably destroy the ecosystem's functional relationships and deprive most indigenous plant and animal species of their basis of existence. However, the drastic loss of species is not limited to tropical regions; it also occurs in Europe. According to estimates by Reichholf (2004), agriculture, with its direct and indirect effects, bears 78% of the responsibility for declining mammals, birds, reptiles, and amphibians in Central Europe. Nitrogen, in particular, has become a "suffocating substance" for biodiversity. The problem of overfertilization is not caused solely by the excessive use of **mineral fertilizers**. In addition, there is a massive increase in the generation of liquid manure from expanded livestock farming.

Although insect living conditions have been adversely affected for some time, insect mortality has only relatively recently come to the public's attention. It took extensive quantification to turn previously held assumptions about the extent of losses into certainties. Globally, approximately 41% of insect species and 22% of vertebrate species are in decline, with annual decline rates of 1.0% and 2.5%, respectively. This is the recent finding of a comprehensive meta-analysis of 73 quantitative studies conducted worldwide based on time series (Sánchez-Bayo et al. 2019). According to this analysis, the extent of global insect mortality is far greater than previously assumed. The main factors identified for changes in insect diversity are human-induced habitat changes, human-induced pollution, a complex of biological causes, and **climate change**. Many entomologists from all corners of the globe have titled a joint publication, "Warning to Humanity!" (Cardoso et al. 2020). Scientists believe that the number of endangered and already extinct species is underestimated to an alarming extent. The researchers have taken an overall view of butterflies, bees, ants, spiders, beetles, and other insect groups. The global analysis of insect occurrences shows a 45% decline in two-thirds of the species studied. Up to half a million species of insects have been lost since the beginning of industrialization due to habitat destruction, the use of ever-increasing amounts of **pesticides** in agriculture, and other human impacts. The authors foresee serious ecosystem consequences if the current negative trend continues unabated. Agriculture is also affected because almost 90% of all flowering plants on earth and three-quarters of crops are pollinated by insects. In addition, essential decomposition processes are of great importance for maintaining soil fertility. Scientists are in broad agreement that it is primarily current agricultural practices that cause negative impacts on the biodiversity of many species groups. However, these cannot be reduced to individual factors. Instead, many influencing factors are involved, which interact and mutually reinforce each other. Given the complexity, it is necessary to consider the system as a whole to get closer to the causes of biodiversity loss (see Fig. 5.11).

Most habitat changes are caused by intensive land management through the simplification of crop rotations and the increase in the size of cultivated plots. This is

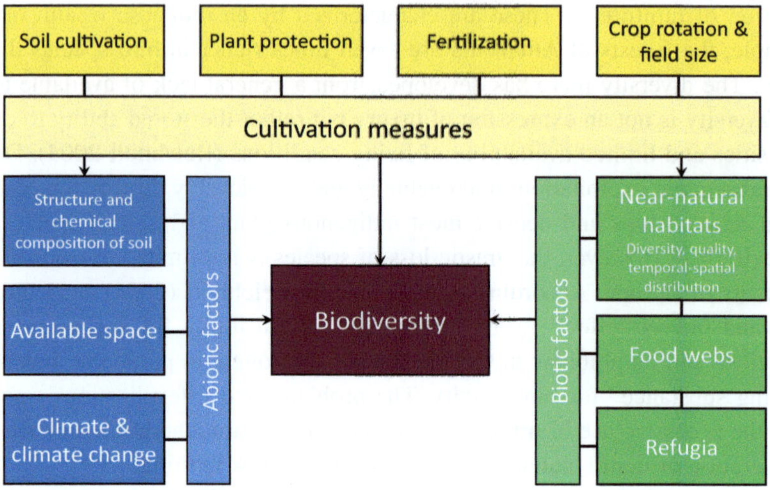

Fig. 5.11 Important factors influencing the biodiversity of agricultural landscapes (Niggli et al. 2020)

accompanied by losses of field margins, which tend to have a diverse vegetation. In addition, there has been a sharp decline in arable flora due to improved weed control. The use of pesticides in intensive agriculture is also a significant factor in biodiversity loss. Habitat changes lead to habitat losses and reduce nutrient supply. Terrestrial habitat changes are also caused by tillage. Changes in soil microhabitats disrupt food webs. This affects the nutritional basis, especially for long-lived organisms such as springtails, mites, and earthworms (Tsiafouli et al. 2015). The agricultural landscape's removal of aquatic and terrestrial habitats (spawning grounds, hibernation sites, loss of migration routes, etc.) is responsible for the decline of amphibian and reptile populations, which are affected by the loss of non-productive marginal habitats (Baker et al. 2013). The ecological crisis in the agricultural landscape has reached an extent that endangers the functioning of the surrounding ecosystems. The ever-increasing biodiversity loss also has consequences for the human population, not only in the adverse effects on food but also for recreation and psychological well-being. Only a change of direction towards a more nature-friendly way of doing business can still avert serious consequences. The National Academy of Sciences concludes this after a surprisingly critical analysis concerning biodiversity (National Academy of Sciences Leopoldina 2020). The scientists, therefore, call for a radical change in European agricultural policy.

5.3.2 Pollutants in the Atmosphere

In agricultural production processes, various residual substances are produced that take on a gaseous state of aggregation by which they escape from the operating system.

The primary concern is ammonia (NH_3), a strongly pungent-smelling, colorless, water-soluble, and toxic gas that emits in the air. It derives first and foremost from the excrement of the animals. It is the portion of the ingested feed that is not utilized by the digestive system and is excreted as feces and urine. In the feces, the nitrogen is organically bound in various chemical compounds. In the urine, the nitrogen is present as urea. Due to the ubiquitously present enzyme urease, urea is converted into ammonia at different rates depending on the boundary conditions (temperature, pH value, air exchange rates, etc.).

Farm animals are the first living systems to come into contact with ammonia. The animals are kept on slatted floors in barns constructed according to labor-economical and cost-saving aspects. These can transfer the excrement to slurry pits below the standing and moving areas. For the farm animals, this means they are permanently confronted with the outgassing of their excrements. The result is irritation of the mucous membranes of the respiratory tract and the eyes, often the starting point for further inflammatory processes. Forced ventilation releases the air from the stables into the immediate environment. A fattening house with a thousand pigs blows about three and a half tons of toxic ammonia into the air every year (Umweltbundesamt 2020). In addition, the air carries fine dust and bioaerosols composed of airborne microorganisms and decay products, which include the dreaded endotoxins. In addition, there are odorous substances that hurt residents' quality of life. A scientific lung study among residents of intensive livestock facilities found that the lung function of people living closer to such stables was impaired (Radon et al. 2005).

However, ammonia does not only have local harmful effects. It reacts with other atmospheric components and can be transported over long distances before being deposited again. Ammonia deposition can contribute to the acidification of soils and waters and **eutrophication** (nutrient enrichment). Acidification of soils and nutrient oversupply of natural and near-natural ecosystems (such as peatlands, rough grasslands, and water bodies) lead to changes in the species composition of biotic communities, which can also result in the extinction of individual species.

Ammonia is not the only reactive nitrogen compound released due to material conversion processes in agriculture. There are also nitrogen oxides and nitrous oxide, which, like ammonia, escape in gaseous form and play an essential role as **greenhouse gases**. Other products of microbial decomposition processes are nitrate and ammonium. However, they do not leave the operating system in gaseous form but are released via **surface waters** and **groundwater**. The initial substrate of the gaseous and water-soluble emissions of the reactive nitrogen compounds are the residual substances that are not present in organically bound form and are therefore washed out or escape in gaseous form.

Which of the two emission paths is followed depends on various physical and chemical initial constellations, which are also influenced by the management measures. Measures that contribute to reducing gaseous emissions in the barn, such as lowering the pH value in the slurry using acid, lead to more nitrogen being brought to the field via the farm manure. There, the nitrogen can be reabsorbed by the plants and fed into an on-farm process of **recycling**) or, in the case of **overfertilization**, can be washed out. The internal **nitrogen cycle** is the constant transfer and biochemical conversion of nitrogen by microorganisms in the soil, plants, and animals in the form of aggregated **biomass**. In the nitrogen cycle, reactive nitrogen alternates between organic bonding in the form of microbiological, plant, and animal biomass and inorganic compounds that can be reintegrated into biomass. Excess, non-organically bound (inorganic) N compounds diffuse into the environment in gaseous or water-soluble form. The proportions of excess inorganic N compounds originating from agriculture compared to other emission sources are shown in Table 5.1.

While agriculture is only responsible for a small proportion (approx. 8%) of nitrogen oxides (NO_x) in the total emissions of inorganic N compounds, the other reactive N compounds are predominantly released by processes in agriculture. Action against the primary polluters is more difficult because the nitrate values in groundwater, continuously recorded at defined measuring points, do not indicate where the inputs originate. Although the polluter-pays principle is an overriding principle in European legislation, it is not applied in agriculture. However, if the international commitments to reduce greenhouse gases are to be met, agriculture cannot avoid contributing and implementing efficient emission reduction measures.

Greenhouse gases released from agriculture are nitrous oxide (N_2O) and methane (CH_4). Nitrous oxide is around 300 times more harmful to the climate than carbon dioxide (CO_2); methane is about 25 times more harmful than CO_2. The primary sources of nitrous oxide released from agriculture are nitrogenous fertilizers, livestock manure, atmospheric N deposition, reactive N compounds in soils from plant residues, and biological N fixation of legumes. These give rise to nitrous oxide due to nitrification and denitrification processes. In addition, especially on converted moorland soils and on

Table 5.1 Share of the essential N-compounds (nitrogen oxides, ammonia, nitrous oxide) and emitter groups in the mean annual total emissions to air and surface water in gigagrams N/year (2005–2010). *Source* Umweltbundesamt 2015

	Air			Water		
	NO_X	NH_3	N_2O	NO_3^-/NH_4^+	Total [Gg N a^{-1}]	Share [%]
Agriculture	33	435	88	424	980	63
Traffic	192	13	2	207	13	13
Industry/Energy	166	15	27	10	218	14
Households/Wastewater treatment plants/Surface runoff*	21	1	6	135	163	10
Total [Gg N a^{-1}]	412	464	123	569	1568	100
Share [%]	26	30	8	36	100	

* Includes urban systems and total surface runoff, as no split between agricultural and other land is currently possible. Air emissions include combustion plants in households

grassland, high emissions of greenhouse gases are to be expected after fertilization due to the high humus content (Harris et al. 2021). All measures to reduce nitrous oxide emissions are based on the efficient use of fertilizers. This means reducing overfertilization and promoting the organic binding of Nr in microbial, plant, and animal **biomass**.

Methane is formed during microbial fermentation processes released under anaerobic conditions in the rumen of ruminants or during the decomposition processes of excrements in the slurry. In addition, methane is formed by wastewater and sewage sludge treatment and sewage sludge utilization in agriculture. It must also be assumed that significant amounts of methane escape into the environment when methane is produced from plant biomass and excrement for energy generation. Reducing livestock numbers of ruminants and influences via feeding are generally considered efficient measures to reduce methane emissions. However, it should be taken into account that the methane releases resulting from the fermentation of plants and, thus, renewable biomass in the rumen of cows by a recycling process should be assessed differently from methane releases resulting from industrial processes (Jackson et al. 2020).

The reduction of greenhouse gases from agriculture has been sluggish. Often, calculations do not even consider that agriculture draws on extensive resources in the upstream and downstream sectors. This includes, for example, the energy-intensive production of mineral fertilizers, which has to be charged to agriculture instead of the energy industry. While other sectors have considerably reduced greenhouse gas emissions, the lack of success in reducing emissions in the agricultural sector is becoming an increasing problem. The share of greenhouse gases from agriculture is successively increasing compared to other industries, thus increasing the pressure on agriculture to contribute to climate protection.

While other sectors have various technical options to reduce greenhouse gas emissions, these options are limited in the agricultural industry. It is true that filtration plants can filter out Nr compounds or that slurry treatment can reduce Nr emissions. Besides the considerable financial expenditure for end-of-pipe technologies, they do not lead to a solution to the problem but only to a shift. If emissions are reduced in one section of the process chain, they may be all the greater elsewhere. Thus, the release of Nr compounds in agriculture should not be based on technical innovations but on modifying biochemical processes. Nr emissions occur as inorganic intermediates. If these are fixed again in organic bonding ratios, they can be prevented from escaping into the environment. Thus, the extent to which reactive nitrogen compounds are released is determined by the quantities produced and the interaction of processes in the respective context. For example, not the absolute amount of fertilizer is decisive, but the quantities available to the plants during the vegetation phase. The goal of maximum performance in plant production can only be achieved if the supply of nutrients available to the plants is kept as high as possible. Top performance can only be achieved by **overfertilization**. Maximum availability in the vegetation phase is bought with a considerable surplus in the other phases, in which a relevant proportion of the soluble nutrients is lost through leaching processes into the surface and groundwater.

5.3.3 Pollutants in the Hydrosphere

By far, the most significant shares of water pollution come from nitrogen (N) and phosphorus (P) compounds. In contrast, carbon (C) compounds are usually efficiently degraded by microbes in the water bodies. The importance of N and P compounds emerges from the quantitative ratio that characterizes the biological-ecological conditions in water bodies (Reichholf 2004). The 106 C to 16 N to 1 P ratio characterizes favorable water conditions. Accordingly, **phosphorus** is the factor that significantly determines water quality. When carbon and nitrogen contents deviate from the quantitative ratio, a surplus or deficiency situation arises concerning phosphorus.

The most significant anthropogenic impairments of surface water quality result from excessive releases of N and P through fertilization or disposal of waste and residual materials from livestock farming. In Europe, attempts to limit the excess and thus the discharge of inorganic N compounds through legal regulations started with the Nitrates Directive (Monteny 2001). Increased nitrate contents in **groundwater** and the associated consumer risk prompted the EU Commission to take countermeasures as early as 1991. They set a maximum permissible limit of 50 mg/liter of groundwater. All European countries should comply with this limit value at the measuring points. However, the implementation of the legal requirements is sobering. Even 30 years after the legal requirement came into force, the limit value continues to be exceeded in many regions, and in some cases, still significantly so (European Commission 2021).

After repeated warnings, the EU Commission has initiated infringement proceedings against those countries that have not yet made any substantial improvements. The European Court of Justice has already rejected complaints by the countries concerned against the procedure. Water companies have to take extensive and costly measures to treat drinking water, which they are forced to charge to consumers, even though the cause and the reasons for this lie in the current form of land management and economic conditions. The nitrate problem shows that, despite a legally fixed limit value, there is insufficient success in limiting groundwater contamination, for which agricultural enterprises are primarily responsible. On the one hand, this has something to do with the political unwillingness to take consistent action against the interests of the agricultural industry. These are not only loudly voiced by the lobbyists but also associated with protest actions and demonstrations of power.

The global application rates of **phosphorus** via fertilizer application currently amount to 14.2 tetra grams of P per year (MacDonald et al. 2011). High P outputs are concentrated in a few world regions with intensive land management. A better global distribution of phosphorus application and improved phosphorus supply to P-poor soils would reduce release levels and increase global crop yields. Along with nitrogen, phosphorus compound discharges tend to cause poor overall water body condition, which leads to eutrophication effects.

The consequences include excessive algal growth. Some algae produce a toxin that is dangerous to fish and humans. In particular, algal growth deprives the water of oxygen, which other organisms in the water then lack (Rabalais 2002). Another cause for concern is the presence of persistent chemicals, **pesticides**, and pharmaceuticals. The leading cause of the high nutrient load in groundwater, rivers, lakes, and coastal and marine waters is the excessive utilization of means of production in agriculture to increase productivity without appropriately considering the adverse side effects on the environment.

5.3.4 Changes in the Lithosphere

The lithosphere is the outermost layer of the Earth's body and includes the Earth's crust and the outermost part of the mantle. The earth's crust is virtually the solid shell of the planet and is, on average, 35 km thick. The uppermost and most fertile soil horizon is the thin layer of "topsoil." In temperate latitudes, it is about 20–30 cm thin; however, it can vary significantly from region to region. Soils, in general, and topsoil are the basis of life for plants, animals, and humans. Their availability is limited and further decimated by erosion, sealing, input of toxic substances, and other depletion. In contrast to deeper soil horizons, topsoil contains, in addition to the main mineral components (fine sand, silt, and clay), a high proportion of nutrients (especially nitrogen) and organic matter (**humus**) as well as a large number of soil organisms (edaphon). Humus is part of the total soil organic matter. The proportion and composition of humus are mainly subject to the influence and activity of soil organisms. Their metabolism continuously contributes to humus' build-up, transformation, or decomposition. Humus consists of a multitude of complex organic compounds. These differ considerably in their enzymatic degradability by microorganisms. While low-molecular carbohydrates and proteins are rapidly decomposed, fibrous compounds such as cellulose, hemicellulose, and lignin are slowly degraded. The enzymatic breakdown of organic fragments releases simple inorganic components (= mineralization) on which the plants feed.

Improper human land use, for example, removal of protective vegetation through overgrazing or deforestation, can cause excessive erosion of soils. Loss of topsoil is particularly problematic. Persistent soil erosion initially results in a deterioration of soil quality (soil degradation). Since 1945, the area affected by soil degradation has totaled more than 1.2 billion hectares worldwide—equivalent to the combined land area of China and India (Montgomery 2012). Degradation can lead to the complete loss of the agricultural usefulness of the soil (soil devastation). It is estimated that the loss of topsoil due to erosion worldwide amounts to about 23–26 billion tons annually. This is equivalent to an annual loss of not quite one percent of the arable land. If they are not prevented from doing so, components of the soils can be removed particularly easily by leaching and wind. Plant root penetration reduces the erosive effects of wind, rain, and surface runoff. The lower the soil cover, the greater the risk of erosion. Protection against rain erosion is based on the effect that water droplets are slowed down by hitting the soil-covering layer and thus have less kinetic energy when hitting the soil. A high degree of ground cover also slows the wind immediately above the soil surface, mitigating wind erosion. Although soils can reform or regenerate, this process is exceedingly slow.

Viewed globally, the current development of soils must be classified as worrying: The already largely cultivated arable land is increasingly affected by soil damage due to erosion, salinization, desertification, contamination, etc. (Amelung et al. 2020). In addition, there are problems such as the growth of cities (sealing). With the further increase in the world's population, it can be assumed that the amount of land available per earth inhabitant will halve between 2012 and 2050. In Europe, about 16% of the total European soil area (~1,000,000 km^2) is threatened by (water) erosion (Europäische Union 2011).

The capacity of soils to accumulate residues and pollutants from agricultural production is limited. Accordingly, substances not organically bound by microbes, plants, or animals and remaining in the operating system are emitted into the environment via soil displacements towards ground or surface waters or air. On the other hand, the topsoil layer of European soils alone contains an estimated 75 billion tons of carbon, making it a vast carbon store.

In terrestrial ecosystems, carbon is present in organic compounds in the living biomass and the soils' **humus**. This opens up a diverse scope for increasing the carbon stocks in the soil and forming a carbon sink with an increase in root biomass and humus contents. Once again, the ambivalence of agricultural production processes regarding their advantages and disadvantages for overriding community concerns becomes apparent. Maintenance or an increase in humus content and root biomass cannot be ensured without a carefully balanced crop rotation between humus-producing and humus-consuming crops and without providing organic fertilizer materials such as farmyard manure. This, in turn, contradicts a strategy that relies solely on productivity increases.

Derivations

The above explanations clearly show the extent to which agricultural production processes have been causing considerable damage for many decades. The harmful effects affect all environmental spheres and, in their totality, unfold a destructive potential that has not yet penetrated public awareness. There are many reasons for this. On the one hand, the debates have generally focused on single aspects, such as CO_2 emissions. At the same time, the equivalent effects of reactive N compounds, which originate primarily from agriculture, have not been sufficiently recognized. The production of large quantities of food at low cost is still the primary objective of primary producers. They are supported in this by agricultural policy and by various interest groups (see Chap. 6). Both the exit pathways and the respective amounts of emissions are very diffuse and highly variable between individual farms. A comprehensive balancing of individual farms' input and output quantities would be required to arrive at reliable estimates. Given the complexities involved, it is relatively easy for farmers and other stakeholders to largely ignore the undesirable harmful effects. Those responsible have so far hardly reacted to the extent of the detrimental effects. The problems of nitrate levels in groundwater and the only minimal reduction of greenhouse gases from agriculture prove how little positive progress has been made.

Pollutant discharges from farm systems into the environment result from complex interactions between on-farm and external factors. The outcome of the interactions is highly context-dependent and thus not generalizable. Accordingly, no improvements can be expected through normative means, such as regulatory measures. The limits of resilience have often been exceeded locally, regionally, and globally. This applies to the destruction of biodiversity and the overloading of soils, air, and surface waters with excessive discharges, especially of reactive nitrogen from agriculture (Steffen et al. 2015).

While conflicts between particular and public welfare interests also affect other sectors, the degree of complexity of the production processes and the heterogeneity of the individual farm conditions are nowhere greater. In no other industry are the residual and waste materials at the same time means of production for subsequent production processes in a continuous internal **material cycle**. This cycle process, in which the nutrients are transformed from the soil via the plants to the animals and the unused feed materials are returned to the soil via the excrement, predestines agriculture in a particular way for the **recycling** of nutrients in an agroecosystem. Farm animals, as an integrated part of agroecosystems, play a key role in **reducing** the amount of mineral fertilizer, reusing and **recycling** plant biomass for food production, and **rebinding** reactive nitrogen and carbohydrates in animal-sourced products and feces, which exceed the emission of methane from the rumen and ammonium from the urine, provided that the farm management makes use of appropriate manure management.

Those who still expect a solution to the roughly sketched environmental problems from technical innovations misjudge or ignore the core conflict between the goals to increase productivity and simultaneously enclose undesired side effects. While the agricultural market system demands further productivity increases, the economic system does not provide substantial incentives for environmentally sound behavior. On the contrary, those who endeavor to reduce pollutant emissions through various measures and additional expenditures without this being economically viable are placed at a competitive disadvantage. On the other hand, those who save themselves the effort, costs, and trouble of reducing substance discharges gain a competitive advantage at the expense of their competitors, but above all, at the expense of the public welfare interest.

The extent of damage varies greatly from farm to farm. It is the extensive abrogation of the polluter pays principle which, in addition to the insufficient implementation of economic incentives, stands in the way of an urgently needed reorientation and an effective reduction in the impairment of public interests. The farm represents the reference system for assessing the environmental pressures emanating from a holding. At the same time, the farm is the reference system for the quantity of products sold. From the community's perspective, the relationship between the amount of material imported into the operating system and that which leaves the operating system again as sales products and as pollutants is, above all, decisive. Thus, resource use efficiency concerning food production and the reduction of substance discharges into the environment is addressed as the decisive parameter and target, representing the common intersection between the public welfare interests and the particular interests of the primary producers. Simultaneously, the conflicts between economic and ecological requirements can be brought to a viable solution.

References

Amelung W, Bossio D, de Vries W, Kögel-Knabner I, Lehmann J, Amundson R, Bol R, Collins C, Lal R, Leifeld J, Minasny B, Pan G, Paustian K, Rumpel C, Sanderman J, van Groenigen JW, Mooney S, van Wesemael B, Wander M, Chabbi A (2020) Towards a global-scale soil climate mitigation strategy. Nat Commun 11:1–10. https://doi.org/10.1038/s41467-020-18887-7

Baker NJ, Bancroft BA, Garcia TS (2013) A meta-analysis of the effects of pesticides and fertilizers on survival and growth of amphibians. Sci Total Environ 449:150–156. https://doi.org/10.1016/j.scitotenv.2013.01.056

Battye W, Aneja VP, Schlesinger WH (2017) Is nitrogen the next carbon? Earth's Future 5:894–904. https://doi.org/10.1002/2017EF000592

Birmingham CL, Muller JL, Palepu A, Spinelli JJ, Anis AH (1999) The cost of obesity in Canada. CMAJ 160:483–488

Blume L, Hoischen-Taubner S, Sundrum A (2021) Alfalfa-a regional protein source for all farm animals. J Sustain Organic Agric Syst 71:1–13. https://doi.org/10.3220/LBF1615894157000

van Bruggen AHC, Finckh MR, He M, Ritsema CJ, Harkes P, Knuth D, Geissen V (2021) Indirect effects of the herbicide glyphosate on plant, animal, and human health through its effects on microbial communities. Front Environ Sci 9. https://doi.org/10.3389/fenvs.2021.763917

Cardoso P, Barton PS, Birkhofer K, Chichorro F, Deacon C, Fartmann T, Fukushima CS, Gaigher R, Habel JC, Hallmann CA, Hill MJ, Hochkirch A, Kwak ML, Mammola S, Ari Noriega J, Orfinger AB, Pedraza F, Pryke JS, Roque FO, Settele J, Simaika JP, Stork NE, Suhling F, Vorster C, Samways MJ (2020) Scientists' warning to humanity on insect extinctions. Biol Cons 242:108426. https://doi.org/10.1016/j.biocon.2020.108426

Cassell DK, Gleaves DH (2000) The encyclopedia of obesity and eating disorders, 2nd edn. Infobase Publishing; Facts On File, New York

Ehlers E (2008) Das Anthropozän: Die Erde im Zeitalter des Menschen. Wiss. Buchges, Darmstadt

Engels J (2015) Die „Mineralstoffwende" in der Pflanzenproduktion – eine gemeinsame Aufgabe für die Pflanzenernährung und der VDLUFA. In: Verband Deutscher Landwirtschaftlicher Untersuchungs- und Forschungsanstalten (ed) 127. VDLUFA-Kongress: VDLUFA-Verlag, Darmstadt, p 16

Erisman JW, Sutton MA, Galloway JN, Klimont Z, Winiwarter W (2008) How a century of ammonia synthesis changed the world. Nat Geosci 1:636–639. https://doi.org/10.1038/ngeo325

Europäische Union (2011) Boden: Der verborgene Teil des Klimazyklus. Amt für Veröff. der Europ. Union, Luxemburg

European Commission (2021) Report from the Commission to the Counsil and the European Parliament on the implementation of Council Directive 91/676/EEC concerning the protection of waters against pollution caused by nitrates from agricultural sources based on Member State reports for the period 2016–2019, Brussels

Food and Agriculture Organization of the United Nations (2000) Food insecurity: when people live with hunger and fear starvation. FAO, Rome

Galloway JN, Aber JD, Erisman JW, Seitzinger SP, Howarth RW, Cowling EB, Cosby BJ (2003) The nitrogen cascade. Bioscience 53:341–356. https://doi.org/10.1641/0006-3568(2003)053[0341:TNC]2.0.CO;2

van Grinsven HJM, Holland M, Jacobsen BH, Klimont Z, Sutton MA, Jaap Willems W (2013) Costs and benefits of nitrogen for Europe and implications for mitigation. Environ Sci Technol 47:3571–3579. https://doi.org/10.1021/es303804g

Harris E, Diaz-Pines E, Stoll E, Schloter M, Schulz S, Duffner C, Li K, Moore KL, Ingrisch J, Reinthaler D, Li K, Zechmeister-Boltenstern S, Glatzel S, Brüggemann N, Bahn M (2021) Denitrifying pathways dominate nitrous oxide emissions from managed grassland during drought and rewetting. Science Advances 7:eabb7118. https://doi.org/10.1126/sciadv.abb7118

Hilgers J, Heger H (2017) Auswertung belegt: Auseinanderwachsen kostet 46.000 Euro. In: Erzeugerring Westfalen eG (ed) Jahresbericht 2017, pp 82–85

Isermeyer F (1992) Optimaler Stickstoffeinsatz in der Landwirtschaft aus betriebswirtschaftlicher und volkswirtschaftlicher Sicht. Landbauforschung Völkenrode, Sonderheft 1992:5–20

Jackson RB, Saunois M, Bousquet P, Canadell JG, Poulter B, Stavert AR, Bergamaschi P, Niwa Y, Segers A, Tsuruta A (2020) Increasing anthropogenic methane emissions arise equally from agricultural and fossil fuel sources. Environ Res Lett 15:71002. https://doi.org/10.1088/1748-9326/ab9ed2

MacDonald GK, Bennett EM, Potter PA, Ramankutty N (2011) Agronomic phosphorus imbalances across the world's croplands. Proc Natl Acad Sci 108:3086–3091. https://doi.org/10.1073/pnas.1010808108

Millennium Ecosystem Assessment (2005) Ecosystems and human well-being: synthesis; a report of the millennium ecosystem assessment. Island Press, Washington, DC, The Millennium Ecosystem Assessment series

Monteny GJ (2001) The EU Nitrates directive: a European approach to combat water pollution from agriculture. ScientificWorldJournal 1(Suppl 2):927–935. https://doi.org/10.1100/tsw.2001.377

Montgomery DR (2012) Dirt: the erosion of civilizations. Univ of California Press; University of California Press, Berkeley

Mottet A, de Haan C, Falcucci A, Tempio G, Opio C, Gerber P (2017) Livestock: on our plates or eating at our table? A new analysis of the feed/food debate. Glob Food Sec 14:1–8. https://doi.org/10.1016/j.gfs.2017.01.001

National Academy of Sciences Leopoldina (2020) Biodiversität und Management von Agrarlandschaften: Umfassendes Handeln ist jetzt wichtig. Schriftenreihe zur wissenschaftsbasierten Politikberatung, Halle (Saale), München, Mainz

Ng M, Fleming T, Robinson M, Thomson B et al (2014) Global, regional, and national prevalence of overweight and obesity in children and adults during 1980–2013: a systematic analysis for the Global Burden of Disease Study 2013. The Lancet 384:766–781. https://doi.org/10.1016/S0140-6736(14)60460-8

Niggli U, Riedel J, Brühl C, Liess M, Schulz R, Altenburger R, Märländer B, Bokelmann W, Heß J, Reineke A (2020) Pflanzenschutz und Biodiversität in Agrarökosystemen. Berichte über Landwirtschaft – Zeitschrift für Agrarpolitik und Landwirtschaft, Aktuelle Beiträge. Ber Landwirtsch 98:1–39. https://doi.org/10.12767/buel.v98i1.272

Olsthoorn CS, Fong NP (1998) The anthropogenic nitrogen cycle in the Netherlands. Nutr Cycl Agroecosyst 52:269–276. https://doi.org/10.1023/A:1009724024770

Pastorelli H, van Milgen J, Lovatto P, Montagne L (2012) Meta-analysis of feed intake and growth responses of growing pigs after a sanitary challenge. Animal 6:952–961. https://doi.org/10.1017/S175173111100228X

Piekenbrock D (2018) Definition: Grenznutzen. https://wirtschaftslexikon.gabler.de/definition/grenznutzen-32826/version-256361. Accessed 19 July 2021

Rabalais NN (2002) Nitrogen in aquatic ecosystems. Ambio 31:102–112. https://doi.org/10.1579/0044-7447-31.2.102

Radon K, Schulze A, van Strien R, Ehrenstein V, Praml G, Nowak D (2005) Atemwegsgesundheit und Allergiestatus bei jungen Erwachsenen in ländlichen Regionen Niedersachsens (Prevalence of respiratory symptoms and diseases in neighbours of large-scale farming in Northern Germany). Pneumologie 59:897–900. https://doi.org/10.1055/s-2005-915572

Reichholf JH (2004) Der Tanz um das goldene Kalb: Der Ökokolonialismus Europas, 1st edn. Wagenbach, Berlin

Sánchez-Bayo F, Wyckhuys KAG, Wyckhuys KA (2019) Worldwide decline of the entomofauna: a review of its drivers. Biol Cons 232:8–27. https://doi.org/10.1016/j.biocon.2019.01.020

Schrödinger E (1992) What is life?: The physical aspect of the living cell. Cambridge Univ. Press, Cambridge

Scientific advisory board of BMELF (1993) Reduzierung der Stickstoffemissionen der Landwirtschaft: Gutachten des Wissenschaftlichen Beirats beim Bundesministerium für Ernährung, Landwirtschaft und Forsten. Schriftenreihe des Bundesministers für Ernährung, Landwirtschaft und Forsten, Vol 423. Landwirtschaftsverl., Münster

Smil V (2000) Feeding the world: a challenge for the twenty-first century. MIT Press, Cambridge Mass. u. a.

Smil V (2002) The earth's biosphere: evolution, dynamics, and change, 1st edn. MIT Press, Cambridge, Mass

Smil V (2012) Nitrogen cycle and world food production. World Agric 2012:9–13

Steffen W, Richardson K, Rockström J, Cornell SE, Fetzer I, Bennett EM, Biggs R, Carpenter SR, Vries W de, Wit CA de, Folke C, Gerten D, Heinke J, Mace GM, Persson LM, Ramanathan V, Reyers B, Sörlin S (2015) Sustainability. Planetary boundaries: guiding human development on a changing planet. Science 347:1259855. https://doi.org/10.1126/science.1259855

Tamminga S (2003) Pollution due to nutrient losses and its control in European animal production. Livest Prod Sci 84:101–111. https://doi.org/10.1016/j.livprodsci.2003.09.008

Taube F (2016) Umwelt-und Klimawirkungen in der Landwirtschaft. In: Deutsche Landwirtschafts-Gesellschaft (ed) Moderne Landwirtschaft zwischen Anspruch und Wirklichkeit: Eine kritische Analyse : DLG-Wintertagung 2016, 11.–13. Januar 2016 in München, vol 110. DLG-Verlag, Frankfurt am Main, pp 13–38

Tsiafouli MA, Thébault E, Sgardelis SP, Ruiter PC de, van der Putten WH, Birkhofer K, Hemerik L, Vries FT de, Bardgett RD, Brady MV, Bjornlund L, Jørgensen HB, Christensen S, Hertefeldt T d', Hotes S, Gera Hol WH, Frouz J, Liiri M, Mortimer SR, Setälä H, Tzanopoulos J, Uteseny K, Pižl V, Stary J, Wolters V, Hedlund K (2015) Intensive agriculture reduces soil biodiversity across Europe. Global Change Biol 21:973–985. https://doi.org/10.1111/gcb.12752

Umweltbundesamt (2020) Ammoniak. https://www.umweltbundesamt.de/themen/luft/luftschadstoffe-im-ueberblick/ammoniak. Accessed 19 July 2021

West G (2019) Scale: Die universalen Gesetze des Lebens von Organismen, Städten und Unternehmen, 1st edn. CH Beck, München

6 Stakeholders

Abstract

Heavyweight particular interests often oppose the realization of public welfare interests. The list of interest groups involved in food production is long. It is not only primary producers and their representatives but also the processing industry and the food retailers who want to see their interests safeguarded. The same applies to consumers, their representatives, and NGOs, who have a decisive influence on public discourse through their impact on the media. A specific justification cannot be denied to all interests. However, since the respective interest groups sometimes pursue conflicting goals, a regulating body is indispensable to balance interests. Concerning this task, the market fails just as much as politics. The latter has so far refused to meet the challenge of representing the interests of the common good and putting excessive particular interests in their place. The limits of particular interests run where they run counter to the interests of the common good. Accordingly, the background and motives of specific interests and the lines of conflict with the interests of the common good must be illuminated. Transparency is the obvious and urgently needed strategy to pave the way for the common good interests and to reconcile vested interests with overriding objectives. Using various examples, it is shown that regulatory processes only work iteratively. Accordingly, state institutions are called upon to take flanking measures with a visible regulatory hand suitable for balancing particular interests and the common good.

The production of food concerns all people; after all, all people depend on the availability of sufficient food, without which they cannot survive. At the same time, all people are affected by the consequences of food production processes, both economically and ecologically. However, the extent to which people are affected and the perception

A. Sundrum, *Public Welfare-Oriented Production of Food*,
https://doi.org/10.1007/978-3-662-69040-6_6

and feeling of being affected vary considerably between individuals and different population groups. If one were to conduct a survey of all population groups and ask the question: What form of agricultural food production would you like to see, then presumably the vast majority of respondents would agree with the objective: Sufficient availability of food at reasonable prices which are of high **product quality** and are associated only with a low level of undesirable side effects in their production and distribution. Based on the premise of a broad consensus regarding the stated objective, the question arises as to why it has not been possible in the past to achieve substantial progress toward the objective. Only partial goals have been achieved: sufficient availability for those who can afford it and a comparatively low price. Even the current market prices for food are still considered too high for those on meager incomes and those dependent on social assistance. Thus, it depends very much on the perspective from which the price level is viewed. The same applies to the level of product and **process quality** of food.

As shown in the previous chapters using various examples, monetary and qualitative objectives are interwoven in a network of interactions and conflicting goals. The coexistence of synergistic and antagonistic effects, whose characteristics also depend primarily on the respective context, creates a confusion that is difficult to penetrate intellectually and hardly understandable to outsiders. In this complex structure of effects, various groups involved in different ways in the production and distribution of food compete for power, influence, and, not least, money. Interest groups can be understood as a loose association of people who pursue common interests. They can join together to form associations to assert their interests more efficiently. In the following, interest groups are addressed as groups of people with similar interests and organized associations.

The assessments made here of the interests of various groups are not without problems. Among other things, information and partial aspects are selected and classified in a necessarily incomplete way. Nor can it be avoided that the weighting of partial aspects encourages an intentional interpretation. If a look behind the scenes is nevertheless ventured here, then this procedure is justified because it is not a final but a preliminary assessment. It marks the first attempt to reflect on interest groups' effectiveness and promote a substantive debate. It is also argued that ignoring the interests involved would counter the aim pursued here far more than a necessarily incomplete reproduction of information. Finally, it is not possible to discuss **public welfare interests** and the possibilities and limits of realization without addressing the particular interests that **conflict** with public welfare interests. This does not exclude the possibility of being congruent in partial areas.

It is acknowledged and respected in advance that each grouping has reasons to pursue specific self-interests from its perspective. This does not mean, however, that the reasons put forward necessarily stand up to critical reflection in terms of the stringency of the argumentation and prove plausible about achieving self-imposed or higher-level goals. The following explanations are guided by the intention of gaining a better understanding of the possible driving forces of the various groupings in the sense of increased comprehensibility before, in a further step, the particular interests are placed on the interests of the common good.

6.1 Primary Producer

Primary producers produce the raw goods sent for further processing as food for human consumption and feed for livestock. Before food reaches the **consumers'** plate, many intermediate steps are taken on the long trail from "farm to fork," and many interests are involved. Other stakeholders are left out only in the case of on-farm feed and direct food sales to consumers. Generally, primary producers decide what is grown, how it is grown, and how it is used within the farm system. At the same time, it is in their hands in which way and to which extent measures for the protection of common goods are taken or renounced. Thus, they directly influence the degree of impairments of the environment, climate, nature, and last but not least, animal health and welfare of farm animals. However, they do not work self-sufficiently but are highly dependent on what the market demands and what the retailers are willing to pay for the raw goods primary producers have to sell to make a living. Correspondingly, primary producers find themselves in an ambivalent dual role. They not only play an active role but are also victims of developments that can impair their interest in securing their economic existence. Due to this ambivalence, deciding which advantages and disadvantages outweigh which side in the respective situation is impossible without well-founded data collection and assessment.

While primary producers play a significant role with regard to both the positive and negative effects of food production, they have little influence on shaping the economic conditions in which production processes take place. Consumers mainly obtain their food from food retailers. In contrast, direct marketing is of marginal importance, i.e., the direct transfer of products, information, and money between primary producers and customers at weekly markets and farm shops. As a counter-model to the prevailing forms of marketing, direct marketing exemplifies what the former lacks. In direct marketing, the direct contact between primary producers and customers results in considering which products are produced, offered, and demanded at which prices and which additional information. In the case of direct marketing, the product range is tailored to customer requirements and is adapted when these changes. Conversely, customers respond to the additional information provided by primary producers, e.g., the background of the production processes. This makes it easier for customers to understand why what is being harvested at any given time is available and why certain products can or must be offered at a lower or higher price. Direct contact, mutual communication, and knowledge of the respective backgrounds are the basis for a relationship of trust between producer and consumer. Suppose customers are cheated by poor product quality or high prices perceived as overpriced without sufficient justification. In that case, this will be sanctioned by a reluctance to buy and a loss of customers. This provides an effective corrective. Positive and negative feedback creates mutual understanding and enables us to consider the interests of both sides.

The outlined forms of balancing interests in direct marketing address what is lacking in the prevailing forms of marketing. Food retailers have taken over communication with consumers. Turned positively, this saves primary producers a lot of effort and time, allowing them to focus entirely on production processes. They also no longer have to adapt to developments in demand in terms of product quantities and quality. If the demand for specific raw goods declines, their production must not be cut back immediately. Product quantities that exceed regional and national demand can be exported. Some farms even use low market prices to expand their production capacities. The logic is that larger quantities are needed for the same operating profit if the profit margin per product unit falls. On the other hand, the expansion of production capacities is linked to the consideration of being better positioned vis-à-vis competitors if market prices should rise again. Farms acting in this way are making a bet on future market developments. There have always been farms that have weathered a period of low market prices better than others. However, suppose farms continue to produce unchanged or expand production capacities despite an oversupply of raw goods. In that case, those who voluntarily reduce their product quantities cause a price-stabilizing effect but behave as "fools" regarding their interests. In the face of a global oversupply of food, the result is a pronounced cut-throat competition, i.e., expanding farmers drive other farmers and, with a time lag, possibly themselves to ruin.

The prevailing forms of marketing lack the prerequisites for a reciprocal relationship between primary producers and consumers. This deprives the former of the necessity to adapt to changing consumer wishes, to inform consumers about changes in the product range, and thus to influence the interplay between supply and demand. At the same time, it makes it easier for both sides to devote themselves entirely to their interests without considering the other stakeholder's interests. Since the primary producers cannot influence the prices paid to them by the buyers, corresponding to the official price quotations, the only thing left for them to do is to minimize further the cost of producing the raw goods. This is more or less inevitably accompanied by compromises in terms of the quality of the products and the production processes. On the other hand, consumers, unaware of the production processes, want to purchase goods of the highest possible quality for as little money as possible. They cannot judge to what extent the goods are of high quality. They are dependent on the data and information offered to them by the retailer. However, the processing food industry and food retailers are less interested in a factual and truthful exchange of information than in an incentive to buy promoted by advertising statements. What remains an inescapable reality is the price consumers pay to retailers and the price received by producers from buyers of raw goods. Everything else is mainly speculative and part of mental associations as long as independent institutions have no qualitative assessments that could provide reliable statements and, thus, orientation.

Instead of direct exchange, as in the case of direct marketing, "communication" between primary producers and consumers now occurs via **standards**. These are defined by compliance with minimum requirements for individual characteristics of products or production processes. The interest groups that establish the standards link this with a qualitative statement and equate them with **"quality standards."** However, standards are not self-explanatory; they merely classify facts based on individual aspects into specific categories that help interpret and evaluate perceptual content. Setting standards aims to suggest a qualitative differentiation in the interest of those who define them. In this way, assigning sales products to standards as categorical pigeonholes reduces confusion in the product range. However, contrary to what is suggested by the term **"Quality standard,"** standards are not synonymous with a robust qualitative assessment. The definition of standards corresponds to the trade requirements for comparability or standardization of raw goods for sale. For example, there is a requirement that fruits and vegetables should have a uniform size, format, and appearance. To meet these requirements, considerable amounts of **pesticides** are used. In addition, requirements for standardization contribute to a drastic genetic impoverishment of the varieties on offer. Standardizations follow the trade's logic and interests regarding logistical, marketing, and cost-saving requirements. However, as explained in more detail in Sect. 6.4, standardizations do not follow the logic of a qualitative differentiation of products and production processes.

Even if standards do not justify comprehensive qualitative statements, they are nevertheless a guideline for action for primary producers. They specify which specific requirements must be met for products purchased by the processing industry or even provided with a price premium. They also determine whether products may be provided with a sales-promoting label (e.g., eco-label). For primary producers, it is no less important that the competitive situation, geared towards cost leadership, forces them to achieve the minimum standards requirements with as little effort and production costs as possible. Just as a showjumping horse does not jump higher to clear the hurdle, many primary producers do not do more than they have to achieve the price quotation linked to the standards. This way, the economic framework conditions promote undercutting competition ("race to the bottom"). Efforts to continuously reduce production costs result in a drop in the quality level of raw goods. In favorable terms, standards defined by minimum requirements, which are restricted to single traits, prevent a drop below the level achieved by complying with the minimum requirements.

Primary producers lack an inherent incentive to strive for a higher quality of their products and the production process if the additional expenditure required for this is not appropriately rewarded. On the other hand, the prescribed standards do not prevent primary producers from selling inferior products that meet the minimum requirements despite a minor quality. This is the case, for example, with products derived from diseased animals. The **proof** that the products come from healthy animals is usually not part of the minimum standards requirements. Therefore, products from sick animals can be sold at the same price as those from healthy animals. As recently as the 1990s, there was a legal requirement to classify the meat of diseased slaughtered animals in the "conditionally fit for human consumption" category. As it was still fit for consumption, it was sold at a discount on separate markets. With the removal of this differentiation, primary producers lack a vital incentive to keep the proportion of low-quality products of animal origin as low as possible. It can be assumed that primary producers have a fundamental interest in maintaining the proportion of diseased livestock low, which often results in reduced production output and further **failure costs**. However, the extent of the reduced output and failure costs are not directly identifiable and can only be measured with additional data acquisition and evaluation. Since these extra expenses are usually not incurred in practice, reduced output due to illness, despite its considerable economic disadvantages, is only of limited relevance to operational management (Habel et al. 2021).

Driven by market-based competition, primary producers are forced to increase productivity continuously. These can often only be achieved by following specific production paths. In the beginning, many farms were encouraged to specialize by the economic successes they were enjoying. However, this also led to increased **path dependencies,** which increasingly restricted the scope for action of primary producers. The concept of path dependency originates from an analytical concept in social sciences (Ackermann 2001). The decision favoring a specific option is followed by a stable phase in which the development can be maintained and further developed on the path once taken due to positive feedback effects. The consequence is that in such a stable phase, deviations from direction are hardly possible as they become increasingly costly. The remarkable thing about stable, path-dependent processes is that decisions tend to be no longer questioned. Self-corrections, fundamental changes, or even leaving the chosen path become increasingly difficult. This is especially true if the investments made so far have not yet been written off. The externally determined market prices and the path dependencies have maneuvered the primary producers into a tightly-knit dependency relationship. Accordingly, few opportunities exist, and little willingness to take on new challenges. This applies to changing consumer demands, such as social demands, for more significant consideration of environmental and animal welfare concerns.

Accordingly, primary producers are seldom interested in assessing the negative impacts of the harmful effects they are responsible for. Many farmers cannot afford, either from an economic or a psychological point of view, to be aware of the true extent of the undesirable side effects and the resulting potential consequences. It is, therefore, understandable if farmers do not want to blame themselves for having made the wrong decisions in the past and for having "backed the wrong horse" or taken the wrong path. While economic developments are positive for some primary producers, the situation is becoming increasingly precarious for most farmers. The attribute "precarious" accentuates that living conditions are complex and existentially threatened. The problem is fragile because it is increasingly difficult for those involved to make the right decisions. Many farmers do not know how to get out of this challenging situation. Sticking to the status quo seems much more manageable to many farmers than the uncertainty that would necessarily be associated with change. The situation is also precarious because farmers defend the economic system. Consequently, they are likely to become victims through their actions against criticism. On the one hand, primary producers and their political representatives are pushing to expand export-oriented agriculture. The export orientation provides the processing industries extended opportunities for the surplus quantities produced that cannot be sold domestically. Since export orientation makes the price level on the world market the yardstick, it ensures low purchase prices for the retail trade and low sales prices for the primary producers. Due to the broad global interchangeability of raw goods, the market system ensures that no shortages arise, which could lead to a relevant increase in purchase and sales prices.

The precarious situation could explain why many primary producers turn a blind eye to the fact that they contribute to flooding the markets with raw goods and enable trade buyers to depress market prices. Most farmers cannot adjust production volumes to demand to avert a price collapse. Because of the path dependency, many farms have little choice but to continue producing even when market prices do not cover production costs. The only thing that keeps them going is the hope that milk prices will rise and the knowledge that there is no turning back if they give up the business. Some farms hold out longer than others, but ultimately only until over-indebtedness, no longer supported by the banks, leaves them no choice but to go bankrupt.

Without state subsidies, most farms would be unable to keep in business. Although farmers have already been subsidized for many years with considerable sums of public money, agricultural lobbyists keep attracting attention by demanding politicians to increase subsidies and prevent any state regulations associated with increased production costs, thus reducing competitiveness. Hence, it is not surprising that it is precisely the representatives of lobby organizations who are very popular, who fend off social demands for change with particular verve, and who reject all considerations of regulatory intervention by the state. Political control approaches to green farming practices under the conditions of free-market and globalized competition for resources, sales markets, and profits are seen as restrictions on economic freedom of action and thus as encroachments on property rights related to fundamental rights. If the farmers' lobby association has its way, the existing subsidy payments should remain untouched and even be increased. Otherwise, politics should stay out of the regulation of production processes. It is argued that state regulatory measures put domestic producers at a disadvantage compared to competitors abroad, who are not subject to such requirements. On the other hand, state aid in the form of substantial direct payments, aid to social security funds, or tax exemptions is not classified as a competitive advantage, although it undoubtedly is. On the contrary, granting public aid is recorded as a justified payment for services rendered in compliance with legal requirements.

While competitors in many countries can pursue their economic interests largely without restriction, primary producers in other countries are more or less constrained by legislation. From the perspective of an economically liberal attitude, this must inevitably appear to primary producers to be unfair, an unfair distortion of competition, and an obstacle to free enterprise. At the same time, the reference to the supposed disadvantages vis-à-vis competitors in other countries distracts attention from the unfair conditions of competition at home. Many farmers gain economic advantages at the expense of others. However, taking advantage takes place behind a great veil of non-transparency. In a liberal economic system, success alone counts, not the possibly dubious ways it was achieved. Because diligence and one's ability to perform can hardly be deciphered, the result is used without further ado to indicate a legitimate achievement. This makes the question of the background to accomplishing the success superfluous.

As long as the Farmers' Association can wring the highest possible subsidy payments out of politics with an influential agricultural lobby, its representatives can credibly claim to represent the interests of all association members. However, the interests of farmers have become much more differentiated, making cohesion among farmers more difficult. While many farmers, for example, are pleased about the state subsidies for cultivating renewable raw materials such as biogas, this runs counter to the interests of many dairy farmers. For example, the cultivation of plant biomass for biogas production has driven up rental prices, making livestock production in the relevant regions unprofitable for those farms that rely on low rental prices. Conflicts between different interests also break out in other places. Often, the promotion of some is at the expense of the interests of other farmers. The fact that primary producers are in a situation of competition with each other, not only at a global level but also at the local, regional, and national level, which threatens their very existence, is concealed by the fact that the associations are working towards common goals (increasing subsidies and resisting regulation). The greater the discontent, the more critical it is for the association representatives to locate the opponents outside their ranks to close their ranks more tightly in defense against adversity from third parties. Those inside the circled wagons build on the argument that one can only successfully fend off attacks from the outside if one shows unity. After all, this has also been used to alleviate internal conflicts. In this way, it has been possible to integrate the organic farming groups, which initially started as a protest and alternative movement and as a counter-model to conventional agriculture, in such a way that the original antagonisms—although still present—have become secondary.

Facing numerous problem areas, a wide variety of positions are held among primary producers to rationalize situations in which they feel trapped and to consider how to escape the unbearable loss of control and powerlessness. Maintaining control is an elementary basic need; feeling able to act is a vital elixir of life. Rationalization involves reinterpreting phenomena. This is mainly observed in people who perceive a situation as threatening without having any options to change it. To avoid a threatening loss of **coherence**, i.e., the inner coherence of the self-model and the subjective suffering associated with it, the situation is reinterpreted to appear halfway logical and bearable for the person concerned (Jost et al. 2004).

Understanding the unique challenges faced by farmers is a complex task. A metaphor often used within the profession helps to shed light on these difficulties, even if it doesn't claim to represent all farmers. In a public statement, a farmer bravely compared his struggles to those of a national football team coach, both facing a barrage of opinions and criticisms from the public (Ackermann 2017). It is implied that the national coach is confronted with the opinions of millions of citizens who claim to know something about football and, therefore, presume to know better than the leading man. Like the national coach, many farmers see themselves as being also harassed by a large number of critics who think they know how farmers should do their job. From the perspective of farmers, they know best what to do. After all, they have learned their profession and see the farm situation and the farm animals living in it best.

The associative comparison drawn here is revealing in many respects. On the surface, it rejects criticism and justifies one's practice. However, not only similarities but also differences become apparent. In professional sports, teams and their respective coaches are judged based on athletic performance. If these fail to materialize, the coaches lose their position more or less quickly. Sporting success is measured in the context of competition with other teams according to clearly defined rules monitored by referees. At the end of the season, the teams are ranked according to their performance. The ranking reflects the level of performance that the teams, together with the coaching staff, were able to deliver. The supporters' satisfaction correlates with the sporting performance of their team and, in the medium to long term, with their economic success.

The comparison with the peculiarities of sport shows what is lacking in agriculture. Whereas the fascination of sport, at least on the pitch, lies in its transparency and manageability, there is little evidence of this in agriculture. This does not occur when it comes to a neutral and independent assessment of the socially relevant services farms provide. While economic success stands in the fore, the social benefits or harmful effects associated with the production processes are of secondary importance. By claiming to know best what is right and wrong, farmers make their assessments the yardstick for their actions (self-referentiality). Suppose one's economic success and evaluation based on one's standards are what counts first and foremost. In that case, this has nothing to do with fair competition conducted according to transparent rules that apply equally to all. It has been widely left to the individual agricultural players to decide how they achieve economic success. What remains unconsidered mainly is the extent of the undesirable side effects accompanying this. Agricultural production methods lack **consistent** rules and regulations, as they exist in sports and other economic sectors. The "good agricultural practice" guideline does not meet this requirement. According to EU Regulation (EC) No 1750/1999, Article 28, "good agricultural practice" is merely the ordinary standard of management that a responsible farmer would apply in the region concerned.

In the occasional perusal of the agricultural trade press, the author has not encountered any reference that could be interpreted as a desire for an improved set of regulations. On the other hand, there is no shortage of positions that castigate possible state requirements as unjustified interference in exercising business. Many positions fear an unnecessary increase in bureaucracy and warn that farmers' competitiveness vis-à-vis foreign competitors will be adversely affected. When an agricultural economist once took the liberty of pointing out weaknesses in the agricultural system in a renowned daily newspaper and called for a significant expansion of controls on farms, he was subjected to a veritable shitstorm in social media (Grossarth 2017). What is remarkable about the reaction to this plea is not only the resistance itself but also that the student council of the Faculty of Agricultural Sciences initiated it. Consequently, it is not just a matter of individual voices opposing regulatory measures but of resistance to any form of regulation broadly anchored in the profession, which has considerable resonance even in agricultural science student circles.

When the Ministries of Agriculture in some European countries, under pressure from the EU Commission, felt compelled to tighten the requirements of the Fertilizer Ordinance, they were confronted with massive protests from farmers. They took their resistance to the streets and blocked important traffic arteries with their giant tractors to emphasize enforcing their interests. In contrast to the usual strike actions, e.g., by public service employees who increase the pressure on employers in connection with demands for higher wages, free entrepreneurs and employers demonstrate this because they see their economic existence threatened. While the negative consequences of the emissions of pollutants escaping from farms are largely negated, the demonstrators complain that society does not value their work enough and that too little respect is shown to them as a professional group. They feel marginalized instead of appreciated for their achievements in providing the population with cheap food.

The protest, however, is not directed against the economic system that exposes them to ruinous cut-throat competition. Nor is it directed against the professional colleagues who are further fueling competition by expanding production capacities, but against a policy that imposes higher legal requirements on farmers to take account of overriding interests in the protection of common goods (such as groundwater quality). Here, a national agricultural policy, forced by the EU Commission to impose requirements on the polluters, collides with the demands of a professional group that does not want to see its freedom of action restricted. With the protest actions, the demonstrators reveal, on the one hand, the hardships to which they see themselves exposed. On the other hand, the demonstration of power expressed using tank-like tractors is, in reality, a sign of powerlessness. While politicians are pushing hard to accommodate the farmers' demands because of fears that the protests will spread, from another perspective, the demonstrations appear to be a last stand against the inevitable. After all, even farmers will not be able to ignore legal requirements in the long run, which were already decided decades ago but have not yet been consistently implemented in practice and not enforced by the executive.

Derivations

The undercutting competition on price and the orientation towards a few partial aspects, which prescribe **"Quality standard"** in the form of **minimum requirements**, has set in motion a spiral of declining product and process quality. By striving for cost leadership, primary producers lack the needed resources for quality production. At the same time, the market rewards those who externalize and disadvantages those who try to meet their qualitative demands with additional expenditure without being reimbursed for the extra services on the market. In this way, the primary producers' room for maneuver is increasingly restricted. The desolate economic situation, the downward spiral in quality production, and its adverse effects on public welfare interests are inherent in the system. Most primary producers have lost touch with consumer demands and social concerns and are thrown back on their perspective and the assessment of facts according to self-referential yardsticks. Lack of room for maneuver causes many primary producers to flee the victim role. This also helps them to ignore their share of responsibility for internal and external undesirable developments.

The tendency towards isolation goes hand in hand with a reluctance to disclose internal conditions. The same applies to evaluating the service of the operating system regarding the protection of the environment and farm animals towards impairments assessed by independent institutions and assessed on an inter-farm basis. This will likely apply to those primary producers who fear a comparatively poor assessment. On the other hand, those primary producers who have already established a qualitative lead should have a strong interest in making their qualitative performance visible to outsiders and advocate fair competition through an inter-farm comparison of socially relevant services. This would allow them to escape the ruinous undercutting competition and secure their economic existence through quality production and the associated generation of additional prices. However, powerful farming lobbies support a general resistance of primary producers against any form of state regulation related to distinguishing between farms according to their socially relevant services. Consequently, all farms are considered "jointly liable" for the **agricultural crisis**.

Farmers, as the owners of the farm animals, are not just legally responsible but also profoundly dedicated to providing for their needs. The farm system is the reference unit for allocating available resources, suited to enable more animals to sustain and live without impairments. However, farm managers are diverse in their skills and the availability of relevant resources. This also applies to their perception of health problems and the expertise needed to deal with them appropriately. Additionally, they are limited to different degrees in their financial flexibility. Trapped in a competition primarily aligned to cost leadership and facing more or less extended phases of market prices that are not covering their total production costs, limitations in the availability of resources are causing trade-offs between aims. All efforts to reduce health and welfare problems must be balanced concerning the farms' economic viability. Thus, protecting farm animals against impairments is not just a duty but a profoundly ingrained part of the overall performance of farm management. As long as the farm management cannot expect a return on investment, it cannot be expected to operate against its interests.

Although it is evident that farms differ considerably in the extent to which they farm at the expense of externalities, distinguishing between farms would undermine the farmers' organizations' self-image of representing the "unity of diversity." This could result in some farms no longer receiving the same support as those that can reconcile food production with protective effects. Underperformers have no interest in farming lobbies that deviate from their previous position. Conversely, this means that the farming lobbies protect, in particular, those who harm public interests. Differentiation of the production performance of individual farms about socially relevant services would make it possible to "separate the wheat from the chaff." This would not abolish competition but would orient it towards new target values. At the same time, a set of rules would have to specify which measures are permitted and prohibited and how transgressions of the set of rules would be punished. With such a set of rules, however, the representatives of the farmers' associations would lose their current supremacy based on the demonstration of strength through unity. Accordingly, the association representatives perceive the differentiation of farms as a threat to their claim to representation, and, therefore, the claim is fought with all available means. Finally, the interests of those who benefit from the current situation and their power to assert them can be seen as the main obstacle to finding solutions, at least for farms equipped with initial conditions to produce raw commodities according to public interests.

6.2 Agricultural Policy

Given the dramatic supply situation during and after the Second World War, national agricultural policy was primarily geared toward securing food for the population. It seemed to be a logical consequence of supporting farmers in their ability to increase food production by providing adequate economic incentives. When the foundations of the European Communities were laid, the basic idea of Common Agricultural Policy (GAP) played a central role. The objectives of CAP were proclaimed in 1957 in the Treaty of Rome (EEC Treaty) (see also Sect. 2.4). The objectives were set out as follows: (a) participation of agriculture in the development of the national economy; (b) best possible supply of the population with foodstuffs; (c) compensation for the natural and economic disadvantages of agriculture; (d) increase in productivity; (e) harmonization of the social situation of those working in agriculture with that of comparable occupational groups. For decades, agricultural policy has been the most communitarian policy area of the European Union. The main decisions on the form of agrarian policy are taken at the EU level in Brussels. This applies particularly to the agricultural market and price policy, as well as the direct income payments to farmers originally introduced to compensate for price reductions (Weingarten 2010). Therefore, national agricultural policy focuses on the national and subnational implementation of EU requirements. In a few agricultural areas, such as soil protection and agricultural social policy, responsibilities continue to lie with the European member states.

Since the food supply situation of the population was already improving at the beginning of the 1960s, the primary objective of the CAP was for a long time to raise the prices of agricultural products through market organization measures. The aim was to improve farmers' incomes and bring them into line with other sectors of the economy. This was achieved through adequate external protection, minimum producer prices, and state purchases to support prices. The knock-on effects of government incentives to increase product volumes and productivity were the much-cited "milk and wine lakes" and "butter, cereal and olive mountains." To cope with the surpluses, the sale of products produced in the EU on the world market was supported by export subsidies. In this way, third-world agricultural markets were flooded with subsidized, cheap goods, which had a massive impact.

The escalating agricultural expenditures created political pressure for action. In 1992, the EU Commission pushed through a reform of European agricultural policy. It included a first step away from an income-oriented price policy towards a market-oriented agrarian policy. Until the reform, export subsidies and other market support measures accounted for over 90% of EU agricultural spending. Although the state purchase of surpluses had significantly stimulated overproduction and had a trade-distorting effect, the CAP reacted with changes in the economic framework conditions only after a considerable time lag. State price support was reduced and compensated for by introducing direct payments to primary producers. While direct payments were initially still linked to production, since 2005, they have been primarily decoupled from production. The fact that the CAP is the only policy area the Community fully finances explains why the European Union's budget is still heavily weighted towards agriculture. In the 1993–99 legislative period, agricultural expenditure still accounted for 46.7% of the total EU budget. It decreased to 36.3% in 2014–20 and is still 30.7% from 2021 to 2027 (Hemmerling and Pascher 2019). In 2020, direct payments accounted for 72% of EU agricultural expenditures, totaling 57 billion euros, 5% of agricultural market expenditures, and around 23% of rural development. Without direct payments from the EU budget, many farms would no longer be viable today.

Decoupling direct payments from production processes was initially expected to lead to a stronger orientation of agricultural production towards market demand. According to Roth (2007), the market economy is essentially about the efficient allocation of **resources** and the avoidance of waste. In market-based competition, the winner will be the supplier who accurately anticipates customer demands and can meet them at a reasonable price.

> *"Just as in sports, it is incumbent on the organizers of the competition—in this case, the politicians—to determine the rules of the game in such a way that they effectively prevent undesirable strategies and unfair practices. If the game's rules are inadequately determined or compliance with them is not ensured, this is not a shortcoming of the idea of competition, but its defective implementation."*

The theory also states that one can only remain successful in the market economy in the long term if the offer is constantly adapted to changing conditions, i.e., reacts to changing customer wishes, production possibilities, and resource shortages. As clear and straightforward as the theory and the distribution of roles might appear, politicians have failed to implement the ideas in practice.

The market economy system was also fully functional in agriculture regarding the supply of low-priced raw materials. Competition for cost leadership has meant that market prices in relation to people's incomes have never been lower than they are today. However, it may be doubted whether, according to free-market theory, the competition winners also offer the goods that customers want. There are doubts about whether fewer resources were needed to produce cheaper goods. According to the above reference, agricultural production involves far more complex processes than are found in sports. Here, the winners are those who run the fastest, jump or throw the farthest, or those teams who score the most points in a particular competition. The rules are fixed and usually apply at the international level. Furthermore, it is regulated who checks the observance of the rules and how and how to proceed in case of rule violations. The fact that spectators are also aware of the rules makes the sport a very transparent event, not only for the players but also for the spectators.

The clarity of the regulatory framework in sport contrasts with the lack of clarity in the regulations governing agricultural production processes. Although there are some rules here, too, these are essentially limited to compliance with minimum standards laid down by regulatory policy, which only affect a few sub-sectors. Furthermore, the control institutions that monitor compliance with the minimum requirements lack the necessary resources. Nor do they have the power to punish violations of the rules effectively. Moreover, compliance with minimum requirements is not synonymous with compliance with fair competition. In sports, care is taken to ensure that the achievements obtained and measured are not achieved by unfair means and not at the expense of competitors. The agricultural industry is a long way from this. To achieve this, the adverse effects of production processes on animal health and welfare, consumer protection, environmental protection, nature conservation, and climate protection would have to be recorded on a farm-by-farm basis. Only then would it be possible to punish and effectively restrict practices that cause above-average harmful effects.

In the nature of European legislation, changes and additions to the legal framework for agriculture take a very long time to be accepted by the majority of member states. Although national governments can initiate their legislative projects, these must not fall below the level of the European minimum requirements. Accordingly, this option is rarely used. Instead, agricultural policy-makers often warn against national go-it-alone measures with excessive regulations. On the one hand, there is the fear of putting domestic primary producers at a disadvantage vis-à-vis their competitors in the competition for cost leadership, thus causing the displeasure of lobbyists from the agricultural industry. On the other hand, introducing rules and regulations does not usually win many votes, even among the electorate. This is more likely achieved by promising to dismantle rules that are perceived as too bureaucratic.

The current agricultural policy of intervening in farm processes with as little regulation as possible is supported by agrarian lobbyists and agricultural economics in several respects. The orientation of production processes towards increased productivity through specialization, intensification, and concentration has been a guiding principle of agricultural economics for decades (Maréchal et al. 2008). A fundamental economic liberal attitude characterizes its theories and positions. According to this, it should be left primarily to the market to find solutions to the problems that arise. This position is occupied even when market failures are apparent, particularly concerning detrimental impacts on public welfare interests. However, as long as the focus remains on increasing productivity and the conditions of competition are aligned with existing minimum legal requirements, there is little reason for lobby groups to doubt the belief in the solution potential of markets. After all, productivity has increased significantly in recent decades. However, focusing on productivity as an indicator of success comes at the price of largely ignoring the negative consequences of production processes. This also applies to the causal relationships outlined in Sect. 5.3, in which the pursuit of cost leadership primarily fuels the negative consequences. **Cost leadership** is based on minimalist expenditure for all those expenses that do not directly affect production performance. The protection of livestock and the environment is usually accompanied by additional spending. Those who reduce these expenses to a minimum and externalize their advantage to the detriment of the community's goods gain a competitive advantage. **External effects** are the uncompensated effects of economic decisions on uninvolved parties, i.e., effects for which no one pays or receives compensation (Mankiw and Wagner 2004). Therefore, they are not usually included in the decision-making calculations of those who cause them. In economic terms, they constitute a form of market failure and can, according to theory, make government intervention necessary.

Agricultural policy continues to rely on an export strategy to sell domestically raw goods that are not in demand. With the certainty of global population growth and the expectation of an accompanying increase in purchasing power, growing sales markets are seen in the future against global climate change, which offers economic profit opportunities for domestic producers. Farmers who once sold their regional products on regional markets at a regional price level will become more and more participants in a world market. However, those farms that profit most from the export orientation of the agricultural economy are often also among the farms responsible for considerable pollutant discharges into the environment. Many individual farm constellations have a veritable conflict of interest between export and public welfare orientations. Of course, there are significant differences between individual farms. Therefore, it is necessary to draw up a separate farm assessment of pollutant discharges in relation to production performance and make a cross-farm comparison. However, such a differentiation of performance between farms is not what is envisaged by farmers' associations or agricultural policy, except for what has already been implemented in The Netherlands (Daatselaar et al. 2015). The priority of securing food for the world's population is argued to ward off counterarguments. From the perspective of agricultural policy, efforts to secure existing markets and open up new ones should be continued and, if necessary, intensified. The effects this will have on feeding the world's population are not detailed.

If market adaptation means that primary producers should increasingly align with customers' wishes, then the CAP cannot be attested to success here. European consumer surveys have attached the most significant importance to livestock farming regarding animal health and welfare for many years (European Commission 2023). More than eight in ten Europeans (84%) believe that the welfare of farmed animals in their country should be better protected than it is now. Yet little has changed for the better in this area in the past. Irrespective of the negative consequences of the prevailing production methods on customer wishes, what can be produced at the cheapest continues to be made. This is not necessarily what consumers understand by high-quality products. On the other hand, the means of production (including fertilizers, feed, and plant protection products) and the costs of housing conditions and rents are becoming increasingly expensive. If the possibility of disposing of the surplus residues into the environment is restricted, this also makes the disposal of waste materials more expensive. Contrary to the theory of the "free" market economy, market conditions have not led to a continuous adaptation to changing requirements but have instead increasingly narrowed the scope for action of primary producers and, thus, the possibilities for adapting to consumer wishes.

For farms that are no longer competitive under the prevailing conditions of undercutting competition, the economic system of the "free" market economy knows only one "way out": the voluntary or involuntary closure of the business. Farms that want to stay in business have to grow; since the number of animals per farm is seldom accompanied by a corresponding increase in the amount of land available for farming, many farmers no longer know what to do with the excrement of their livestock. The inevitable consequence is an increase in the discharge of substances that cannot be used on the farm via **surface water** and airborne **emissions** into the environment. The complexity of the consequential effects has been described in various scientific reports, for instance, in the special report "Environmental Problems of Agriculture," (SRU 1985), published as early as 1985. The report attracted much attention among experts but failed to have any discernible impact on agricultural policy. If, more than 35 years later, the first changes in agricultural policy are now becoming apparent, this is not due to the ability of agricultural policy to understand the worsening problem. In the European Union, the EU Commission has repeatedly sued several member states, and the European Court of Justice (ECJ) has ruled in favor of the EU Commission. The ECJ ruled on 21 June 2018 that some member states do not comply with the legal requirements of the European Community, which had already been laid down in 1991 in the Nitrates Directive (91/676/EEC). Now, agricultural policy faces legal consequences in the form of substantial fines. This example shows that even agrarian policy does not always comply with the legal requirements, and higher authorities must urge them to do so.

Agricultural policy is now, with a considerable time lag, falling victim to the decades-long failure to introduce effective regulations to protect **groundwater** on its feet. Without the ECJ, agricultural policy would still not be obliged to effectively limit the discharge of **reactive N-compounds** from agricultural production processes. Without the ECJ, the issue would not be on the political agenda; without the ruling, governments would not have decided to tighten legal requirements. For this, however, agricultural policy is now reaping angry protests from farmers, who feel that they are being bullied and that the tightening of the legal regulations threatens their economic existence. Despite the ECJ ruling, governments are trying to keep the burden on primary producers as low as possible. For a long time, politicians considered it reasonable to point out that they have succeeded in freeing agriculture from impending environmental regulations. Then, when the ECJ successfully sued the German government for failing to comply with environmental regulations, ex-Federal Minister of Agriculture Schmidt was quoted in 2020 as saying, "*Now we're in deep shit.*" (Fried and Wernicke 2020). This attribution was supplemented at the same event by a farmer to the effect that many farmers would soon no longer know "where to take their shit from livestock farming" because of the impending legal regulation. The politicians promised the farmers demonstrating on site that they would work towards changes in the respective rules. The example reveals to which degree politicians feel obliged to emphasize the common ground with their clientele at the expense of overriding public interests and the community's environmental protection strategy.

The publicly presented representation of agribusiness interests by politics masks another influencing factor that is not immediately apparent but no less compelling: definitional power. The source of legitimacy from which this power draws is the supposed moderation between different interests and the orientation towards what is feasible. In agricultural policy, the coining of terms and setting standards through defining minimum requirements has a long tradition. For instance, by linking husbandry systems with the term "animal welfare," **"quality standards"** are defined. In this context, it does not matter that there is no functional or scientifically reliable connection between compliance with legally defined minimum requirements and the welfare of farm animals (Sundrum 2018). An analogous approach is applied to the issue of **"animal health."** Regulation (EU) 2016/429 of the European Parliament and of the Council of 9 March 2016 is called "animal health law." This title gives the impression that "animal health" is comprehensively regulated. Instead, its content almost exclusively addresses the issue of dealing with epizootic diseases. On the 208 pages of the regulation, many things are controlled, but not the handling of production diseases, which, as the main problem in livestock farming, affects not only the health of farm animals and thus the protection against pain, suffering, and damage but also **consumer protection** as well. Article 4 of the Regulation contains 56 definitions; however, there is no definition of the central term "animal health." However, "health status" is defined. According to the Regulation, this means: "*the status with regard to the listed diseases relevant to a particular listed animal species.*"

A similar approach is taken in agricultural policy with **"environmental compatibility."** Implementing specific environmental, species, and water protection measures is equated with an environmentally compatible production method. Compensation payments are promised by the political authorities solely for the implementation of specific individual measures, irrespective of whether and to what extent the measures in the respective context contribute to the achievement of the envisaged objectives and regardless of whether other harmful effects are caused by the farms at the same time, The same applies when compliance with generally binding minimum requirements concerning the quantities and manner of application of fertilizers is equated with "good professional practice." Without performance reviews, minimum requirements mutate into an end in themselves. This view and approach are maintained, although the current ruling of the European Court of Justice has established by the highest court that the previous regulations did not achieve the specified objective, namely compliance with the nitrate limit value in the groundwater at the numerous measuring points.

Instead of reflecting on strategies to achieve objectives and reviewing the effectiveness of measures in the respective context, the agricultural policy debate revolves primarily around the question of what burdens farmers can bear due to government regulation. The focus on reasonableness distracts from the core question of the primary causes; simultaneously, the main polluters are taken out of the picture. Even though the Treaty of Nice (2001) on European environmental policy explicitly prescribes the polluter-pays principle, it is only applied in isolated cases in agriculture. The fact that farms are involved in discharging pollutants into the environment to very different degrees suggests that the discharges should be assessed on a farm-by-farm basis. Only in this way could it be possible to identify farms that are mainly responsible for pollutant effects. However, agricultural policy, supported by the agricultural trade associations, still does not see any reason to enforce the polluter pays principle. The unreasonableness of bureaucratic expenditure is one of several arguments with which agrarian policy continues to draw a protective wall around the polluters of harmful effects. At the same time, agricultural policy fails to answer how the challenges, including compliance with the emission reduction targets for NO_x, particulate matter, and ammonia in the air set out in the EU NEC Directive, are to be met in the future.

A scientific study outlines the reasons for the failure of agricultural policy to meet the requirements of an environmental policy as follows:

> *"Agricultural policy operates based on an economy-oriented mode of governance and the principles of the capitalist mode of production. The continuity of the involvement of sectoral actors (e.g., heads of departments in ministries, presidents of associations) with the design of agricultural policy measures, the dependence of ministerial bureaucracies on the primary cycle of profit maximization within the framework of the fiscal state, bureaucratic self-interest (in competences and resources), the ideological congruence between ministerial bureaucracies on the one hand and the dominant business associations on the other through common university education and socialization, the Association of Chambers of Agriculture (VLK) as representatives of agricultural self-administration, agricultural bureaucracies, and agricultural faculties, and the "symbolic" (re-)production of the dominant, agricultural sectoral ideology by scientific agricultural economics lead to the maintenance of existing power relations between the policy sectors and the blocking of the implementation of EU environmental protection."* (Kaufer 2015)

Derivations

Politics, in general, is nothing more than the regulation of the affairs of a community through binding decisions based on power. Like other affairs, agricultural policy is challenged by various and partly conflicting goals. On the one hand, the supply of the population with inexpensive food should be guaranteed. On the other hand, farmers' income is to be adjusted to the income development in different sectors. Last but not least, the policy is obligated to protect and develop the common good, here to protect the common good from impairments caused by agricultural production processes. While the first objective has been over-achieved, the second and third objectives are far off. Agricultural policy has failed to reconcile different sub-objectives by over-achieving one objective and under-achieving others. Like a sorcerer's apprentice, agricultural policy has unleashed market forces without having the power or the means to rein them in for the good of the community and the good of farmers. While structures, processes, and content should be used to achieve specific goals in government and society, the goals are usually formulated in general terms and leave much room for self-referential interpretation. The same applies to the paths to be taken to achieve them.

Politicians never tire of presenting themselves as pioneers of a social or eco-social market economy. Agricultural policy cannot protect the citizens and the common good against the externalities and adverse side effects of producing agricultural commodities. It is also evident that agricultural policy has failed to establish fair competition between competitors. They cannot even withhold public money from primary producers, who affect public interests particularly severely. The "free" market economy prevails where social and ecological goals are only vaguely formulated, where their contours are lost in nebulous terms, where (if at all) only minimum standards are defined, and where it is not checked whether these are purposeful in the respective context, where no comprehensive control is established, and where no substantial sanctions prevent the actors from acting at the expense of the interests of the common good.

In this, the actors align themselves with their respective economic self-interests. This is, *per se,* neither morally reprehensible nor nonsensical. What primary producers decide and implement in their operating systems remains private as long as it does not affect the interests of the common good. However, the extent of the harmful effects of agriculture has long since reached a dimension that urgently requires counter-regulation. It is a matter of fairness that regulations should be applied first and foremost to those farms that are conspicuous for their high levels of pollutant emissions and high rates of animal welfare-related diseases. Those farms run particularly counter to the interests of the common good. The refusal of agricultural policy to use regulatory measures to establish an assessment of individual farms based on uniform yardsticks against the particular interests of the primary producers and other stakeholders involved is tantamount to refusing to orient and shape the agricultural economy in the sense of an eco-social market economy.

In the global markets, it is not primarily food that is short in supply, but above all, market prices that people experiencing poverty can afford and simultaneously cover the production costs. The assumption that a food shortage to be expected in the future could already be remedied by providing more of the same is a fundamental fallacy. Instead, food is subject to progressive devaluation as it increases. Thus, primary producers are essentially dependent on a gigantic devaluation machinery. The hopes for a cost-covering increase in world market prices are based on a disproportionate increase in demand. However, this can only be expected if production capacities of a relevant magnitude are lost, as is when droughts occur or when pandemic animal diseases spread. The outbreak of African swine fever in some Asian countries showed the extent to which unaffected countries could benefit—at least temporarily—from the sudden onset of shortages.

Given the worldwide expansion of production capacities and new product lines (see the following Sect. 6.3), the hope of rising world market prices is becoming increasingly remote. The continuing pressure to minimize costs pursued on an individual farm basis, primarily by increasing production output and capacity (economies of scale), further fuels predatory competition. Opposite of a small number of winners stands a large number of losers. Of the latter, many have already surrendered to the seemingly inevitable fate. Others take their displeasure and protest to the streets, even if they cannot change the economic framework conditions and the rules of competition in a "free" market economy.

Worried that the protest movements will continue to expand, agricultural policy is shying away from drastic decisions and delegating the overdue disputes about a strategic reorientation to various commissions. Their recommendations can then be interpreted as they see fit. Through this procedure, the political actors believe that they are gaining time. At the same time, there is a growing danger that they will only be able to react to developments that threaten to break out all the more radically as crises.

6.3 Processing Industry

For some time now, the processing of raw goods into food has been outsourced from primary agricultural production and taken over by a specialized trade. This division of labor probably began with the grinding of **grain** in mills that could harness the energy of wind or water power. The division of labor was successively extended to processing other raw vegetable products. The same applies to the processing of milk and the slaughtering and processing of meat. As long as there is prosperous cooperation and neither side feels it is being taken advantage of, the division of labor between primary production and further processing represents a classic win–win situation for both sides. However, this does not protect against attempts by one side to gain advantages at the expense of the other. Accordingly, it is of great importance for prosperous cooperation that verifiable parameters are available, i.e., that the quantities delivered can be quantified as precisely as possible and the qualitative properties of the raw materials can be assessed understandably to take both aspects into account in the remuneration appropriately.

The market economy processes of specialization, intensification, and concentration explained in the context of primary production did not stop at the processing industry. With the increasing improvement of transport and the reduction of transport costs, the catchment areas for the supply of raw materials expanded, and simultaneously, the competitive situation. Almost every community had a flour mill, dairy, or slaughterhouse for a long time. However, local and regional structures gradually disintegrated and were largely destroyed by a rapidly increasing concentration of the processing industry. For example, while there were still around 19,000 mills in Germany in 1950, only 196 were counted in 2020. In the same period, dairies fell from around 3400 to 155. This also had far-reaching consequences for the interaction between primary producers and the processing industry. The fewer buyers there were for the primary producers' raw materials, the more monopoly-like structures and one-sided dependency relationships were established. The changes in the power and dependency relationships run counter to a fair distribution of the value added that can be achieved with primary production and further processing. These developments are not only the result of technical progress, which drastically increased the scale of the necessary investments and thus accelerated the concentration processes. Legislators and local authorities also played their part in this development.

In the case of **meat** processing, the BSE crisis, which originated in Great Britain in the 1990s, resulted in a comprehensive reorganization of food law at the EU level. This, among other things, meant that from 2009, even smaller slaughterhouses had to meet the increased requirements of EU approval. This entailed extensive investment in structural changes to ensure hygienic conditions. Many slaughterhouse operators took these requirements as a reason to cease operations (Fink-Kessler 2020). In addition, the enlargement of the EU made it possible to employ workers under contract for work (Posted Workers Act) in the slaughtering industry. As a result, contracts for work and services, especially with Eastern European workers, became a business model, leading to reduced labor costs in the slaughter industry and further concentration processes. The larger the slaughter companies, the better investments could be absorbed through economic economies of scale. In addition, large slaughter companies began to buy up minor cutting and processing companies to combine several links in the process chain in one hand. In this way, the growing discount meat markets could be better served. In some cases, food retailers even set up their meat plants. Added to this were technological developments such as modified atmosphere film packaging, which promoted the entry of discounters into the fresh meat business.

Against the concentration processes and the economies of scale that can be achieved with them, the communal slaughterhouses that have been predominant in the past had no chance of holding their own in the competition for low processing costs. Many butcher shops now get meat that can be purchased cheaply and cut and portioned in the desired manner. While the slaughtering and processing processes have been concentrated in large companies, the tasks of food monitoring have not been handed over to higher-level specialist authorities but have been municipalized. The primary responsibility for food monitoring thus lies with the district offices. At the same time, the state ministries have only a few rights of intervention, and the federal government has hardly any powers or possibilities of intervention in food monitoring. The shift of responsibility to the municipal level means that the prices that slaughter companies have to pay to the district offices for live animal and meat inspection vary greatly. According to the Butchers' Association, the big players in the industry pay considerably lower fees per pig than a butcher's business (Fink-Kessler 2020). In addition, various local authorities subsidize the large meat companies by foregoing relevant fee income to maintain jobs and income from trade tax. On the other hand, they contribute significantly to distorting competition between the slaughter companies.

With the expansion of processing capacities, the large slaughter companies also began serving export markets within the EU and third countries. Again, the large slaughterhouses had structural advantages that allowed them to continue growing at small-scale operations' expense. Furthermore, they successfully match the heterogeneous nature of carcasses with country-specific consumption habits and demand for specific products.

Technological developments make it possible to record the differences in carcass composition in detail and cut the cuts with a view to **standardization** and adaptation to the market's needs. The development of export markets made it possible to cut sales that were classified as inferior in the country of production profitably. A frequently cited example is pigs' feet, which have no market value in European countries but are in great demand in China (Klepper 2020). The same applies to belly pork, which is in great demand in some Eastern European countries, while there is hardly any demand for these cuts in other countries. Those who were able to sell pigs' feet, ears, and heads to China at reasonable prices realized an additional profit (the so-called "fifth quarter") and, at the same time, with an underlying mixed calculation, were able to offer the meat to the processing industry more cheaply than other slaughter companies. The resounding success of this business model has benefited mainly the slaughterhouse but least of all the primary producers. They must continue to be satisfied with the price quotations the market sets for them, irrespective of whether these cover costs. The original win–win situation has turned into a win-lose situation.

The business model of the processing industry is based primarily on the economic advantages that arise from cost-minimizing economies of scale and increased sales. Raw materials are transported over long distances to utilize the technological capacities of processing as fully as possible. This is possible because transport costs are subordinate to the economic calculation for the delivery of raw materials and the distribution of processed products. While primary producers cannot pass rising production costs to the processing industry, the latter occupies a better market and power position to countervail price fluctuations.

The purchased raw materials are differentiated primarily based on quantitative characteristics. In the case of meat, this is the categorization of the carcasses based on the meat-fat ratio. In the EU, this takes place within the framework of the EUROP classification anchored in law. This has no reference to the sensory **product quality** of the meat. Also, the reference to the operational context in which the raw materials were produced is primarily canceled. In the case of defective raw materials, a legally anchored system of traceability is intended to ensure that, if necessary, the source farms of defective raw materials can be identified and recourse can be taken against them. In this way, the entrepreneurial risk of further processing is significantly limited. The business model is based on realizing high unit sales, achieved primarily through high technology. Accordingly, there is little interest in a qualitative differentiation of the raw materials insofar as this is associated with additional logistical expenses not additionally rewarded by the market.

The degradation of animal-derived products into raw materials for the processing industry is particularly evident in the case of **raw milk.** The milk, which is milked from the udder of the cows, consists of proteins, fatty acids, carbohydrate lactose, minerals, and trace elements. The daily milking of the individual cows flows into large tanks from which the mixture is collected by tanker trucks at regular intervals. Milk from different farms is mixed in the tank trucks, so the raw material loses its reference to the producing farm. This reference is only relevant in case of damage. If contaminants are discovered in the milk samples taken at each collection, the polluter is liable for the damage caused to the dairy due to the affected batch being discarded. While the polluter pays principle is excluded concerning damage to the public interest caused by primary production, it nevertheless applies to the processing industry. By taking various measures to record the origin of raw materials, companies protect themselves and hold themselves harmless in the event of damage.

In the dairies, the raw milk is pasteurized to kill the pathogenic germs present in the milk. In addition, the milk cells, which consist primarily of immune cells that accumulate in the milk during the inflammatory processes that frequently occur, are centrifuged. In further steps, the milk is separated into individual groups of substances. Depending on the milk product to which the initial components will be further processed, individual components are mixed to form a predefined standard product. The only thing that the milk for sale and other milk products have in common with the original composition of the milk secretion formed in the mammary gland of a cow is the chemical starting materials. It is not far from separating raw milk into individual groups of substances and combining them to form a standard product to mixing chemical groups of substances originating from sources other than the udder of a dairy cow. The original substrates are replaced, for example, by plant substance groups and formed into a product that has a similar composition to milk and is marketed as a substitute for milk. In this way, whole milk has been replaced for decades by so-called milk replacers, which can be produced more cheaply than whole milk.

The first experiences with substitutes for milk from dairy cows were made with soy milk. In the meantime, oat milk is **competing** with soy milk. Oat milk suppliers are succeeding, not least because of the bad image attached to soybeans due to the problematic production conditions in the countries of origin. Many consumers associate soybeans with the cutting down of rainforests in South America. Oat milk producers are going on the offensive against competing products, including milk from dairy cows. One company, with the support of NGOs, has launched a petition to the German Parliament and campaigns in various European countries to the effect that in the future, all food products should state how many greenhouse gases their production has caused. The calculation behind this is obvious. In a direct comparison, oat milk comes off considerably better than cow's milk, provided that only those criteria appear relevant to the producers of the substitute products.

In addition to oats, peas are also used as a raw material for milk substitutes. These have a significantly higher protein content and, as a legume, are also associated with various advantages in terms of **sustainability** during cultivation. For example, the world's largest food company, Nestlée, is aggressively promoting a pea-based milk substitute. With this product, the company wants to participate in the emerging boom for vegan products. The business model promises big profits, as production, processing, and marketing are all on the same hand. It is speculated that new markets can be opened, detached from the agricultural production conditions and the ever-worsening image of animal-sourced food. The acceptance of these products is exceptionally high among consumers, who criticize the conditions in livestock production and therefore welcome substitute products that are produced without the undesirable side effects of livestock farming and have a supposedly better environmental balance. Accordingly, the substitute products are advertised with the reference to non-animal source materials. In addition, the conventional production processes are aggressively denounced. Nestlé's advertising message to the target group of millennials (those born around the year 2000) is: "*We want to take on the role of the helping superhero for our customers.*" (Wenzel 2021).

In this way, the suppliers of vegan products are attempting to place themselves at the forefront of a movement that opposes the previous production structures that have been branded as destructive. By producing substitute products, the players in the market distance themselves from the harmful effects of conventional production processes. They do not want to be associated with the negative consequences of agro-industrial production but to belong to the "good guys" without renouncing industrial processes and the advantages of a "free" market for those who have market power at the expense of those who don't. At the same time, they are avoiding their share of responsibility for the misguided developments in the agricultural economy, of which food corporations undoubtedly have a considerable share. A growing clientele wants to eat products that have as little as possible to do with the abuses in livestock farming and the negative implications for farm animals, the environment, and the climate (see Sect. 6.5). This is particularly true for meat products. While meat sales have been declining for years, there are clear signs of a boom in substitute products, which are forecast to have high growth rates in the future. The processing industry cannot ignore this trend but is challenged to adapt to changes in consumer demands at an early stage. In comprehensive sections of the population, there is a growing need to influence adverse developments closely associated with livestock farming through their consumer behavior. Correspondingly, the proportion of consumers who ate meat more or less is decreasing considerably, and the proportion of so-called flexitarians is increasing. Behind this is a large group of consumers striving to reduce **meat consumption** for various reasons. Although meat consumption is deeply rooted in Western food culture, the conditions under which the meat is produced are questioned by more and more consumers. This conflict situation promotes a search for alternatives.

A meat alternative is a product that can replace the consumption of meat. Five product groups are generally distinguished, each designed to appeal to different target groups and with distinct advantages and disadvantages. These include Cultured Meat, Insects, Microorganisms, and Plant-based and Pure Plant-based products. Cultured Meat is grown from animal cells and produced synthetically on an industrial scale. Just like insect-based products, Cultured Meat is of animal origin. Accordingly, neither is, in fact, an alternative to an animal-free or vegan diet. However, they open up options for "meat production" that avoid or significantly reduce the disadvantages and weaknesses of industrial livestock farming concerning consumer, animal, environmental, and climate protection. Consumer surveys in various countries suggest a high acceptance of Cultured Meat among the younger generations, especially when these products are advertised with benefits they supposedly contain for their health and the protection of the community (Bryant and Barnett 2020). To increase the acceptance of Cultured Meat, the authors explicitly recommend that advertising strategies address it as a solution to problems in livestock farming and a way to reduce meat consumption.

By adding additives, laboratory-grown meat is also expected to offer new options for adapting products more specifically to the industry's interests and consumers' specific needs. New sales opportunities are luring, promising high growth rates and economic success. However, alternative products are not free of problems either. They are usually subject to a high degree of processing using various additives whose effects on health have often not been subjected to thorough testing. Accordingly, attempts to imitate meat with vegetable fats, synthetic proteins, and artificial flavorings cannot be considered harmless to health without further ado. Nevertheless, there are various possible applications to replace "cheap meat" with substitute products, especially in convenience products, where the focus is not on enjoyment and quality.

In December 2020, Singapore became the first country in the world to release lab-grown chicken for consumption. The company that makes the nuggets predicts nothing less than the end of meat production as we know it. The company expects that by 2040, only 40% of all meat products will come from farm animals, and therefore, it anticipates gigantic new revenue opportunities for financial investors (Blage 2021). In the meantime, several large-scale projects have been launched, aiming to cultivate chicken, pork, beef, shrimp, and tuna. The term "cultured meat" promises a high level of acceptance among consumers because it holds out the prospect of solutions to the very problems for which livestock farming has fallen into disrepute among many consumers: animal suffering, environmental destruction, global warming, loss of biodiversity, waste of resources and antibiotic resistance, to name just the main aspects. On the other hand, producing substitute products requires enormous investment resources, corresponding starting substrates (e.g., calf serum), and large amounts of energy. Obviously, for the proponents of the substitute products, the balance favors the cultivated products against the previous production processes. However, this ignores the embedding of agricultural production processes in natural material cycles and the use of biological resources to produce foodstuffs, which do not require a high level of investment or external energy and chemical inputs.

Plant-based products contain elements extracted from plants, such as proteins, oils, fibers, and minerals, recombining in specific mixing ratios and structures. Since plant proteins, fats, and fibers do not have the same properties as raw materials of animal origin, the greatest challenge for plant-based products is to mimic the texture of meat. Accordingly, hybrid products are also being considered, combining plant and animal source materials to reduce the meat content without losing too many structural properties.

Concerned about losing market share, various meat processing companies have already started including hybrid, vegetarian, or vegan product ranges in their assortments. The growth rates are considerable and lead to increased investment in these areas. At the same time, the willingness to invest in traditional product lines is decreasing. Producing alternatives proves much more evident for the meat industry than initially assumed. Many production processes can be adopted and adapted. What changes, first and foremost, are the raw materials.

Despite the still comparatively low shares, it has long since ceased to be a question of principle, but only a question of time, when and to what extent products that can replace the consumption of animal-sourced food will noticeably compete with conventional raw goods of animal origin. No other area in the food sector is growing as rapidly as that for substitute products. According to market analysis, Europe and North America will reach "peak meat" by 2035, perhaps as early as 2025 (Witte et al. 2021). From that point on—according to the forecast—the consumption of conventional meat, milk, egg, and fish products will irreversibly decline and be compensated by substitute products. Assuming that the technology for producing substitute products develops even faster and, at the same time, animal products become more expensive due to regulatory requirements, thus making plant-based products more attractive, the market share could even rise to 22% by 2035. The market for substitute products is mainly driven by innovation and technology. At the same time, considerable profit options offer incentives for investors.

Derivations

For the processing industry, it is essential to have access to as large a pool of raw materials as possible to utilize their processing capacities. Where the raw goods come from is initially of secondary importance since they have been **standardized** on a global scale through far-reaching processes of genetics and feeding, which have made them highly interchangeable. The almost unlimited import possibilities ensure that raw materials for the processing industry only become scarce in exceptional situations. As long as there are no shortages, they are immune to paying significantly higher purchase prices beyond the usual price fluctuations on the market. On the other hand, the processing industry is quite open to processing regional products and ensuring traceability. However, this is only true when the additional logistical costs are within reasonable limits. After all, the focus is not on promoting regional material cycles and value chains but on generating windfall profits.

New industries are joining the existing processing industries of agriculturally produced raw materials. These also produce foodstuffs, although they no longer rely solely on raw materials from agricultural production but also on products from the laboratory. In this way, companies are decoupling food production from the individual farm and regional agrarian context and increasingly from agricultural production processes. This development can be seen as a logical continuation of the previous methods of agro-industrial production. The past processes of specialization, mechanization, and intensification of production have already produced a high degree of standardization of raw materials. At the same time, the differences in product characteristics caused by biological variation and heterogeneous production conditions have become less and less critical. Primary producers have been degraded to raw material suppliers and forced into a one-sided dependency relationship, making them largely powerless. However, the processing industry companies refuse to share responsibility for the damaging effects caused by primary production due to undercutting competition, even though they de facto shape the economic framework conditions that rob primary producers of their room for maneuver.

These do not have the efficacy to resist the price dictates of those who take the sales products from the primary producers. The dominant business model of the processing industry is not threatened by the suppliers but by other competing companies. Discouraged by the harmful excesses of the processes in primary production and further processing, more and more consumers are turning to various substitute products. These promise relief from the growing reservations and unease that creep over young people, particularly concerning the prevailing livestock production methods. At the same time, suppliers of substitute products are striving to come as close as possible to previous consumption habits. A breakthrough in the substitute products can be expected, especially if they taste, smell, and look exactly like the original. Consequently, the current business practice will come under pressure. It does not require a prophetic gift to predict that products of animal origin, which are characterized by a below-average level concerning relevant features of product and **process quality** (see Sect. 9.3), will be less in demand in the future and will have to deal with the negative image and competition from various substitute products.

The trend towards higher-quality products will also be driven by those consumers who do not want to switch to substitute products and will, therefore, increasingly demand high-quality products of animal origin. At the same time, some consumers will demand that more extraordinary accounts be taken of aspects of consumer and animal protection as well as environmental, nature, and climate protection in production and processing procedures. In the future, it will no longer be sufficient to propagate a higher level of quality through advertising slogans alone. Instead, there will be a need for proof of quality that can withstand critical scrutiny by independent experts. The larger the jungle of product diversity becomes and the more it expands due to substitute products, the greater the need for reliable statements that provide orientation. This is the only way to enable consumers to purchase products that promote a diet for themselves, their families, and for the good of the community that offers more benefits than it causes harm. Currently, the processing industry appears ill-prepared to meet the changing demands of consumers.

6.4 Food Retailers

Food retailing acts as an interface between many supplier companies and a heterogeneous group of consumers. It decides what selection of products is offered to them and where they are placed with what sales prospects in an increasingly large and confusing assortment. By making **aliments** available to consumers and significantly shaping purchasing behavior through the presentation and pricing of products, the food retail sector directly influences the upstream sectors of the processing industry and, indirectly, the primary agricultural production processes. While consumers in Western countries have quickly become accustomed to the sprawling assortment of apparent variety in the product range, a supermarket must seem like the veritable land of milk and honey to people from third countries. There is no lack of anything here. A feel-good oasis is offered that is pleasantly different from other life situations. This is true even for those who cannot afford products that are highly priced from their perspective. In this way, people retain the desire for products they hope to be able to afford in the future. In Germany, approximately 9.5% of consumer spending goes on food. This puts this country in one of the last places in the EU. Only in Austria, Luxembourg, and Ireland is the figure even lower. The highest share is in Eastern Europe. This does not mean that everyone there buys the more expensive organic products. Instead, in Romania and Lithuania, incomes are significantly lower, so the population inevitably has to spend more on food.

Retailers lure their customers with advertising messages whose truthfulness they do not have to prove and for which they do not have to assume any responsibility. This also applies to the undesirable side effects caused by the products during their production and further processing and transport to the supermarket shelf. At the same time, the shopping temples are kept free of any indications of possible problems and conflict areas that could evoke negative associations among consumers. All that remains is the 'agony of choice' many perceive as luxurious. The **marketing** saves the customers from having to think about the background of the production or even from having to ward off a guilty conscience. Removing the products for sale from the original context of production provides a feel-good oasis of carefree consumption. The only real thing is the price customers must pay at the checkout. The value they have acquired in return is left to the individual customer's value concepts and associations.

The grocery business model is based on high sales rates, which correlate closely with profits. Increases in sales are most likely to succeed when customers are enticed with low-price offers or supposedly favorable price-performance ratios and when they are tied to the retail chain via **brand programs**. As a consequence, lure offers are repeatedly launched via advertisements. Individual products are advertised with low prices that differ from the usual price and the competition's prices. The low prices encourage some consumers to go on a real bargain hunt. All the retailers advertise products of animal origin, in particular as special offers. Despite the low prices, the advertising messages simultaneously suggest to consumers that even the cheapest discount offers the **"best product quality"** and "maximum meat enjoyment."

The special offers are cross-financed by mixed calculations so that the prices at the point of sale do not necessarily directly relate to production costs. Instead, they are geared to what consumers are willing to spend on specifically advertised products. It is not the production costs but the customers' **willingness to pay** compared with the expected sales that determine the price quotations decisively. In addition, various marketing tricks influence the desire to pay off the consumer clientele targeted by the product. These include specific packaging formats and private labels, which make it difficult or impossible to compare products and make a reliable assessment of the price-performance ratio.

A significant portion of the real costs incurred in producing food is not included in the consumer price. "*Prices lie*" because they do not reflect the real costs of the production processes and the consequential costs to the community (Engelsman and Geier 2018). The main issue is external costs, i.e., costs that have arisen as undesirable side effects in the production, transport, and processing processes. External costs include, among other things, poor working conditions and low wages, expenses for the treatment of polluted groundwater, and the consequential costs of greenhouse gas emissions or the development of antibiotic resistance. A significant proportion of external costs are imposed on future generations, who have to live with environmental degradation and a climate crisis for which they are not responsible.

In Germany, a discounter has dared a highly unusual experiment in 2020 (Gaugler 2020). For eight selected conventionally and organically produced private label products, scientists had included the so-called "true costs" of the effects of nitrogen, climate gases, energy, and land use changes on the sales price. Based on the exemplary evaluation, the sales price of the eight conventionally produced foods (apple, banana, potato, tomato, mozzarella, Gouda cheese, milk, and mixed minced meat) would have to increase by around 62% per kilogram. For organic alternatives, the increase was only around 35%. Considering consumption habits, the average increase in the price paid at the checkout was 52% (conventional) and 32% (organic). With this experiment, a discounter admitted for the first time that the prices do not contain reliable information about the "true costs." The food is too cheap to cover all the expenses associated with the production processes. The error is embedded in the economic system. Correcting the systemic error would mean one thing for consumers: a significant price increase. This example illustrates the fundamental conflict between the particular interests of consumers and food retailers on the one hand and the interests of primary producers in cost-covering prices and the public welfare interests in a significant reduction in the harmful effects of production processes on the other.

> *"We must come to make the consequential costs of our consumption visible. Only then can customers decide on the shelf. As a company in a competitive market, we are undoubtedly* ***part of the problem****. However, I believe we can become part of the solution with this step,"* said the Divisional Board Member for Retail Germany at REWE Group. (quoted from Gaugler 2020).

This remarkable admission is not shared by all those responsible in the retail sector. The head of purchasing a big discounter sees his primary task as establishing a bulwark against price increases; he says it is a matter of defending the discounter's brand essence: "*high quality at the lowest possible price*" (Kläsgen 2021). How the discounter defines **quality** is the trade secret of the food supplier. No less significant is the fact that the discounter sees itself as the price leader and does not want to let anyone take the butter off its bread in this role. With its purchasing volume, the company sees itself in a position to convince or make obdurate suppliers compliant. Equipped with immense market power, the discounters, in competition with the supermarkets, dictate to the suppliers and, thus, the primary producers the framework within which food can be produced, processed, and traded. In doing so, they are supported by the consent of many consumers, who approve of the business model with their purchasing behavior. The negative consequences of this business model for the community are largely ignored.

One does not offend the buyers in the food retail sector if one assumes that they are fundamentally concerned with purchasing raw goods as cheaply as possible. Whether the primary producers can survive economically on the market prices and cover the direct costs is just as little considered in the purchasing decisions of the food retail buyers as it is by the consumers at the 'point of sale.' Instead, the anonymity and interchangeability of raw goods enable the food retailers to play off the supplier companies and, thus, the primary producers against each other worldwide to depress purchase prices further. When the Federal Association of German Dairy Farmers attempted in 2008 to draw attention to the plight of milk prices that were far too low and did not cover costs and to force the food retailers to pay higher prices by stopping deliveries, the retailers let the strikers go nowhere. A representative of the retail association made the following comments at the time:

> *"If some farmers don't deliver, others will step in. (…) There is too much milk, so this strike can't achieve anything. It fizzles out; it is absurd—if necessary, the milk will come from abroad."* (Postel 2008)

That's how it turned out. When individual politicians took the desolate situation of farmers as an opportunity to demand a minimum price for animal-sourced products, this suggestion was rejected by the food retailers' association. The association president criticized:

> *"Some politicians have lost the regulatory compass that focuses on the benefits of the social market economy and the goal of creating prosperity for all."*

The fact that the association of food retailers styles itself as a representative of a "social market economy" in the face of the grocery trade's pricing policy testifies to its self-confidence vis-à-vis politics and self-image. When the Federal Minister of Agriculture of Germany dared to criticize the pricing policy of the food retail sector and accused the food retail sector of unfair trading practices, the heads of the major German retail chains felt compelled to write a public letter of protest to the Federal Chancellor complaining about the statements made by the Minister of Agriculture. The media did not disclose how the Chancellor reacted to the protest letter. Representatives of the retail trade had already made it clear in advance what was particularly important to them. On a poster campaign by a big supermarket equipped with media celebrities, one slogan, in particular, was put in the foreground: "*Food deserves a price: the lowest*" (Gassmann 2020). However, this campaign brought the company fierce protest and opposition, and not only from farmers. Finally, responsible persons felt compelled to distance themselves from the action.

In addition to the dictate of low prices, it is also of great interest to the food retailer not to be held liable for its products. On the one hand, this applies to individual batches that are contaminated or otherwise defective. Accordingly, it was important for the trade to introduce traceability of products back to the producer, who can be held liable if the worst comes to the worst. This also applies to all other undesirable side effects in the production process. One does not want to be associated with the harmful effects massively forced by the food retail trade price dumping. On the other hand, they like to promote their image with good deeds. By introducing husbandry labeling, retail companies are staging themselves as the ones making a relevant financial contribution to improving farm animal welfare. However, this does not prevent companies from pursuing merciless purchasing practices that deprive primary producers of cost-covering prices. Even though livestock farmers bear primary responsibility for the welfare of their farm animals, their room for maneuver is excessively restricted by the pricing policy of the food retail sector. Under the economic conditions in which livestock farmers try to organize their financial survival, animal welfare efforts involving additional expenditure are primarily doomed to failure. Farmers do not have the necessary income or are not compelled by the state or the market to offer farm animals the degree of protection from biotic and abiotic stressors and to provide the living conditions that would be necessary for the animals to adapt to the living conditions and to live without severe impairments.

Consumers usually do not know enough about the hardships of primary producers and farm animals. When entering the supermarket, customers enter a make-believe world. Everything is in abundance as if there is no shortage of food elsewhere. The viewer's eyes are unaware of how the products were produced, how the farm animals lived beforehand, and that a not inconsiderable proportion of the goods on offer end up in the rubbish later on (Kreutzberger and Thurn 2011). Everything in the shops radiates freshness. At the meat counter, the fresh meat is illuminated with specific light so that it appears redder than it is. Milk cartons have pictures of cows in the pasture, yet the cows from which the milk in the carton comes may never have seen a pasture. The same goes for images of pigs in straw that have never been rooted in it. Some products are decorated with an emblem indicating regional production, even though only a few subcomponents come from the region or where the area takes on national proportions without defined borders. However, the scope for interpretation is extensive when assessing what additional benefits might be associated with regional production. Why should a product be produced in a region, however geographically delimited and marketed regionally, be qualitatively better than the products of the same company that are exported and have to compete in the global market? Consumers try to assess the quality of products utilizing the region's image or using individual characteristics associated with the area. In the purchasing situation, the indication of origin replaces the further examination of information about the factors to be evaluated (Banik et al. 2007).

Discounters and supermarkets usually offer many price variants of the same product, such as milk. To what extent the price differences are associatively or factually related to differences in product characteristics is not apparent to outsiders. Neither is an expensive product automatically of high quality nor is an inexpensive product necessarily of inferior quality. As a rule, the customer has neither sufficient information nor competence to assess and differentiate between product qualities adequately. This applies all the more to the assessment of the production processes underlying the product. Whether the products of animal origin come from animals that were doing well or whether the animal populations were confronted with high rates of disease and corresponding pain, suffering, and damage can neither be seen on the product nor inferred from the product information. The same applies to the release of pollutants and greenhouse gases released during the production processes. Instead, the superficially considerable variation in product properties and the undesirable consequences of the production processes are broken down into a few individual aspects. As explained using the example of farming methods, these particular aspects are used to identify so-called **"Quality standards."** These are intended to suggest a level of quality, which is misleading only because quality is not defined based on individual characteristics but by the totality of all attributes of a product, system, or process. Explanations of the definition and assessment of product and process quality are given in Sect. 9.1.

Standardization is a unifying way to describe something. It thus belongs to one of the various strategies of complexity reduction. Standards are considered a kind of guideline. In the original meaning of the word, a **standard** is a kind of rallying point around which people gather—similar to the standard, a field emblem hoisted on a pole in earlier times and thus became the symbol of a unit of troops. Suppose one wants to orient oneself in the complex tangle of interactions between biological, technical, socio-economic, and cultural food production processes; one cannot avoid a **reduction of complexity**. Reducing complexity by utilizing standards is a simplifying option. Still, it is not the only one, and, as will be discussed in more detail in Sect. 9.2, it is not the best choice for dealing with complexity and the resulting differences in quality. However, the use of standards, in particular, has become established in food production because they are particularly well-suited as a marketing tool. Therefore, it is not surprising that standardization is propagated in particular by agricultural marketing and economics. They are suitable for both customer acquisition and customer retention. **"Quality standards"** can be advertised, and corresponding products can be singled out from the anonymity of the masses to achieve a competitive advantage through product differentiation. In addition, quality standards are suitable for positively influencing a brand's reputation and a company's image. At the same time, standards provide customers who wish to better assess the product and process quality of foodstuffs with a supposedly suitable criterion for the purchase decision without overloading them with information.

Legal "minimum standards" usually mark the starting and anchor point from which production methods and the products produced in this way are delineated. An increase in **minimum requirements** is equated with a higher standard. The simple message is that the higher the minimum requirements, the higher the **"Quality standard."** Various brand programs, also those used to identify organic products, are based on this cognitive link. Standardization reduces the existing variation in the characteristics to a manageable number of standards. Even if a different impression is conveyed to the customer, standardization is the opposite of differentiation of product and process qualities. Instead, goods or processes are standardized based on standards to simplify marketing and reduce manufacturing, information, transaction, and distribution costs.

A distinction is made between de facto, institutional, and legislative standardization (Genschel 1995). Factual standardization occurs when the choice of a standard is left to the market participants. Institutional standardization occurs within the framework of recognized international organizations such as the International Organization for Standardization (ISO) or the German Institute for Standardization (DIN). Legislative standardization is when the state regulates certain specifications in binding legislation. In the food industry, de facto standardization prevails. Facts are created by highlighting individual characteristics that define new standards as increased minimum requirements. Through standardization, branded products are produced which are oriented toward the expectations of specific customers (Wollny 2008). Competitive advantages can be generated through market differentiation based on branded products, quality seals, or labels that are based on standardization. Ecological management represents a legislative standardization. However, it is based on de facto standardization developed by organic farming associations to differentiate themselves from conventional products. The standards developed by privately organized associations were transferred in slightly modified form to European regulation. This regulates the increased minimum requirements for the production processes and the labeling and certification of the products throughout Europe (Sundrum 2005). In this way, the EU Commission aimed to protect the "organic" production method from distortions of competition by free riders who label sales products with similar terms without fulfilling the increased minimum requirements.

Two cornerstones of standardization have been set with the legally defined conventional and organic minimum requirements regarding production conditions. This has created an intermediate space that the food retail sector increasingly uses to set its own de facto standards. Among other things, the food retailing sector is using the results of various studies financed with public funds to find out which information consumers react to in a particular way and associate them with positive associations. For example, the pasture keeping of cows is highly rated by consumers, and it offers options to establish new products, such as conventional and organic milk. While pasture milk production has only been economically viable for a few agricultural dairy farms, the food retail sector will surely gain a better image by offering pasture milk.

> *"A topic like pasture milk "peps up" retail brands and brings many image benefits, but costs relatively little. Pasture milk products can thus be raised in price almost to the level of much advertised branded products."* (Spiller and Schulze 2019). *With its rules, regulations, and standards, the* food retail sector is *"increasingly taking over the role of politics as a pacemaker for agriculture—in both the conventional and organic sectors."*

Other examples of mandatory certification standards as a performance requirement are GMO-free milk for private labels, the ban on year-round tethering, and the mandatory "husbandry compass" for meat products. Increasingly, advertising messages are enriched with quality labels. Extensive studies precede the introduction of quality seals and labels to determine consumers' willingness to pay for specific labels and claims. The focus is on the question of to what extent this willingness to pay can be increased (Schulze et al. 2021). Numerous research groups in agricultural marketing have tried hard for decades to identify the determinants influencing purchasing decisions. The findings are expected to support agribusiness stakeholders in promoting the efficiency of marketing strategies using labels (Grebitus et al. 2009). In situations where consumers are uncertain—which is the rule rather than the exception for food products—many consumers rely on quality cues communicated in advertising and packaging. These include price in particular, but the point of sale characteristics are also important. Above all, however, consumers try to deduce the quality of products from brand names, seals of approval, and certificates with which they are familiar (Iop et al. 2006). This makes it easier for them to make a purchase decision. Accordingly, products from branded meat programs or meat products bearing quality labels are considered higher quality (Bredahl and Poulsen 2002). Information about the production process (e.g., organic production, free-range farming) or regionality is also relevant for quality assessments from the consumer's perspective. This also applies if there are no direct biological relationships.

For example, in agricultural marketing, "pastured beef" is seen as a product with promising potential for market differentiation. Consumer surveys revealed that such labeling is associated with high-quality product attributes (Schulz et al. 2021). The term can evoke positive associations in many consumers. Furthermore, the respondents were asked to assign a monetary value to their preferences to classify their willingness to pay. However, the interviews remained open on whether and to what extent this is associated with better meat quality or a higher level of welfare in the case of cattle that are temporarily kept on pasture. At the same time, consumers were left ignorant that a considerable variation in product and **process quality** between individual grazing animals and farms that practice grazing fattening exists (Schulz et al. 2021). From a scientific perspective, the significant variation contradicts any attempt to make an overriding quality statement by emphasizing a single characteristic, such as the temporary grazing of beef cattle. Therefore, the effort to generate statements on willingness to pay in this way exposes the intentionality of the approach, which does not aim at qualitative improvement but at harnessing consumers' associations to increase willingness to pay and, thus, the profit margin.

Analogous associations also operate in the case of pork. For example, many consumers rate meat from pigs raised under organic production conditions or in free-range husbandry significantly higher in terms of its enjoyment value (Scholderer et al. 2004). The product is refined in a quasi-psychological way by the associations and appreciations that the references to the husbandry conditions evoke in the consumers. What changed husbandry conditions mean for the farm animals is—if at all—of secondary importance for marketing. After all, the importance of product information for consumers is at stake and not the question of what influence different living conditions have on the quality of the products and the welfare of the animals.

Derivations

The food retail industry likes to invoke the fundamental economic law according to which the relationship between supply and demand determines the price. Even if this reference has its justification, it mainly distracts from the fact that the supply of goods is influenced to a large extent by market strategies or standardization. This also applies to demand and the decision to buy, which the food retailing industry tries to steer in a targeted manner in the desired direction with advertising measures. Neuromarketing is a marketing field that has long since established itself. It uses the findings of brain research and psychology to optimize the effectiveness of advertising and harnesses this knowledge for purchase incentives. Before new or newly labelled products are launched, they are tested in preliminary studies and on test markets to see whether the labeling proves effective in purchasing decisions and willingness to pay. The comprehensive knowledge that the food retailers have acquired about consumer behavior has an effect at the point of sale and influences the profile of requirements that the trade places on the products and the agricultural production processes. Supply and demand are, therefore, not independent variables in a free play of market forces but are subject to the manipulative efforts of actors who have the appropriate means to influence supply and demand in their interests. Since it does not matter to the food retailers which articles are on the shelves, as long as the turnover and the margin are proper, they can act highly flexibly. In contrast, the upstream processing companies and the primary producers depend highly on the production and economic path. While changes to standards are merely a "stroke of the pen" for the food retail sector, they result in severe changes in other areas.

Economic inferiority, lack of legal knowledge, susceptibility to subversive advertising messages, and a vast array of goods are some of the aspects that also put consumers at a disadvantage vis-à-vis product providers and give rise to the model of the consumer in need of protection. Consumers have neither the expertise nor the resources to obtain comprehensive information on goods and services and to assess the relationship between performance and prices. Consequently, product suppliers and consumers have an immense power and information imbalance. In contrast, the food industry assumes the following:

"Guiding principle of the ***responsible consumer*** *who thinks critically about his behavior and needs in a competent and informed way. The consumer has personal responsibility for his purchasing decisions. The state creates the framework conditions for a balanced market that protects consumer and producer interests; it ensures health protection. The consumer should also be protected against misleading and deception but not patronized. The sovereignty of the consumer ensures competition and a needs-oriented product range. The consumer can (individually or in an organized way) express his interests and influence the market by his actions."* (Bundesvereinigung der Deutschen Ernährungsindustrie 2019)

By emphasizing consumers' sovereignty and personal responsibility and the state's regulatory role, the food industry reveals its interest in distracting from its responsibility for a valuable food supply. As long as consumers buy the offered goods, the food industry can claim to serve only consumers' needs as a service provider. The role of the service provider provides the food industry with a variety of argumentation aids to legitimize its actions and to shift the responsibility for undesirable developments onto the behavior of consumers, producers, or politicians. Given the existing balance of power, there can, however, be no talk of a balanced market and a balance of power.

Under the eyes of agricultural policy, the food retail sector has created a strong position of power to exploit both vis-à-vis suppliers and customers. On the one hand, cheap goods are offered at dumping prices. At the same time, higher-priced goods associated with **"Quality standards"** are offered without the price difference that correlates with the qualitative differences of the products. The food retail sector does not wait for the state to prescribe the framework conditions for doing business but defines the standards it considers purposeful. The labelling of self-defined farming methods is only one example of the power of definition of the food retailers, which they use because of an inactive agricultural policy. Customers do not know what level of product and process quality is offered to them for the respective purchase price. From a moral perspective, this constitutes "unfair competition". Whether this also stands up to legal scrutiny cannot be judged here.

Supermarkets are illusion buildings. The **marketing** of the food retail industry continually provides new projection surfaces to which customers can dock their associations to indulge in the illusion that they are doing something "good" for themselves and the common good by buying specific products. Accordingly, marketing is primarily concerned with avoiding cognitive dissonance among customers. Marketing has one primary purpose: to persuade people to buy. For the sake of sound bites and at the expense of **information,** complex issues are communicated with buzzwords. These neither reflect the respective quality level nor provide information about the external effects caused during production and, thus, the costs caused for the community. Therefore, supermarkets are not places of enlightenment, from which one goes out wiser than one goes in. Here, customers pursue their particular interests by using a comprehensive and rich range of food products at different price levels and with diverse advertising messages.

The food retail industry has no original interest in a qualitative differentiation of the products and the underlying production processes based on valid assessment criteria and standards. It is much easier to generate profitable sales by maintaining illusory worlds than by elaborate differentiation, which would result in considerable additional costs for logistics, infrastructure, and distribution of the goods. A differentiation of products based on standardization is carried out based on easily comprehensible individual characteristics, which only pretend to be a qualitative differentiation but de facto represent pseudo-qualities. With the standardization of products, the food retail sector does not offer customers any orientation. On the contrary, standards are primarily responsible for the market's confusion and lack of transparency. Standards facilitate and simplify trade in goods by sorting them into prefabricated boxes and, at the same time, into mental pigeonholes. In contrast, however, reality is characterized by much variation. The quality attributes aggregated and determining value in products and production processes rank on an extensive continuum from very low to very high levels of quality. Therefore, an evidence-based assessment of the actual level of product and process qualities requires a different approach (see Chap. 9).

6.5 Consumers

Individual eating behaviors result from a lifelong sociocultural learning process similar to language learning (Gahagan 2012). The central learning principle here is learning by observing others. This is supplemented by operant conditioning through positive taste experiences. The primary needs of hunger and satiety are overridden by secondary needs in the context of food intake. In adulthood, most purchasing decisions are dominated by the following motives: enjoyment and taste, convenience, and price. Health motives also represent relevant determinants of food choice for part of the population. However, for most consumers, the choice of products is primarily determined by habits that have developed over the years. Purchasing decisions are guided by preconceptions, current needs, financial resources, and opinion-forming processes, which are also influenced by advertising campaigns or media coverage.

The current megatrend in dining choices is called individualization. People no longer want to be lumped together. The concern of getting lost in the big crowd drives the desire for a personal identity. According to (Ellrott 2016), we eat "*to define, stage and optimize ourselves.*" The pluralization of eating styles ranges from the so-called Stone Age diet (Paleo diet), in which everything is supposed to be as pristine as possible, to functional food, the foods optimized through targeted chemical interventions. Vegetarianism and veganism are particularly popular. Some people are also concerned with sustainability or animal welfare. However, the focus is increasingly on self-dramatization and group affiliation. Given the loss of traditional order systems, many people ask themselves: Who do I want to be? How do I want to be perceived by others? The question of what and how much I eat contributes significantly to forming an identity because the eating style is easily changeable even with a limited budget. Food is an excellent digital tattoo that can signal a sense of belonging. With the snapshot of a perfectly arranged plate or via social media, in which certain nutritional styles are discussed, one can define and stage oneself excellently. Using one's eating style, one can feel that one belongs and, at the same time, set oneself apart from others, thus generating **individuality**. A better humanity could be expressed with the affiliation to vegetarianism or veganism, for example, self-discipline, commitment, consideration, responsibility, and animal protection. This makes it attractive to profess it. What is bought and rejected provides at least a partial answer to who I am and who I want to be. The trend toward individualization underscores that there is no such thing as "the consumer."

Results of a nationally representative survey in Austria revealed how food meanings relate to a wide range of food consumption patterns (Kokkoris and Stavrova 2021). Four factors were identified that explained 45% of the total variance. The first factor describes health-conscious eating behavior patterns. They reflect a genuine concern about healthy eating, such as following a nutritious diet, cooking and eating fresh food (salads, fruits, vegetables, regional food), or seeking medical/expert advice about nutrition. The second factor describes discerning eating behavior patterns. They capture a rather highbrow taste and a preference for upscale market offerings, such as buying organic foods, shopping at small stores or markets, caring less about price and more about quality, or avoiding meat products. The third factor corresponds to behaviors driven by pleasure-seeking or convenience, such as consuming junk food, ready-made meals, eating on the go, or overeating. Finally, the fourth factor describes functional eating behavior patterns. They express a rational approach to eating with an emphasis on knowledge (e.g., checking information on product labels) and careful planning (e.g., shopping with a shopping list and avoiding impulse purchases) as well as some affinity with technology (e.g., shopping online) and food innovations (e.g., consuming dietary supplements and products enriched with nutrients).

Food retail companies are happy to embrace the trend towards differentiation because it allows them to respond to the pronounced competitive situation in the retail sector. Whereas in the past, consumers were trained to pay primary attention to the (low) price, food retailers are increasingly trying to offer higher-priced products, combined with the suggestion that these higher-priced goods are also associated with higher additional benefits. However, the opacity of the pricing and the manipulative orientation of the advertising messages are likely to reinforce many customers' essential skeptical attitudes toward food retailers. Customers are hardly offered opportunities to base their food purchase decisions on the best price-performance ratio. Without reliable information, many consumers are primarily guided by price as the relevant reference point. This means they can do the least wrong and minimize the risk of being overcharged. What is not valid for food is true for price: it is directly comparable. Direct price comparisons have not been around for very long. It only became established at the end of the nineteenth century with the first department stores. Before that, prices were primarily negotiated individually. The department stores democratized shopping. Everyone could see what quantity of product they were getting for what price and, in this way, compare prices. The ability to compare prices was a significant factor in making department stores a resounding success.

Currently, however, the trend is moving back toward individualized pricing. Self-learning algorithms, electronic loyalty cards, and apps enable retailers to return toward target group-oriented and individualized pricing and willingness to pay. If the price that individuals pay for the same product varies, this increasingly invalidates the theory about the market mechanism, according to which supply and demand determine the price. The price no longer marks a point of collective equilibrium but the sum of what group of consumers are willing to pay for food products. While fluctuating prices make the situation even more confusing for consumers, they are a paradise for traders. In the case of food, they not only lack the expertise to validly assess **product qualities** and qualitative performance on the part of primary producers in a valid manner. They also cannot evaluate whether they have just saved or paid on top when buying. It then proves to be entirely illusory to assess the price-performance ratio reliably.

Marketing has always been geared towards making it palatable for consumers to purchase products that they would not have fallen for on their own. For Trentmann (2017), marketing has always had a slightly cynical undercurrent. "*Stinginess is hornier than ever,*" writes, for example, the industry magazine *Der Handel* in an August 2017 issue, adding that Germany is the discount republic par excellence. This is especially true for food. Bargain prices are supposed to make people's heads spin, turn off their rationale, and tempt them to make purchases they hadn't planned at all. Bargains have, therefore, become a very effective and seemingly indispensable marketing tool: Customers who are made to feel like they're getting a good deal are happy to come back. They often count it as a personal achievement if they have managed to snag large quantities of products at low prices. This gives bargain hunters a sense of satisfaction and participation that would not be available at higher prices. Retailers' priority goal is to lure customers into stores with promotional messages and low-price offers. Once they have crossed the threshold, customers will find something to tempt them to buy (Häusel 2013).

Since it is already difficult enough to make a purchase decision given the large selection of products, it is essential to avoid complex weighing processes between qualitative aspects inherent in the product and the price to be paid for it. This is all the more true for areas of conflict related to production processes. Consumers should not have to worry or even have a guilty conscience. After all, the consumer is not a *homo oeconomicus.* For purchasing decisions to be truly rational, the customer must be fully informed of all his options and their consequences. They would then have to be able to derive their preferences from this and put them in a stable order. In contrast, purchase decisions, except for high-priced products or services, rarely rely on extensive reflection and consideration processes. Instead, decision-making and behavioral patterns have become established with which the decision-making process can be implemented with as little mental effort as possible. The metaphor "the agony of choice" expresses the fact that when there is an overabundance of purchase options, a stressful situation requires mental effort. The orientation one must acquire when entering a supermarket for the first time is imprinted on the mind when repeatedly visiting the same supermarket. It facilitates the spatial orientation and the decision for specific products, which we expect to find in a place we know. It helps us minimize the expenditure for searching and deciding.

It has been known for some time that consumers find a wide range of products attractive but increasingly challenging to decide. A scientific study has shown what happens in the brain when people are "spoilt for choice" (Reutskaja et al. 2018). Brain activity in certain areas is always highest when the preferred average number of choices is available. If the choice becomes too large, the effort required to make the decision exceeds the benefit, activity decreases, and the "agony of choice" results. As the range of options increases, the benefits of more excellent choices, i.e., the likelihood that an even better option will be there, diminish. Conversely, the deciding costs always increase: it takes more time because one cannot remember all the possibilities. Thus, comparison becomes more difficult. At some point, the cognitive costs exceed the benefits of making a significant choice (the effort exceeds the benefit). We are then demotivated and dissatisfied with the decision. Consequently, the production processes (see Sect. 5.2) and the cognitive processes of decision-making are subject to a marginal utility function. In decision-making, branded products are particularly well suited to facilitate orientation. Well-known brands can't be that bad; customer brains think they have a good reputation, and ultimately, a large group of customers can't be wrong.

As far as possible, shopping should be characterized by a flair of effortlessness to transport what is wanted in the lightly rolling trolleys through the wide aisles. Since even this effort is perceived as an effort by some contemporaries, more and more people are resorting to delivery services that offer ordering from the sofa and delivery to the doorstep. We can quickly get used to conveniences and take them more or less for granted. Accordingly, a feeling of well-being no longer arises with every purchase. Instead, there is a growing tendency to complain if not all specific products are available or if we must wait a little longer at the checkout. Accordingly, in the supermarket, it is avoided that products are sometimes not available. The full range of perishable goods is still available shortly before closing time. It is more critical for supermarkets not to scare away customers than to accept that some food must be thrown away because it is no longer fresh. The fear is too great that customers will look for other shopping opportunities if they have the repeated experience of standing in front of an empty shelf. Customers are given the feeling that they are kings and can autonomously decide what they consider a good choice based on self-referential judgment. This is why they can sometimes react sensitively when they are told by politicians what they should prefer to consume. When Green politicians in Germany brought up the idea of a "veggie day" offered in public canteens in 2013, the party and its proposal were met with ridicule and scorn. The party was presented with the receipt in the subsequent elections. The conclusion is that many consumers do not want to be patronized.

Consumers have long since become accustomed to meat's ubiquitous and inexpensive availability. It is bought in portions and increasingly packaged. For years, the trend has been to purchase valuable cuts: Fillets, brisket, and cutlets. Pieces of meat that require elaborate preparation can hardly be sold in food retail anymore. However, suppose only a part of the entire carcass is demanded by the consumer as meat. In that case, additional production costs can only be passed on to the valuable cuts, which makes them disproportionately more expensive. The fact that the products are components of previously living and sentient animals is largely ignored by many consumers (Nocella et al. 2014). This also applies to the production conditions under which the animals are kept.

These assessments are contrasted by current consumer surveys, which show that the most essential task of agriculture is to produce healthy and high-quality **food**. Half of the people surveyed in Germany gave this answer in a survey that allowed only one citation (infratest dimap 2020). 29% of respondents considered it a priority for agriculture to ensure animal welfare; 16% saw protecting the environment and climate as a priority task.

In contrast, only 4% of respondents thought that the most crucial task of agriculture was to produce low-priced food. With such a basic mood among the population, it should be easy to meet consumer demands with corresponding high-quality product offers without fearing that the price increases associated with qualitative improvements will be a significant obstacle to purchasing. However, how do you get customers to pay higher prices for food when food has been communicated through price for decades and continues to lure customers into the shops with low-price offers? To change consumer behavior in the long term, it will be necessary to communicate to customers credibly and understandably the additional benefits associated with a higher-priced product. In the past, organic products, in particular, have managed to be perceived by consumers worldwide. This is particularly true in Western countries, where they are classified as higher-quality food products that justify a premium price. They are marketed under a separate label and have been able to steal certain groups of buyers away from the conventional market. These consumers want to eat healthily and attach importance to foods whose ecological footprint is as small as possible. However, the organic market continues to be a niche market.

Given the discrepancy between the results of consumer surveys and the niche existence of organic products considered to be of higher quality, it is often assumed that most consumers continue to be only conditionally willing to pay higher prices for food. Consumer research refers to the differences (inconsistencies) between what people—as citizens—think and express and what they—as consumers—do as the consumer-citizen gap (Schulze et al. 2023). Many citizens wish for higher animal welfare standards and stricter **animal protection** in agriculture, prefer pasture farming over stall farming, and find organic livestock farming advantageous. Nevertheless, the market share of such foods is significantly lower than the attitude scores in scientific surveys, and the public debates around these issues would suggest. The supposedly low willingness to pay for supposedly high-quality products identified by consumer research with a self-referential approach is exploited by other interest groups to divert attention from their deficits, using consumers as a target for criticism. These include, in particular, the primary producers, who never tire of blaming consumers and their unwillingness to pay higher prices for their plight (van der Ploeg 2020).

The agricultural system has created a primary production geared to **cost leadership**. Consumers can hardly be blamed for a behavior suggested by marketing strategies, which does not differ from the behavior of food retailers towards the primary producers, i.e., wanting to buy the goods as cheaply as possible. From the consumer's perspective, one could conclude that consumers behave following the market. They have learned to pay attention to the price, especially the price difference. Suppose something is offered at a higher price. In that case, many consumers legitimately ask whether the extra price is also accompanied by adequate added value and additional benefits, for example, for their health and well-being, or for that of farm animals, a reduction in environmental pollution, or an improvement in nature and climate protection. Consumers rely on marketing claims to match the difference in price with the difference in added value. However, a large proportion of consumers have learned to distrust them. On the other hand, they cannot check the quality. The lack of clarity in the product range and the inability to understand the relationship between price and performance make it seem plausible from this perspective that consumers tend to hold back on higher-priced products. Without reliable quality assessments, lower-priced products still promise the best value for money. If consumers are invited to buy higher-priced products, they should also be allowed to check the truth of the promised benefits (Di Domenico et al. 2022). However, food retailers generally have no interest in informing customers about the true price-performance ratio. It is all the more critical that no efforts are discernible on the part of the scientific discipline of agricultural marketing to bring light into the darkness, i.e., to inform consumers about the predominant strategies of food retailers.

A valid assessment of the price-performance ratio, i.e., the relationship between the qualitative properties of the products and the production processes and the price that appears to be justified, lies outside the consumers' competence to assess. Therefore, it is unsurprising that consumers' assessments are often not in accordance with a professionally experienced quality assessment. This is particularly true for meat products (Grunert et al. 2004). When consumers make a purchase decision, they form their quality expectations self-referentially, i.e., based on their perceptions and patterns of judgment shaped by previous experience. In this sense, consumers' quality assessment is distinctly subjective. When making a purchase decision, already-known information interacts with new external information (Stankevich 2017). In this way, a mixture of different associations arises, which are evoked by specific product characteristics in combination with cues such as the price, advertising claims, and packaging, and compared with the credibility of the claims. While food retailers are best informed about the consumer decision-making process, consumers are seldom aware of what's happening in food marketing and production.

While food production and processing processes have developed rapidly through rationalization, mechanization, and intensification, large parts of the population still have the "old" image of agriculture. It is no coincidence that these images have taken root in many people's minds. The majority of the population no longer knows agriculture and livestock farming from their own experience; they only know it through the media and the advertising messages of suppliers. Advertising brochures and packaging suggest rural agricultural idylls and "happy farm animals": dairy cows on green meadows, preferably in alpine landscapes, pigs reared on straw with plenty of fresh air, and poultry kept outdoors. In reality, the production of raw goods of animal origin takes place away from the direct perception of consumers and primarily from the point of view of cost savings. The real production conditions contrast with the images disseminated by marketing. However, agriculture, processing, and trade have little interest in informing consumers about the real production conditions and showing "real" pictures of modern animal husbandry on farms, during transport, or at slaughter.

It is, therefore, not surprising that media reports about abuses in so-called "factory farming" have a disturbing effect on many consumers. Recurring scandalous reports and images of unhygienic conditions, substandard products, animal diseases, and suffering creatures have cracked perceptions of high-quality and safe food. They are further confused about environmental pollution and climate change links with livestock production. As a result, products of animal origin have experienced a significant loss of esteem. Images of animals subjected to obvious pain, suffering, and harm plunge many consumers into a moral dilemma. Accordingly, consumer reactions to reports of abuses vary widely. They range from complete ignorance of the abuses to complete rejection of products of animal origin. By changing consumption behavior, more consumers wish to distance themselves from the adverse effects of livestock production by increasingly favoring plant-based over animal-sourced food. However, the majority of consumers are trying to rationalize in the face of this ambivalent situation and are trying to come to terms with the contradictions between the glossy images of the advertising industry and reality, between their interests in high-quality products at the lowest possible prices and their shared responsibility for the abuses in primary production. Since consumers are not in a position to judge the qualitative differences in the range of products on offer, many reach for the low-priced products in case of doubt. Accordingly, a large proportion of products of animal origin continue to be bought in discount stores. On the other hand, consumer uncertainty is resulting in an increased desire for high-quality products (Brečić et al. 2017). If individual quality aspects are identifiable, consumers are willing to pay more. This is, among other things, the case with organic, fair trade, and regional foods. However, given a product range that is almost impossible to track and a flood of quality seals and labels, each advertised with different benefits, market transparency, and comprehensibility are in danger of being lost in the overflow of information.

Apart from the insufficient assessment skills of most consumers concerning the quality of food and healthy diets, it has to be considered that food prices are a relevant factor for healthy food choices. Food purchasing is financially challenging for persons with meager incomes. While fresh-cooked plant-based diets are less pricy than the unhealthy, highly processed standard omnivore diet, diets with reduced carbohydrate content are considerably more expensive, limiting their use for people with low income (Kabisch et al. 2021). Therefore, organizations for consumer protection claim that high-quality food has to be available for all citizens. The quality of food would have to be ensured by the primary producers and guaranteed by responsible authorities without consumers being expected to pay a higher price. After all, people pay taxes and can expect something in return. This fundamental attitude is reinforced by various protest groups, as well as by trade unions and social politicians, when they repeatedly articulate the claim that everyone, regardless of income, should be able to afford high-quality food. This claim brands any food price increase as anti-social. After all, farmers already receive a lot of money from taxpayers. The argument that food prices must be kept low so that everyone can afford them does not stand up to scrutiny from a common good perspective. Trying to solve social inequality with a low food price policy ignores all the harmful effects that the previous pricing policy of the food retail sector has contributed to. It also ignores that many primary producers belong to the precariat. A strategy that leads to the external costs of food production being priced in and creating social compensation for those in need would be a far better agricultural and social policy. The fact that the argument of the unreasonableness of higher food prices for the less well-off is still being used again and again by various groups and also by the media as a 'killer argument' testifies to the low willingness to abstract from one's interests and to undertake a change of perspective concerning overriding public welfare interests.

Derivations

The food retail sector does not offer consumers reliable information that could enable them to make purchasing decisions based on a reliable price-performance ratio. Just like retailers, consumers also pursue their interests. However, there are considerable differences in the possibilities of exerting influence. The strategy of reducing complexity using standardization gives the food retailing sector enormous power to define and thus shape the market, which it knows how to exploit for its benefit. With the help of **"quality standards,"** consumers are thrown back on the patterns of perception they had previously been asked about by marketing. Ultimately, however, they are left in the dark regarding the context in which food production is embedded. Since they cannot assess how bad or good the goods they have bought at dumping prices or as high-priced "quality goods," disorientation prevails.

In principle, most consumers cannot be denied an interest in a qualitative differentiation of the product and an appropriate price-performance ratio. The need to purchase a supposedly high-quality product at a low price makes them susceptible to manipulative marketing. Consumers' purchasing behavior will only change if convincing alternatives are available that enable consumers to create a reliable assessment of the actual price-performance ratio of products. This would require an intersubjectively valid evaluation not limited to single aspects but based on a comprehensive quality assessment. The prerequisite for this would be an institution authorized by the state and independent of the interests of the food retail sector, dedicated to clarifying the complex issues involved in primary food production, processing, and marketing.

6.6 Non-governmental Organizations

Food production is a complex process involving primary producers, processing, trade, and the significant influence of non-governmental organizations (NGOs). These organizations, often referred to as NGOs, were initially defined by the United Nations (UN) as representatives of civil society participating in the UN's political processes. Today, the term 'NGO' encompasses a wide range of non-governmental associations advocating for socially relevant issues, including those related to food production (Götz 2008). NGOs are thus associations of interest that have come about through civil society. However, they are not legitimized by a public mandate. Thus, they are private organizations that engage in specific goals in the public interest or declared so by the NGOs. About their contribution to the pursuit of public welfare interests, a critical look at individual activities is taken below, without claiming to do justice to the NGOs in their entirety. Possible benefits and disadvantages concerning public interests are discussed in the case of animal rights activists and environmentalists.

6.6.1 Animal Rights Activists

Where would **animal protection** be today without the commitment of animal rights activists? This question initially appears to be purely rhetorical, as it may evoke the immediate association in some that without animal welfare activists, there would be no **animal welfare**. However, the circumstances in farm animal husbandry are complex, so no simple answer is indicated but a more comprehensive reflection. The following remarks cannot do this comprehensively but are to be understood primarily as an impetus for a change of perspective, which allows a state of affairs to appear differently from another point of view. It is indisputable that the activities of the animal protection associations have repeatedly made the apparent abuses in livestock production a topic of discussion through continuous publicity campaigns and have thus brought them to the population's attention. Thereby, it is, again and again, the pictures of maltreated creatures, with which the medial attention and that of the association members is awakened, and the problems are made attentive. However, over the years and decades, most people have become accustomed to these images. Many seem to shrug them off like the deterrent images on cigarette packets. Those with a strong affinity for meat products are unlikely to be tempted by these images to change their consumption behavior. However, the photos affect young people, resulting in a significant reduction in animal-sourced food (ASF), at least in some Western countries, while the consumption of ASF is further increasing in developing countries (Knap 2022). Considering the changes in consumer behavior, one might argue that animal protection organizations have succeeded in raising awareness and promoting a shift in dietary choices, particularly among the younger generations. But the question remains: Has this shift in consumer behavior translated into a tangible reduction in the suffering of farm animals in animal stables?

Fearing legal objections that animal rights activists could have raised against the caging of farm animals, the EU Commission launched regulations for the supposed protection of animals kept indoors more than 30 years ago, many of which are still valid today. The minimum requirements for single features of housing conditions did not protect farm animals from pain, suffering, and harm. The prevalence of production diseases, being high for decades, speaks for itself. What was protected here was the competitiveness of livestock farmers. The regulations gave and still give livestock farmers the legal certainty they need for the investments they have made. From the perspective of the farm animals, the activities of animal protectionists have failed to protect farm animals from the impositions of industrialized livestock farming and the extent of animal suffering that goes hand in hand with it. Therefore, the work of animal protection organizations concerning farm animal husbandry cannot be interpreted as a success story so far. It could be argued that activists cannot be blamed for not being powerful enough to force the parliament to change legal requirements about farm animals' animal health and welfare. However, they can be blamed for focusing solely on housing conditions and not on whether housing conditions as a means to an end are adequate for the envisaged goal of reducing pain, suffering, and damage to farm animals. While simplifying complexity and distracting public attention from the underlying background of animal health and welfare problems and possible solutions, they prioritize their interests above those of the farm animals they claim to represent. For decades, they have demanded that agricultural policy must enhance the minimum standards of housing conditions. As the enhancement of minimum standards in housing conditions is not suited to improve animal health and welfare substantially, but on the other hand is very costly, society might be fortunate that NGOs do not have the power they wish to have. Nevertheless, the animal welfare movements, with their media-prepared and high-profile criticism of the cramped space conditions of farm animals in stables, have made a significant contribution to the fact that housing conditions, particularly the space allowance per animal, have burned themselves into the public consciousness as a supposed central animal welfare indicator (European Commission 2023).

The lack of change in agricultural practice has prompted the German **Animal Welfare** Association to take action and introduce an animal welfare label based essentially on an increased level of husbandry conditions between the legal minimum requirements and the requirements of organic livestock farming. Equipped with a great deal of media competence and based on a considerable membership, the German Animal Welfare Association felt well-equipped to use its initiative to give new impetus to the improvements in farm animal husbandry that were deemed necessary. However, the results are very sobering. The animal welfare label of the German Animal Welfare Federation, introduced in 2013, has so far spread only very slowly. In 2019, for example, across all animal species (fattening pigs, broilers, laying hens, and dairy cows), only 330 farms were certified by the increased minimum requirements (German Animal Welfare Association 2019). The unsatisfactory result leaves little room for interpretation. Either the German Animal Welfare Association has been unable to ensure that appropriately labeled products are listed in supermarkets and discounters. Or the umbrella organization of approx. Seven hundred forty associations with a considerable number of fee-paying members (more than 800,000) have not succeeded in persuading them to buy products of animal origin, which the association considers to have a higher animal welfare standard. The members have not taken the opportunity to contribute to animal welfare or set an exclamation mark for their purchasing behavior. The indication that there might be a higher proportion of vegetarians among animal rights activists is not convincing. Even with a higher proportion of vegetarians compared to the population as a whole, there are still very many members of animal protection associations who use ASF, but not the higher-priced products marked with the label of the German Animal Welfare Association.

The unwillingness of the association members to incur the additional costs for the purchase of supposedly more animal welfare-friendly and higher-quality animal products makes it clear which target group the animal welfare associations have primarily in mind with their demands. The demands are not mainly directed at their members but at politicians, who are called upon to follow the proposals of the animal protection associations. Thereby, Animal Welfare Associations do not only act selflessly but also concerning the media attention and the feedback to their clientele, that one is anxious to work tirelessly for the welfare of the animals. Finally, one is intensely interested in increasing the number of members and donations to pay for the campaigns and the permanent employees.

The groups involved in animal protection show a broad spectrum regarding membership, staff numbers, and the focus of their respective activities. Public attention is not directly correlated with the number of members. Among animal rights activists, few groups with only a few employees specialize in undercover operations to capture animal welfare-related abuses in stables or slaughterhouses with the camera. The footage is then leaked to the media. By publishing pictures of cruel conditions in livestock stables and slaughterhouses, attention is drawn to existing abuses. At the same time, the pressure on livestock farmers is drastically increased, who have to fear being attacked by camera-armed animal rights activists. However, the undercover activities and the uncovering of abuses alone cannot eliminate them. The spectacular footage aims at individual livestock farmers and stamps them as individual perpetrators. In the process, it is lost sight of the fact that the conditions in farm animal husbandry vary significantly from farm to farm and are nevertheless a systemic consequence of the general conditions in agriculture, which leaves the farm animal owners alone with the challenges. Neither are robust controls carried out nor does a system of incentives and sanctions exist to counteract the **unfair competitive conditions**. In addition, many livestock farmers have lacked the resources for many years to provide livestock with improved living conditions that better protect them from pain, suffering, and harm. Even if the images catch on with receptive people and increase reservations about eating products of animal origin, this does not improve the situation for farm animals. A decrease in domestic ASF consumption only increases the share of exports without causing any changes in the living conditions of domestic farm animals.

Even if the various animal welfare organizations are not in a position to protect farm animals from harm, they do have the effect of ensuring that the concern for animal welfare is and remains present in the public perception. At the same time, however, they are also responsible for how the animal welfare issue is dealt with publicly and which proposed solutions stick with citizens and politicians. The public sees political statements and decisions in the context of the animal welfare issue as purposeful if they meet the demands of animal welfare activists. Thus, animal welfare is what animal welfare activists demand. In this way, animal protection organizations are granted competence by the political and public sides, which animal protectionists can only claim for themselves to a limited extent. After all, they only judge the level of animal welfare based on the minimum standards of husbandry conditions. On the other hand, the welfare conditions do not address the protection of farm animals from pain, suffering, and physical harm as indispensable prerequisites for **welfare**; instead, they follow an **anthropocentric** approach. Of primary importance is what humans, in this case, the members of animal welfare organizations, consider relevant concerning animal welfare and not how the respective living conditions are to be assessed from the perspective of the farm animals. The degree of harm to which farm animals are exposed and which is contrary to their state of physical and mental health can only be assessed in the individual farm context, based on assessments to which degree the individual farm animals can cope with the respective living conditions. However, animal welfare organizations generally do not have insight into the on-farm contexts that determine the extent to which welfare-related abuses occur. In contrast, animal welfare organizations are familiar with the minimum requirements for husbandry conditions, improvements for which they have been calling for decades. Changes to husbandry conditions mutate into an end in themselves if a reference to the specific situation on the individual farms and an overview concerning the conditions for the farm animals' welfare is missing.

Several retail trading companies used the demand of the animal welfare associations for improved husbandry conditions as an excellent opportunity to enhance their image. With the definition of husbandry conditions, it has bypassed agricultural policy and the interests of other stakeholders. It is now trying to distinguish itself from consumers as a supporter of animal welfare. In this way, it saves itself the expense of generating advertising-effective and image-enhancing media attention. At the same time, the trade is supported by the agreement of the animal protection organizations and the approval of many consumers. Animal protection organizations also profit from this, seeing their repeated demands for improved husbandry conditions confirmed. They can interpret this to their clientele as a success of their efforts. It also pleases those consumers who now have the opportunity to contribute to improving the welfare of farm animals and relieve their conscience by buying labeled products.

Once again, the farm animals do not benefit from this approach. Due to the negligible relevance of slightly improved husbandry conditions for their welfare, hardly anything changes. This is all the more true for the high portion of diseased farm animals in the herds (see Sect. 4.1). Due to diseases usually accompanied by pain, suffering, and damage, these animals cannot be certified as being in good health, irrespective of the farming methods used. Many scientific studies on organic livestock production largely agree that, on average, significantly improved husbandry conditions did not substantially reduce production diseases (Åkerfeldt et al. 2021). Just as it is not sufficient for humans to call in an interior designer in the event of illness and associated impairment of welfare, the welfare of farm animals cannot be improved simply by giving them more freedom of movement and providing straw as bedding.

By focusing on the housing conditions, public attention is diverted from production diseases as the core problems of livestock production and the freedom of diseases as an essential prerequisite to animal welfare to a side issue. As a result, the level of production diseases and the associated pain, suffering, and damage will not change for the better for the time being, and the conditions will persist indefinitely. At this point, it should become clear that the commitment of animal rights activists is not necessarily practical but can also turn into the opposite. This is especially true when the focus is not on the immediate welfare of the animals but on the particular interests of those involved. While animal welfare organizations compete with other interest groups for media attention, they have lost sight of the complexity of the animal welfare issue. This is the only way to explain that they are campaigning for the labelling of husbandry systems in public and in the relevant commissions, to which they are consulted as supposed representatives of the interests of farm animals. In this way, they contribute significantly to the indefinite continuation of the suffering of farm animals. Consequently, they are acting against the essential interests of farm animals, which has not yet occurred to them. The focus on their interests closes their eyes to the lack of protection for farm animals and undermines their credibility. When sweeping solutions are propagated, which, on closer inspection, turn out to be sham solutions, animal welfare organizations become the victims of their pronouncements. In this way, animal welfare organizations support not only the interests of the food retailers but also those of the primary producers. By promoting the husbandry systems as a solution, the individual farms are simultaneously prevented from being subjected to an assessment of animal welfare-relevant facts and the animal welfare services the individual farm system provides or does not provide. Given their knock-on effects, pseudo-solutions are even more problematic than admitting to having no solutions. They discourage further efforts to find practical solutions in the respective context, and the abuses persist.

Not all animal welfare organizations and **NGOs** focus exclusively on housing systems. In a joint position paper, some advocate for a nationwide animal health monitoring system to accurately record and combat production-related diseases in farm animals (foodwatch 2018). Healthy animals are classified as a prerequisite for effective animal, environmental, and consumer protection.

> *"Regular checks on the animal health status of the live animal and the management conditions on the holding must be carried out unannounced by specially trained, independent inspectors. An animal health protocol, an action plan for improvements, follow-up inspections, and—where necessary—intensive farm support and advice are required. Existing recording systems ... can be used to evaluate mortalities and exits on a farm-by-farm basis. ... Only through consistent, farm-specific animal health monitoring is it possible to guarantee a sufficient level of animal welfare. This applies to all husbandry methods, as the livestock farmer must prevent diseases on his farm in the best possible way."*

6.6.2 Environmentalists

The question already addressed in the context of animal welfare can be asked analogously about environmental protection: What would environmental protection be without environmentalists? Without the recurring references to undesirable developments, the issue would not have been brought to the public's attention. But have they also initiated measures that have demonstrably reduced environmental pollution? There are justifiable doubts about food production.

The example of Greenpeace being a major environmental organization whose campaigns are also dedicated to the environmental pollution caused by agriculture will be used to examine this question. On its website, Greenpeace presents itself as an international environmental organization that fights with non-violent actions to protect the natural foundations of human life and nature and for justice for all living beings (Greenpeace 2021). It has become known to the public mainly through its spectacular actions and campaigning ability. In agriculture, Greenpeace's actions are primarily directed against factory farming and using specific means of production, such as soybeans or pesticides. These are mainly held responsible for environmental damage that already occurs during cultivation or when they are created and also lead to considerable ecological pollution when they are used on farms.

By naming clearly defined problem areas and clearly demanding that others do what Greenpeace believes must happen to solve the problems, the organization is seen by many citizens as one of those who are on the right side and who, through their sometimes fearless actions, put the fear of God into the big companies in the agricultural industry and the food trade. Above all, however, they are the mouthpiece for simple messages that many people understand. Finally, it is also about increasing the number of supporters. Greenpeace has an impact by exerting pressure on politicians and food retailers by generating media attention. For example, a press release states:

> *"With its purchasing policy, the trade is largely responsible for humans, animals, and the climate being massively damaged by industrial meat production. (…) The concept of supermarkets offering meat at dumping prices to lure consumers into their stores is part of the sick system of cheap meat. (…) We call on the large chains to quickly remove cheap meat from their product range and to pay farmers fairly. Only then can animals be kept in a manner appropriate to their species. (…) In principle, labelling with the farming method creates transparency. But if customers can buy almost exclusively cheap meat, the labelling degenerates into greenwashing."* (Greenpeace 2020)

According to Greenpeace, such statements leave no doubt as to who is primarily responsible for the abuses. But that is not all. The Greenpeace spokeswoman also claims for herself:

> *"We would have a large toolbox to solve the problem, but politicians are dragging out proposals."* (Töwe 2021)

Furthermore, Greenpeace believes that methane emissions from ruminants must be significantly reduced. This is to be achieved by having only about half as many animals in stables and on pastures by then and by halving meat consumption. The fundamental difficulty of addressing the complex interdependencies in the production and distribution of food and bringing about substantial improvements with sweeping accusations and simple proposals for solutions has already been discussed in the previous chapters. Here, the question arises whether the generalizations and deliberate simplifications do not run the risk of having the opposite effect than intended. They possibly torpedo what is most urgently needed, namely a differentiated approach that does not lump everything together but differentiates in a graduated way between those products and production processes associated with a high or low discharge of pollutants in the respective context.

The blanket call to reduce **meat consumption** does nothing to change the pollutant emissions associated with producing meat products. Halving meat consumption cannot be decreed by law. Still, it can only be brought about by changing framework conditions with far-reaching consequences and making meat products significantly more expensive. Anyone who believes that meat consumption is directly related to the extent of meat production is ignoring all the processes between "farm and fork," which are characterized above all by heterogeneity and variability. Meat products that cannot be sold domestically are exported, i.e., the decline in meat consumption does not automatically result in a decrease in meat production. It is also ignored that the extent of pollutant emissions concerning the quantities of food sold varies greatly between farms. A decline in meat consumption does not necessarily mean that those production processes with the highest pollutant effects per unit of product are discontinued first. Also, the methane releases that occur during the fermentation of vegetables and thus renewable biomass in the rumen of cows are to be evaluated entirely differently than the newly added methane releases, as is the case above all with industrial processes (Jackson et al. 2020). Further explanations and justifications are given in Sect. 8.5. No attention is paid to the fact that a general discrediting of livestock farming does not consider the positive ecological effects of livestock farming in the **nutrient cycle** of an agroecosystem. The positive impact on the environment and the conversion of nutrients that are not usable by humans into high-quality foodstuffs are widely ignored. However, the possible benefits of livestock farming cannot be generalized; they have to be envisaged and assessed by appropriate measures adapted to the context of the respective farm systems. While a differentiated approach and assessment of the impacts are needed, NGOs often act with sweeping allocations and striking demands. This way, they can better mobilize supporters and promoters.

Derivation

NGOs have developed over the past decades as a rallying point for people from different sections of the population who see active or passive membership as a way of expressing their protest against undesirable societal developments. In the process, the various NGOs have been and continue to be successful to varying degrees in establishing themselves as influential opinion-forming organizations. In addition to the public, which can be reached through the media and their membership clientele, their activities are directed primarily at politicians, whose decisions are to be influenced. While a wide section sees NGOs of the population as working for the common good, it is often forgotten that NGOs also pursue their interests. These are primarily linked to the desire to increase their influence on political decisions. However, the real currency of NGOs is the attention they can attract from the media as multipliers. Accordingly, NGOs aim to disseminate their messages as effectively as possible to the press and, thus, to the public. In this way, they reach their own and potential new supporters and politicians. The catchier the messages, the greater the emphasis with which they are advocated, and the more spectacular the actions that are resorted to, the greater the attention NGOs receive from the media, the population, and politicians.

Agricultural policy and the agri-food industry form a largely hermetically sealed self-referential system concerned primarily with particular economic interests and completely inadequately with the interests of the common good. Since the legislature and the judiciary have lacked effective regulation to stop undesirable food production developments, NGOs have an essential and indispensable watchdog function. As a counter-movement, they draw attention to grievances and attract media attention to problems. As a rule, the media themselves show little competence in food production. The general public media do virtually no continuous reporting but leave the field to the so-called "specialist media," which are part of the self-referential system. Public media, therefore, like to fall back on information presented to them by NGOs.

As undoubted as the good intentions with which NGOs try to get their messages across to people are, it should no longer be ignored that this approach also has its price. This refers to the damage that signals unambiguity through the generalizations of the criticism presented and the generalizations of supposed proposals for solutions. This has little to do with the diversity of phenomena and the variability of operational conditions, both in terms of the conditions to be criticized and the options for eliminating them. Fueled by self-interest, the quest for media attention is bought by reducing complex issues to simple statements. These do justice neither to the problems nor to possible solutions. Simplifications are not an adequate strategy to deal with complexity and variation, which are caused depending on the respective context and require a differentiated approach.

Just as consumer advocates cannot directly protect consumers from harm associated with food consumption, animal advocates cannot protect farm animals from harm directly associated with production processes. The same is true for environmentalists. They can't protect the environment directly from the emissions that escape from farms. This can only be done by those who are responsible for the processes in the respective farm systems. However, those responsible on-site can only do this if they have resources and framework conditions that enable them to proceed in a targeted manner. Which measures are target-oriented depends on the respective context of the operational circumstances. Therefore, it is necessary to implement specific measures and continuously review whether the implementations are suitable to promote the achievement of objectives or run counter to them. In this sense, sham solutions are not only counterproductive because they do not contribute to accomplishing the goals. Without implementing a performance review, they prevent the search for demonstrably effective solutions in the respective context.

Given the potentially counterproductive consequences, it would be urgently necessary for not only the primary actors in agriculture, the professional associations, and the media but also the NGOs to face up to the debate as to whether the complexity reduction they are pursuing is serving the interests of the common good, or possibly instead harming it because it is distracting them from the actual requirements. Even if the intention is worthy of all honor, NGOs are not immune to causing the opposite of what they intend and thus are not part of the solution but part of the problem. This danger exists, especially when interest groups become prisoners of simplistic explanations they represent to their clientele and the public. The less complex issues are understood, the more individual partial aspects come to the fore, charged with exaggerated meaning. As a result, other elements inevitably fade into the background. Once "under-complex" thought patterns have taken root in people's minds, it is difficult to reform them.

Over-simplification leads can be reflected in the example of the standardization of husbandry systems, which addresses the issue of **animal welfare** and provides a supposed solution. The simplistic message of the animal welfare organizations: "improved husbandry conditions equal more animal welfare," is gratefully taken up by NGOs and consumer centers and communicated to their clientele. The NGOs are not familiar with the complexity of farm animals' physical and mental health, nor of the occurrence of diseases and behavioral disorders and their dependence on the highly variable living conditions that farm animals encounter in livestock. It is not the housing conditions but the individual interactions of the farm animals with the living conditions that determine the extent to which the welfare of farm animals is impaired. The standardization of husbandry systems obscures the complexity of the interactions. The propagation of forms of husbandry suits vested interests in terms of effective public relations far more than dealing with the background of complex processes responsible for impairing farm animals' welfare. However, these fall outside the competence of NGOs. They require professional expertise, which records the individual reactions of animals to the respective living conditions based on ethological, physiological, and pathological findings and classifies to what extent they are detrimental to the welfare of farm animals.

6.7 How Do We Balance Particular and Public Interests?

Modern agricultural production follows economic principles. Investments and food system innovation primarily focus on increasing productivity. Since agricultural produce is traded internationally as commodities, they are subject to global price-focused competition. The resulting economic framework leaves little scope for actions that are not focused on economic efficiency and hampers all those efforts in farm management, from which a financial return appears unlikely. Market prices and production costs govern the framework to which the farm systems' production processes must be adapted when striving for economic viability. The negative consequences of food production processes are an expression and consequence of the misguided functions of an economic system that has produced these effects or has been unable to prevent them. In the economic system, various actors interact with other actors to influence the flow of materials, money, and information to safeguard their interests. The actors are **self-referential** about their interests. They act **anthropo-**

centrically, i.e., they focus on their own human needs. Although people are fundamentally capable of a change of perspective in which they can put themselves in the position of other people with different interests, they make only limited use of this in economic life. Due to the focus on one's interests, the undesired and harmful side-effects outside the immediate sphere of interests are widely disregarded or come into view with a considerable delay. Thus, it sometimes takes a long time before one is willing or compelled to take a closer look at the undesirable developments and to link them to one's activities causally. It is undisputed that individual interest groups are involved in the adverse effects to varying degrees. The allocation of responsibilities is correspondingly complex.

In complex contexts of action, undesirable developments are generally caused by different sub-areas competing for limited available resources. This leads to disruptions if no functioning system of rules has been established to ensure appropriate distribution and **cooperation** concerning a common and overriding **objective**. If, in the flows of materials, money, and information, specific sub-sectors and particular interests are given preferential treatment or made use of, this is usually at the expense of the supply of other sub-sectors. The inevitable consequences are imbalances, which sooner or later impair the functioning of the entire system. Unlike in a living organism, there is no superordinate authority established in an agricultural system to control the interaction of the subsystems and to ensure that, as far as possible, all subsystems are supplied by their respective needs. A supply of materials, money, and information appropriate to need is not only a prerequisite for the functioning of the subsystems. This also applies to the contribution they can make to the functioning of the overall system and the achievement of higher-level goals. Without a local, regional, or national regulating body, it is impossible to strengthen those who can contribute positively to the community and to act against those who act contrary to the interests of the common good. Therefore, promoting the public welfare interests in food production can only succeed if the institutions predestined for this at the respective process levels control the distribution processes. At the national level, the coordinating task falls to agricultural policy. As already outlined in Sect. 6.2, however, it performs this task only to a limited extent.

The principle of equality in community law is related to minimum standards. It does not address the outcomes of implemented standards concerning preventing undesired side effects and realizing common goods. Compliance with the law is equivalent to avoiding impairments related to accommodations below the minimum requirements. On the other hand, implementing living conditions above the minimum is left to the self-referential estimations of stakeholders and market forces where retailers play a dominant role. The heterogeneity in priorities between the involved stakeholders demands finding the least common denominator for the various and partly conflicting interests. It seemed reasonable to propose enhancing minimum standards to continue along the same path.

After all, it is not primarily a matter of putting forward the highest possible proportion of one's interests and defending them against objections from other perspectives but instead of initiating fundamental changes to solve socially relevant problems whose origins are located in the one-sided pursuit of particular interests. A minimal consensus between divergent interest groups may be seen by many as a first step in the frequently cited "right direction." Viewed from a different perspective, however, the hope of small-scale progress is an elementary fallacy. A minimal consensus has little to do with realizing the common good. It pretends to be progress, while in fact, it is primarily to maintain the status quo. Even if it is part of the interests of the common good to take as many interested parties as possible along on the path to new goals, in the pursuit of the interests of the common good, one cannot avoid leaving behind, in a figurative sense, and protecting through social systems those who are not willing or able to change.

In this context, it is helpful to remember the original definition of stakeholders. This definition sees the individual stakeholders as part of a higher-level organization, without whose support the organization would cease to exist (Freeman 1984). As part of the parent organization, the individual stakeholders are interested in ensuring that the organization prospers and does not suffer harm. In this sense, the stakeholders involved mutually influence each other. At the same time, they shape the superordinate organization and are also shaped by it. Such a basic understanding of the structure of effects of sub-groups of a superordinate community of interests results in a mutual conditionality and the necessity to organize a balance between diverging interests. The interests of the involved stakeholders have to be conflated into a common overarching goal. After all, the interest groups sit together in a boat, each fulfilling necessary sub-tasks that keep the boat afloat and can move in the desired direction, i.e., towards a goal. Without agreement on a common goal, the respective activities and the resources deployed can neither be used in a goal-oriented nor efficient manner. The more clearly the common objective can be outlined, the more likely it is to provide all stakeholders with the necessary orientation.

Sustainability interests oriented towards public welfare differ from particular interests; they are not anthropocentric but system-centric. It is not the interests of people that are in focus, but the preservation of the higher-level system, in this case, the preservation of a higher-level community. Thus, it is not the interests of primary producers and their association representatives, the agri-food industry or the retail sector, nor the interests of consumers that are the focus of attention. Instead, they are subordinate to the common good interest. Strategies to sustain communities follow a different logic than pursuing particular interests. A fundamental shift in perspective must be carried out to ensure sustainability or permanence for human communities (Ziegler and Ott) and non-human animals (Boscardin and Bossert 2015). Starting from a superordinate point of view, the question becomes central as to what is conducive to preserving communal, regional, national, and ultimately also European community structures and what is suitable for promoting them. In principle, this is in line with the **principle of subsidiarity**. It states that state institutions should always, and ideally only, intervene in a regulatory way when the possibilities of individual actors, groups, or state institutions alone are not sufficient to solve a community-oriented task. The principle of subsidiarity is an essential concept of federal associations of states, such as the European Union. It is also a central element of the regulatory concept of a social and ecological market economy (Wurzel et al. 2003). However, it has yet to be applied in food production.

In this context, the overall objective of food production is apparent from the perspective of public welfare interest. According to this, the **primary purpose of** government bodies at the various levels should be to promote an agricultural economic system that produces **foodstuffs of high quality**, with a **low level of undesirable side-effects and adverse effects** on consumers, livestock, and the natural environment, and which are offered to consumers **at fair prices** through a variety of markets. Before Chap. 9 goes into more detail on the methodological tools that can be used to approach this objective, several essential prerequisites must first be addressed and reflected upon.

Contrary to what is suggested by the perspectives of the different stakeholders, the first addressee for implementing public welfare-oriented measures is not the agricultural policy but the farm. Each farm is the producer of food and the primary producer of **public goods** To produce these, they depend on framework conditions that enable them to deliver food and public services in a way that covers costs and ensures that the two objectives are well balanced. The elementary prerequisites include suitable nutrients for efficient nutrition of crops and livestock, cost-covering prices, problem-solving *know-how,* staggered payment according to performance, valid procedures for individual and inter-farm assessment, independent and incorruptible performance monitoring, reliable prevention of unfair practices and illegal benefits, etc. Many primary producers do not have sufficient access to these elementary prerequisites. In addition, **cost leadership** competition among farms has massively intensified the antagonistic relationship between the quest for minimizing effort and costs and reducing undesirable and harmful side effects. Since securing economic livelihoods is a top priority for primary producers, the main actors for producing public goods lack the necessary resources and any incentives to behave differently.

From a social and ecological market economy perspective, better agricultural landlords succeed in linking their economic interests with those of the common good, i.e., in internalizing external effects. To achieve this, the economic parameters relevant to the viability of the farm must be reconciled with the requirements of consumer, nature, climate, environmental, and animal protection in the form of the overall performance of the farm management. However, the necessary framework conditions for this are still lacking. After all, primary producers—like other professional groups—cannot be expected to act against their existential interests. Accordingly, only changed economic framework conditions and fair rules can help here, which, on the one hand, create incentives for a reorientation and, on the other hand, prevent **externalization**, i.e., economic activity at the expense of goods for the common good. A balanced relationship between production and protection services can only become a management objective for entrepreneurs and thus bring the entrepreneurial innovation potential to fruition if this results in competitive advantages. This presupposes that economic behavior detrimental to the interests of the common good must no longer be rewarded but punished. Those who do more for the common good deserve to be paid more, and those who spend more money on food have a right to get more quality. At the latest, it becomes clear that this cannot be realized in an economic system of the "free" market economy, which is defined by competition for cost leadership concerning compliance with (enhanced) minimum requirements.

While the primary producers must set the course within the operational boundaries and implement the measures that are suitable for reducing impairments to the detriment of the community to a minimum, it is decided outside the operational systems whether the farms are provided with the necessary resources through the sale of their products and via subsidies. Decisions are made outside the farm boundaries, whether the farms' performance is subject to comparative assessment and control that prevent unfair competition. The unfair conditions of competition maintained by the interest groups involved for various reasons avoid establishing economic framework conditions that protect consumers, farm animals, the environment, and nature. Even though each farm is required to contribute to its livelihood and the interests of the common good, it cannot do so without the support of the other interest groups.

For example, the principal goal of animal health and welfare of farm animals as a public good is to protect them from pain, suffering, and damage, which challenges their well-being and self-preservation. Other stakeholders' interests must be accommodated to safeguard farm animals' interests. The resulting collaboration must consider the interests of livestock farmers who have to protect farm animals from harm by providing appropriate living conditions for the animals to cope with. In addition, consumers must be prepared to pay the required price to produce food from healthy farm animals. The involved market participants and, last but not least, the legislators are responsible for establishing framework conditions that enable fair competition for ASF with outstanding quality. At least in theory, farm animals, livestock farmers, and consumers who adhere to a common understanding of animal health and welfare form a community defined by the common interest in the absence of severe health problems that threaten the fitness and survival of farm animals. At the same time, health problems threaten the productivity of livestock and food safety when derived from diseased animals. Furthermore, food from diseased animals threatens the expectations and associations of consumers concerning the quality and value of food. It thus reduces their willingness to pay premium prices or purchase ASF. The latter is indicated by the dramatic decrease in the consumption of pork and beef and the success of the upcoming production processes of replacement products, which emerge as a new and productive competitor. From this perspective, animal health problems are undermining the business model of livestock production in developed countries, not abruptly but with a high probability in the long run.

Just as the economic viability of an operating system requires generating a positive farm income, farms cannot afford to ignore the elementary public interest in minimizing long-term harmful effects. Otherwise, they risk being deprived of financial and infrastructural support from local, regional, and national community structures. The magic word is **transparency**. At present, the interactions between farms and a society divided into different interest groups are incredibly confusing. The individual groupings appear entangled in a tangled web of interests that is difficult to understand. Often, veritable self-interests are concealed under the guise of generally formulated interests in the common good.

Consequently, the main issue must be to organize a fairer distribution of resources in terms of the interests of the common good. This implies that something must be taken from those who have benefited in excess so far to make more resources available to others. Reorienting the entire agricultural economic system and organizing the transformation processes to realize common good interests is a mammoth task. It is not enough to set up new subsidy pots, e.g., to construct new stables, while other areas and the export orientation of the agricultural sector remain untouched. All interest groups in this endeavor must support primary producers who serve the public interest. This applies to the processing industry, food retailers, and, last but not least, consumers. The latter must be enabled to contribute by buying food whose production is beneficial to the public interest. It is equally important that those who persist in a method of production that is detrimental to the public interest lose the support of the other stakeholders.

In its current constellation and orientation, agricultural policy appears overstretched to contain the various particular interests and subordinate them to the interests of the common good. Last but not least, the expansion of farm subsidies after the end of the Second World War (see Sect. 2.4) has contributed significantly to the fact that agricultural policy today is closely linked to the economic interests of the agricultural and food industries and their representatives. However, the successes achieved in increasing productivity due to intensification have long since exceeded the benefit optima, leading to an exuberant level of undesirable side effects and no longer tolerable consequences. Therefore, a "business as usual" approach is not a political option. The assumption that agricultural systems can self-regulate (by an invisible hand) has proved an elementary fallacy. Just as after the experience of hardship and deprivation during and after the world wars, it was seen as a national and European collective task to increase food production, so now concerted strategies at the national and European levels are once again called for to put a stop to the far-reaching aberrations in food production. Examples from the Netherlands and Denmark show how national agricultural policy can be coordinated with other policy areas, such as environmental and health policy, to take account of overriding national interests.

At the various hierarchically arranged levels, state bodies' task is to ensure a valid assessment of public welfare performance. Additionally, measures of promotion and support, as well as sanctioning misconduct, are required. Further proposals in this regard are made in Chap. 9. Only when there is agreement on the priority objectives of primary production and the methodology for assessing the success of implemented measures will this open up the search for and the debate about the most effective and efficient strategies for achieving common goods. On this basis, mutual concessions and compensation possibilities can then be negotiated between the various interest groups to determine how the pursuit of public welfare interests can be realized in a socially acceptable manner.

In contrast to the previous approach, the public welfare interests in food production should be given normative significance, i.e., significance binding on all actors and pointing the way ahead. **Normativity** can only be effective if it is clearly defined which services benefit the public interest and which are detrimental. On this basis, framework conditions can be implemented to promote achieving the objectives. With the specification of the interests of the common good, the particular interests do not disappear into thin air. They continue to exist but are relativized, i.e., placed in a new relationship with each other and to higher-level objectives. The common good is the yardstick, and individual farms and farming systems are the reference systems against which progress is measured. How difficult it is for actors to reorient depends on the degree of path dependencies to which not only primary producers but also members of other stakeholder groups are subject. Substantial incentives are therefore needed to encourage actors to move away from the paths they have taken and to follow new ones. The more numerous the enterprises are, and the more comprehensively they provide the public good, the more the community benefits. The same applies to how those practices are pushed back that are used to the detriment of the common good. In the absence of substantially changed framework conditions, there is little reason for individual primary producers to leave the paths they have been following and to strive to reconcile self-interest with the interests of the common good.

References

Ackermann R (2001) Pfadabhängigkeit, Institutionen und Regelreform. Univ., Diss., 1999. Die Einheit der Gesellschaftswissenschaften, vol 120. Mohr Siebeck, Tübingen

Ackermann J (2017) DBV-Milchviehtagung. Agrarmanager

Åkerfeldt MP, Gunnarsson S, Bernes G, Blanco-Penedo I (2021) Health and welfare in organic livestock production systems—a systematic mapping of current knowledge. Org Agr 11:105–132. https://doi.org/10.1007/s13165-020-00334-y

Blage J (2021) Ersehnter Klops. Süddeutsche Zeitung

Boscardin L, Bossert L (2015) Sustainable development and non-human animals: why anthropocentric concepts of sustainability are outdated and need to be extended. In: Meisch S, Lundershausen J, Bossert L, Rockoff M (eds) Ethics of science in the research for sustainable development, 1st edn. Nomos; Nomos Verlagsgesellschaft GmbH & Co, KG, Baden-Baden, pp 323–352

Brečić R, Mesić Ž, Cerjak M (2017) Importance of intrinsic and extrinsic quality food characteristics by different consumer segments. Br Food J 119:845–862. https://doi.org/10.1108/BFJ-06-2016-0284

Bryant C, Barnett J (2020) Consumer acceptance of cultured meat: an updated review (2018–2020). Appl Sci 10:5201. https://doi.org/10.3390/app10155201

Bundesvereinigung der Deutschen Ernährungsindustrie (2019) Leitbild Verbraucher. https://www.bve-online.de/themen/verbraucher/leitbild-verbraucher. Accessed 8 Jan 2022

Daatselaar CH, Reijs JR, Oenema J, Doornewaard GJ, Aarts HFM (2015) Variation in nitrogen use efficiencies on Dutch dairy farms. J Sci Food Agric 95:3055–3058. https://doi.org/10.1002/jsfa.7250

Di Domenico G, Premazzi K, Cugini A (2022) "I will pay you more, as long as you are transparent!": an investigation of the pick-your-price participative pricing mechanism. J Bus Res 147:403–419. https://doi.org/10.1016/j.jbusres.2022.04.037

Ellrott T (2016) Sage mir, was du isst…: Ich-Performance. https://www.spiegel.de/spiegelwissen/ernaehrungspsychologie-interview-mit-thomas-ellrott-a-1081565.html

Engelsman V, Geier B (eds) (2018) Die Preise lügen: Warum uns billige Lebensmittel teuer zu stehen kommen. oekom verlag, München
European Commission (2023) Attitudes of Europeans towards animal welfare: special eurobarometer 533 (99.1)
Fink-Kessler A (2020) Regional Schlachten: Jedem Landkreis ein Schlachthaus – oder was brauchen wir wirklich? https://www.bauernstimme.de/fileadmin/Dokumente/Verlag/Bauernstimme.pdf/pdf-20/09-2020.pdf. Accessed 7 Jan 2022
foodwatch (2018) Positionspapier Tiergesundheit. https://www.foodwatch.org/fileadmin/Themen/Tierhaltung/Dokumente/2018-09-03_Positionspapier_Tiergesundheit.pdf. Accessed 7 Jan 2022
Freeman RE (1984) Strategic management: a stakeholder approach. Pitman series in business and public policy, Pitman, Boston, Mass
Fried N, Wernicke C (2020) Mist aus Bayern für die Hauptstadt: CSU-Klausur in Kloster Seeon. https://sz.de/1.4745890
Gahagan S (2012) Development of eating behavior: biology and context. J Dev Behav Pediatr 33:261–271. https://doi.org/10.1097/DBP.0b013e31824a7baa
Gassmann M (2020) „Essen verdient niedrigsten Preis" – Edeka erntet Bauern-Zorn. https://www.welt.de/wirtschaft/article205376911/Essen-verdient-niedrigsten-Preis-Edeka-erntet-Bauern-Zorn.html. Accessed 21 July 2021
Gaugler T (2020) PENNY weist bei ersten Produkten „wahre Verkaufspreise" aus. https://www.penny.de/presse/wahre-verkaufspreise. Accessed 21 July 2021
Genschel P (1995) Standards in der Informationstechnik: Institutioneller Wandel in der internationalen Standardisierung. Schriften des Max-Planck-Instituts für Gesellschaftsforschung, Köln, vol 22. Campus Verlag; Campus-Verl., Frankfurt/Main
German Animal Welfare Association (2019) Tierschutzlabel. https://www.tierschutzbund.de/information/hintergrund/landwirtschaft/tierschutzlabel/. Accessed 28 May 2019
Götz N (2008) Reframing NGOs: the identity of an international relations non-starter. Eur J Int Rel 14:231–258. https://doi.org/10.1177/1354066108089242
Grebitus C, Menapace, L, Bruhn M (2009) What determines the use of brands and seals of approval as extrinsic quality cues in consumers' pork purchase decision? In: Berg H et al (eds) vol 44, 171–182
Greenpeace (2020) Greenpeace-Recherchen: Supermärkte setzen weiter auf Billigfleisch. http://presseportal.greenpeace.de/204443-greenpeace-recherchen-supermarkte-setzen-weiter-auf-billigfleisch
Greenpeace (2021) Greenpeace stellt sich vor: Volle Kraft voraus für die Umwelt. https://www.greenpeace.de/ueber-uns. Accessed 11 Nov 2021
Grossarth J (2017) Ein Aufklärer, den sie Denunziant nennen. https://www.faz.net/aktuell/wirtschaft/unternehmen/tierhaltung-ein-aufklaerer-den-sie-denunziant-nennen-14762951.html. Accessed 7 Jan 2022
Grunert KG, Bredahl L, Brunsø K (2004) Consumer perception of meat quality and implications for product development in the meat sector—a review. Meat Sci 66:259–272. https://doi.org/10.1016/S0309-1740(03)00130-X
Habel J, Uhlig V, Hoischen-Tauber S, Schwabenbauer E-M, Rumphorst T, Ebert L, Möller D, Sundrum A (2021) Income over service life cost—estimation of individual profitability of dairy cows at time of death reveals farm-specific economic trade-offs. Livest Sci 254:104765. https://doi.org/10.1016/j.livsci.2021.104765
Häusel H-G (2013) Emotional Boosting English Version, 1st edn. Haufe Fachbuch, vol 1639. Haufe, Stuttgart
Hemmerling U, Pascher P (2019) Situationsbericht 2019/20: Trends und Fakten zur Landwirtschaft, 1st edn. Deutscher Bauernverband e. V., Berlin

Iop S, Teixeira E, Deliza R (2006) Consumer research: extrinsic variables in food studies. Br Food J 108:894–903. https://doi.org/10.1108/00070700610709940

Jackson RB, Saunois M, Bousquet P, Canadell JG, Poulter B, Stavert AR, Bergamaschi P, Niwa Y, Segers A, Tsuruta A (2020) Increasing anthropogenic methane emissions arise equally from agricultural and fossil fuel sources. Environ Res Lett 15:71002. https://doi.org/10.1088/1748-9326/ab9ed2

Jost JT, Banaji MR, Nosek BA (2004) A decade of system justification theory: accumulated evidence of conscious and unconscious bolstering of the status Quo. Polit Psychol 25:881–919. https://doi.org/10.1111/j.1467-9221.2004.00402.x

Kabisch S, Wenschuh S, Buccellato P, Spranger J, Pfeiffer AFH (2021) Affordability of different isocaloric healthy diets in Germany—an assessment of food prices for seven distinct food patterns. Nutrients 13. https://doi.org/10.3390/nu13093037

Kaufer R (2015) Umsetzung von EU-Umweltschutz in der deutschen Land-und Forstwirtschaft. Dissertation, Georg-August-Universität Göttingen

Kläsgen M (2021) Aldi will die Preistreiber stoppen. Süddeutsche Zeitung

Klepper R (2020) Schweinefüße sehr begehrt: Ein Gespräch über die Landwirtschaft in China, deutsche Export-Chancen und die Corona-Krise. Wissenschaft Erleben 2020:8–9

Knap PW (2022) Drivers of animal source food consumption: a biophysical approach. Front Sustain Food Syst 5. https://doi.org/10.3389/fsufs.2021.732915

Kokkoris MD, Stavrova O (2021) Meaning of food and consumer eating behaviors. Food Qual Prefer 94:104343. https://doi.org/10.1016/j.foodqual.2021.104343

Kreutzberger S, Thurn V (2011) Die Essensvernichter: warum die Hälfte aller Lebensmittel im Müll landet und wer dafür verantwortlich ist. Kiepenheuer & Witsch

Mankiw NG, Wagner A (2004) Grundzüge der Volkswirtschaftslehre, 3rd edn. Schäffer-Poeschel, Stuttgart

Nocella G, Romano D, Stefani G (2014) Consumers' attitudes, trust and willingness to pay for food information. Int J Consum Stud 38:153–165. https://doi.org/10.1111/ijcs.12080

Postel T (2008) "Milch ist Macht": Der Streik der Bauern. https://www.stern.de/wirtschaft/news/-milch-ist-macht--der-streik-der-bauern-3851378.html. Accessed 21 July 2021

Reutskaja E, Lindner A, Nagel R, Andersen RA, Camerer CF (2018) Choice overload reduces neural signatures of choice set value in dorsal striatum and anterior cingulate cortex. Nat Hum Behav 2:925–935. https://doi.org/10.1038/s41562-018-0440-2

Roth SJ (2007) Die Marktwirtschaft – Fairer Wettbewerb um die beste Lösung. https://www.bpb.de/politik/wirtschaft/wirtschaftspolitik/64333/marktwirtschaft. Accessed 7 Jan 2022

Schulz L, Halli K, König S, Sundrum A (2021) Potenziale der Erzeugung von Qualitätsfleisch in der Population Rotes Höhenvieh. Züchtungskunde 2021

Schulze M, Spiller A, Risius A (2021) Do consumers prefer pasture-raised dual-purpose cattle when considering meat products? A hypothetical discrete choice experiment for the case of minced beef. Meat Sci 177:108494. https://doi.org/10.1016/j.meatsci.2021.108494

Schulze M, Sonntag W, von Meyer-Höfer M (2023) Is less more? Investigating citizen and consumer preferences for the future direction of livestock farming policy. J Clean Prod 390:136136. https://doi.org/10.1016/j.jclepro.2023.136136

Spiller A, Schulze M (2019) Der LEH macht die Spielregeln. top agrar 2019:126–129

SRU (1985) Umweltprobleme der Landwirtschaft: Sondergutachten März 1985. Sondergutachten/Der Rat von Sachverständigen für Umweltfragen, vol 1985. Kohlhammer, Stuttgart, Mainz

Stankevich A (2017) Explaining the consumer decision-making process: critical literature review. JIBRM 2:7–14. https://doi.org/10.18775/jibrm.1849-8558.2015.26.3001

Sundrum A (2005) Paradigmenwechsel – Vom ökologischen Landbau zur ökologischen Landwirtschaft. Ökologie & Landbau 1(2005):17–19

Sundrum A (2018) Beurteilung von Tierschutzleistungen in der Nutztierhaltung. Ber. Landwirtsch 96. https://doi.org/10.12767/buel.v96i1.189

Töwe S (2021) Es geht um die Wurst. Süddeutsche Zeitung

Trentmann F (2017) Herrschaft der Dinge: Die Geschichte des Konsums vom 15. Jahrhundert bis heute, 1st edn. Deutsche Verlags-Anstalt, München

van der Ploeg JD (2020) Farmers' upheaval, climate crisis and populism. J Peasant Stud 47:589–605. https://doi.org/10.1080/03066150.2020.1725490

Weingarten P (2010) Agrarpolitik in Deutschland. Parlam Beil Polit Zeitgesch:6–17

Wenzel FT (2021) Nestlé setzt auf veganen Milchersatz. Göttinger Tageblatt

Witte B, Obloj P, Koktenturk S, Morach B, Brigl M, Rogg J, Schulze U, Walker D, von Koeller E, Dehnert N, Grosse-Holz F (2021) Food for thought: the protein transformation. Boston consulting group and blue horizon corporation. Ind Biotechnol 17:125–133. https://doi.org/10.1089/ind.2021.29245.bwi

Wollny VP (2008) Zur globalen Standardisierung von Markennamen im Konsumgütersektor Erfolgspotenziale, Problemfälle und Lösungsansätze//Zur globalen Standardisierung von Markennamen im Konsumgütersektor – Erfolgspotenziale, Problemfälle und Lösungsansätze, 1st edn. Diplomica, Hamburg

Wurzel RK, Jordan A, Zito AR, Brückner L (2003) From high regulatory state to social and ecological market economy? new environmental policy instruments in Germany. Environ Politics 12:115–136. https://doi.org/10.1080/09644010412331308214

7 Scientific Approach to Reality

Abstract

Given the complexity and opacity of procedural processes, as well as the undesirable and harmful side-effects of production processes, it would be the task of the sciences, using their methodological toolbox, to illuminate the interrelationships and the possibilities of realizing the interests of the common good. At present, the sciences are still far from achieving this. The agricultural sciences, in particular, have, to date, played a decisive role in promoting the intensification of production processes. At the same time, the undesired and harmful side effects are given significance at best concerning general issues but not in the context of individual agroecosystems. What has made agricultural science successful in increasing productivity is now becoming a severe obstacle. The self-reinforcing tendencies of specialization have led to the agricultural sciences depriving themselves of their problem-solving competence. Solving problems that arise from the complexity of interrelationships can only be achieved through inter- and transdisciplinary research approaches. However, this is not the way to earn scientific merit. It requires a look behind the scenes to understand how agricultural science has come to be an auxiliary science of rural economics and to become part of the problem. Only when the past role of agricultural science is understood can we reflect on how it can move from being part of the problem to being part of the solution.

The current orientation of the agricultural economy and the **undesirable side-effects** that go hand in hand with it can hardly be understood if the contribution of agricultural sciences to this development is not considered. Suppose the harmful effects are to be contained and an orientation towards public welfare interests is to be promoted. In that case, one cannot avoid reflecting on the scientific processes of knowledge involved in the emergence of the present state of the agricultural economy, which has shaped the

A. Sundrum, *Public Welfare-Oriented Production of Food*,
https://doi.org/10.1007/978-3-662-69040-6_7

patterns of thought. Since the findings of the agricultural sciences have been instrumental in promoting the intensification processes in the production of foodstuffs while mainly ignoring the extent of the undesirable side effects for decades, they are undoubtedly part of the problem. On the other hand, they are part of the solution because the knowledge gained to date about the extent and causes of harmful effects and the proposals for minimizing undesirable developments also result from various sub-fields of agricultural science. In the face of this ambivalence, all attempts to understand the activities of the agricultural sciences as "value-free" to avoid being drawn into socio-political debates or even being held jointly responsible for undesirable developments come to nothing. The image of a value-free **science** may still be maintained in physics, even if Friedrich Dürrematt should have robbed us of this illusion long ago with his comedy 'The Physicists,' written as early as 1961. In no other scientific discipline as in agricultural sciences, the interest in knowledge is so closely linked to the interest in its application. Linked to the interest in applying scientific knowledge is achieving an intention and a goal.

Following the previous ductus from the general to the specific, the various explanatory approaches, the main branches of science in which the agricultural sciences are embedded, and the forms of knowledge produced are roughly outlined. Subsequently, agricultural sciences as an applied science will be distinguished from the natural and technical sciences. Finally, the specific characteristics of the agricultural sciences will be elaborated to clarify what they have achieved in the past and what they cannot do given future challenges without a **paradigm shift**.

The goal of science is to find out the truth about what exists (Chalmers 1999). However, facts are accessible to assess their truth content to a very different degree. Many processes are inaccessible to direct observation, even with diverse and sophisticated methods. Or they are subject to such dynamics that only snapshots of partial aspects are possible. Therefore, a systematic approach is required to gain knowledge accepted within the scientific community and thus claim intersubjective validity. This is distinct from the gains in knowledge that each person believes they have attained. **Scientific knowledge** is rightly called the best possible expertise because it emerges from a process of cognition that is or should and could be continuously subjected to a critical evaluation, provided that the necessary conditions exist. Knowledge is indeed oriented toward truth but can only be measured conclusively against it if someone already possesses it (Ladenthin 2020). In that case, there would be no need for science. From this, Ladenthin concludes that evaluating what "excellent science" is is impossible because the person evaluating would have to know more than the person being assessed. However, this is rarely the case because of the ever-expanding special knowledge. Accordingly, science consists, above all, of questioning science for its validity.

What constitutes scientific activity at its core can be outlined as follows:

> *"Science is a system of knowledge about the essential properties, causal relations, and regularities of nature, technology, society, and thought, fixed in terms of concepts, categories, measures, laws, theories, and hypotheses."*

At the same time, science refers to the systematic process of intersubjectively comprehensible research and cognition in a specific field. This process produces a well-founded, orderly, and assured body of knowledge in the form of a coherent system of propositions, theories, and procedures, which is subjected to strict validity tests and associated with the claim of **objectivity** and supra-personal validity (Carrier 2011). Verifiable laws and regularities can be derived from the results and cognitions gained in this way, and predictions can be made by applying logical considerations. **Scientific methods** of acquiring knowledge can be used in almost all areas of life. However, not all areas of life are accessible in the same way or even of interest to be researched in more detail to make reliable statements about them. As different as the methods of gaining knowledge are, the possibilities of using the findings to solve open questions and problems or to achieve goals set by oneself or by others are also different.

7.1 Scientific Knowledge

Kant prefaced his enlightenment philosophy with three questions: What can we know, what may we hope, and what shall we do? (Schlageter 2013). These questions still provide the basic structure for a rational approach to a constantly changing world. In Kant's time, the rational use of **knowledge** was still reserved for a small circle of people. Today, not least because of manifold technical innovations, dealing with knowledge has developed into a dimension that determines almost all areas of life. We live in a knowledge society. The fact that we can access **information** very quickly does not mean we understand a great deal. The exponential increase in data and information, which act as the knowledge substrate, is becoming an ever-greater challenge. The transformation from data to information and from information to personal knowledge is not a no-brainer. Instead, the path is paved with numerous obstacles and is often not well signposted, so one can quickly lose one's orientation.

Science is a subsystem of social reality. The reputation of science is based on its ability to explain phenomena and facts reliably, in the sense that one can rely on the explanation in theoretical argumentation or practical action and in coping with problems of everyday life. The knowledge gained and processed according to the principles of scientific research can be called scientific knowledge (Mohr 1999). The practical interest in assured knowledge is self-explanatory. Knowledge about an issue can be defined as the capability to solve problems connected with it. This is enabled by connecting information about the issue with cognitive structures relevant to understanding its behavior. To explain a state of affairs in science means to reduce it to general propositions and to be able to trace it back to the system-specific boundary conditions. Assured statements in science are generally made by singular propositions (statements of fact) or general propositions (statements of law). A law is a confident statement that applies to a variety of situations but not necessarily to all situations. Concerning the generalizability of the assured propositions, a distinction can be made in decreasing order of precedence between all propositions, particularly all propositions, theoretical laws, empirical laws, and regularities. Different approaches are available to explain facts: the causal, the functional, and the teleological explanation.

7.1.1 Causal Explanation

Causal explanations concern the lawful sequence of interrelated events. Event A is the cause of effect B if B is brought about by A. The causal principle describes the belief that the same cause always leads to the same effect. The structure of causal explanation is described in the sciences by the Hempel-Oppenheim model (Mohr 1999). It contains three elements: general propositions, statements about the system-specific circumstances (initial and boundary conditions), and a statement about the state of affairs that is to be explained (explanandum) or that is predicted. The quality of the laws determines the quality of an explanation or the precision of a prediction. In contrast to physics or technology, biological processes are rarely governed by laws but rather by regularities. Process outcomes are often overlaid by unforeseen events, which require the biological systems to adapt and put the predictions into perspective. In addition, there are multifactorial processes, which usually represent a structure of effects that can hardly be penetrated in thought and can, therefore, only be reproduced to a limited extent experimentally under standardized boundary conditions. In contrast to monocausal interactions, multifactorial interactions cannot be described with the Hempel-Oppenheim model because the interactions do not follow general laws or theorems.

In addition, the complexity of biological systems places tight limits on simplifying boundary conditions. This leads to a considerable loss of precision in the explanation and has the consequence that no reliable predictions can be made. The precise explanation of an individual event is thus not possible. The Hempel-Oppenheim model fails when dealing with very complex and chaotic systems. Systems behave chaotically when small changes can cause very different effects. This is the case, for example, with developing a disease in an organism. The same cause (pathogenic germ) can cause very different disease symptoms and courses in the animals of a group, even under very standardized initial (e.g., genetics) and boundary conditions (including feeding, husbandry, and hygiene). Since, in reality, neither the past initial conditions nor the current or future boundary conditions are known or can be recorded with sufficient accuracy, this essentially precludes a precise explanation and prediction of the course of processes. In the presence of a large number of animals, regularities can be determined using statistical procedures and statements about the probability of the occurrence of specific reactions. However, these only apply to the respective population, not the individual units from which the population is formed. Suppose no differentiation is made between the population and the individual units regarding explanatory content and predictability. There is a severe error in thinking. In other words, what applies to the population does not necessarily apply to each sub-unit.

7.1.2 Functional Explanation

While causal explanation is primarily used in physics and chemistry, two further forms of explanation are added in biology, which play an indispensable role alongside causal explanation: functional and teleological explanation of facts. They are not sufficiently familiar to scientists who orient themselves to physics as a paradigmatic natural science. Functional explanations are statements about a part's role in a functioning whole. The functional explanation is based on the premise established by evolutionary science that living organisms are optimized and purposeful biological systems. They can sustain and reproduce themselves because they function (Nurse 2020). This means that they contain virtually no useless parts. Each part of the whole serves a particular purpose, often several purposes simultaneously, but in different compartments. This applies, for example, to minerals and hormones. The epistemological principle of isolation introduced by Galileo, i.e., the analysis of ideally singled out individual moments and subsequent assembly in superposition, has worked well in many branches of science and especially in physics, at least until the advent of quantum physics. However, it fails in biology.

"With the concept of the organism, Aristotle gave biology its fundamental principle of knowledge." (Schlageter 2013). For the organism as a whole, it is true that

> *"the parts which are separated from the essence [are; author's inset] to which they belong do not exist at all"* (Aristotle).

The living cannot be disassembled into individual parts without being destroyed simultaneously. Ernst Cassirer (1974) adds 2000 years later:

> *"In a living organism, the parts are not merely adjacent but interact with each other, and all work together toward a common end. No task requires the help and cooperation of almost all the others. ... The concept of an organism cannot be grasped without including in it the moment of purposefulness."*

In this sense, functional explanations are statements about a part's role in a higher-level system, such as an organelle in a cell, an organ in an organism, an animal in a herd, and animal husbandry in an operating farm system. The functional explanation is based on the premise that biological systems have a purpose and that the subunits (sub-systems) perform a function within the whole. A disruption in their ability to function is not without consequences for the ability of the whole to function. At the same time, there is a diversity of variation, which is the prerequisite for **adapting** to changing living conditions. The functionality of subsystems can thus also be explained based on the ability to adapt to changing conditions to maintain the system's functionality as a whole.

7.1.3 Teleological Explanation

Because the processes in a living system are goal-oriented, they are logical concerning achieving the goal. It is assumed that the functional end states and the development processes leading to them are oriented toward optimization. Mohr elaborates on this:

> *"The purposeful function of a part in a whole usually requires a fully developed, functionally mature system. The developmental process that leads to the actual functioning part is usually not accessible to functional explanation. It is at this point that teleological explanation plays an indispensable role. The teleological explanation of a part means that its existence can be understood from the functional end state."* (Mohr 1999)

For example, living organisms are capable of self-regulation. They can alter internal processes to maintain vital functions and adapt the system to new conditions. For instance, during an infection, the immune system, cardiovascular system, and many other bodily functions are put at the service of immune defense. During physical stress, the supply of nutrients to the muscles can be increased via an altered blood flow at the expense of other organs. At the same time, the blood flow is improved via an increase in cardiac output, and processes in different organ systems adapt to the altered requirements. In this sense, organisms are adaptive systems that adapt to changing conditions through feedback and can achieve the goal of self-preservation despite disturbances (target-actual deviations). The decisive factor is that each part of the whole serves the overriding purpose of survivability.

7.2 Branches of Science

According to the classical view, science is divided into natural sciences, social sciences, and humanities. The diversity of the different branches of science makes it understandable that there is no consensus in the scientific community on a general classification of the individual sciences. Accordingly, different degrees of overlap between the various fields of science make clear classifications and categorizations difficult. On the other hand, the engagement and reflection of different scientific institutions with their respective perspectives and methodological spectrums with the same research object would offer an excellent opportunity to understand science more comprehensively than to divide it into individual disciplines that hardly have any points of contact with each other.

7.2.1 Natural Sciences

The natural sciences belong to the empirical sciences. They are characterized primarily by their research object, the inanimate and animate matter. Some natural sciences, such as physics and chemistry, are characterized by a mathematical approach to their research object. They are called exact sciences and have set the gold standard with exactness over long periods, which other disciplines still follow today. This often happens even though physics has long since undergone a paradigm shift due to quantum physics and Heisenberg's uncertainty principle. This principle states that the momentum and the location of a subatomic phenomenon cannot be determined precisely simultaneously. This means that exact and, at the same time, complete descriptions of what happens at the subatomic level are no longer possible in quantum theory. Also, in quantum physics, the observer and his measuring instruments cannot be separated from the phenomenon to be observed. Thus, the assumption of the modern natural sciences that the world is thoroughly calculable and objectifiable can no longer be maintained.

Research in the natural sciences is primarily concerned with questions that can be answered by investigating law-like relationships in nature. The focus is on describing the processes and not attributing meaning or a goal to them. The focus is not on the why but on the how. For example, the question: Why does it rain? does not find its explanation in: so that plants can grow. It is answered in this way: because water evaporates, rises, collects in clouds, and finally condenses, which leads to precipitation. Natural science, therefore, does not answer questions directed toward a purpose or a goal (teleologically) but rather traces the processes under investigation back to natural laws or to already-known facts. Insofar as this succeeds, natural science is ascribed as a descriptive and explanatory character.

Natural scientists observe, measure, and analyze the states and behavior of actors in nature by methods primarily intended to ensure the **reproducibility** of their results and identify regularities. The diversity of natural phenomena has produced many methods by which they can be analyzed, and the mechanisms of action underlying the phenomena can be interpreted and thus thoughtfully penetrated. However, there are two main barriers to the generation of knowledge with this approach. On the one hand, the processes that produce the phenomena are primarily hidden from direct insight. Consequently, breaking open what is concealed by dissecting and cutting it into ever smaller sections with various tools and procedures is one of the elementary activities of the natural sciences. It makes an ever-greater precision of investigation possible and necessary to bring hitherto hidden structures to light. At the same time, the development of magnifying instruments, which now make even structures on the nanoscale visible, has brought to light an almost unmanageable wealth of insights into the micro-areas. New subject areas emerge with each methodological extension and decomposition of the sub-areas into sub-sub-areas. These are interpreted and defined as gaps in our knowledge so that they can be the subject of further investigation. In this way, scientists do not run out of work.

The other barrier to knowledge is the network of relationships and effects between the individual sub-areas and elements that make up the respective phenomena. Identifying the relations connecting the sub-areas is often limited to the quantitative relations during the analysis. At best, this succeeds under experimental setups conducted under strictly standardized conditions and focusing on very few sub-areas. The network of relationships and effects is destroyed through analysis, i.e., by cutting up, separating, and dissecting phenomena. Living systems, which are based on complex networks of effects, in which each part can be both cause and effect, are not accessible in this way. Although recording details are becoming increasingly precise, this does not necessarily mean the whole is understood. In the case of complex biological processes, the natural sciences quickly reach their limits. Consequently, other methodological approaches are needed to get to the bottom of complex interactions.

The natural sciences claim to investigate natural phenomena in a way that is independent of the subject (the scientist) and thus produces **objectivity**. These results are independent of the subject. Since the selection and delimitation of the object of investigation and the choice of the methods used are based on a subjective decision, the ideal of a subject-independent and value-free science cannot be maintained. If questions about the "what for" of natural phenomena are still largely ignored in the natural sciences, they come to the fore at the latest when it comes to making them useful: what are the objects and phenomena of nature useful for and, above all, for whom are the findings about natural phenomena useful? Harnessing natural conditions for human purposes inevitably involves **anthropocentrism**, in which human nature is viewed, understood, and harnessed from the perspective of humans. Nevertheless, human beings are able, by their capacity for **abstraction**, to relativize their view of things and to adopt other perspectives. Through the change of perspective, as is well known, the things considered become relative, i.e., they are placed in relation to different things or contexts and different observer's perspectives. At the same time, statements about an object of observation lose the claim to absolute correctness in the effort to depict **reality**. It can never be ruled out that other insights regarding reality arise from another perspective, which limits the assessment considered correct from the original perspective. From this, it follows that there is not one but always different realities. Different persons who look at a complex state of affairs from different perspectives will only, in exceptional cases, arrive at an identical assessment of reality. The different realities do not exist separately but—insofar as they are directed at the same object—create a greater or lesser degree of common intersection. Positively speaking, the different perspectives broaden the field of vision and enable a more comprehensive view and, thus, more realistic insights into facts. However, this is only true if combining the different insights into a coherent overall picture is possible. In science, this is done by formulating a **theory**. The best and most valid theory is the one in which as many as possible of the insights gained from the different perspectives are combined into a **consistent** (free of contradictions) and coherent (relating things conclusively to one another) construct of ideas.

7.2.2 Social and Economic Sciences

The social sciences investigate phenomena of the social coexistence of people. In this context, the structures and functions of social interdependencies of institutions and systems and their interaction with the action and behavioral processes of individual actors are analyzed theory-based or empirical. Economics is the science of the economy and belongs to the social sciences. It primarily examines the rational use of scarce goods, i.e., goods only available to a limited extent. Economics examines fundamental social relationships and laws about individual economic units (microeconomics) and an economic context (macroeconomics). The objects of knowledge are economic activity, i.e., the planned and efficient use of scarce resources to satisfy the needs of different actors in the best possible way. Business economics deals with the economic interrelationships and regularities of individual companies. It provides insights into the potential of monetary value creation as a function of company structures and processes compared to the market-economy framework. Econometric methods based on mathematical models are used to empirically test economic models and analyze economic phenomena quantitatively. These models are also used to make forecasts for economic developments.

7.2.3 Humanities

The humanities are concerned with human beings' thoughts, actions, and achievements in all their diversity. This includes everyday culture, language, history, law, social, political, and religious life, the arts, the practice of all sciences, and their ways of thinking. The humanities make us aware that historically evolved conditions shape our actions and that assessments from various perspectives are inevitable, although they can never be final. The humanities continue what is at the origin of modern science: the critical examination of notions, ideas, and linguistic images. They attempt to replace dogmatism and ideology with critical discussion and thus create reflective and orienting knowledge. Correspondingly, as early as the nineteenth century, the humanities have concluded that there is no one reality but somewhat different views of reality. Since then, good scientific work in the humanities has involved researchers disclosing and reflecting on the standpoints from which they have developed their considerations. While scholars in various disciplines are very good at generating new knowledge through their practical work, they are not exceptionally skilled at taking a step back from their work and describing, characterizing, and reflecting on their thought processes (Chalmers 1999). Successful sub-fields of the various branches of science have become independent as individual disciplines and have each developed their form of reasoning. As a result, the connection to higher levels of reflection has been weakened or lost altogether. Philosophers' methodological skills and abilities can be used to re-locate themselves in the overall context of the sciences and achieve new connectivity. Part of philosophy is meta-theory for disciplines in the natural, social, and cultural sciences in the form of general theory of science and **epistemology**. Philosophy thus reflects the use of reason in our world of life and various systemic contexts such as economics or politics, as well as in the context of the disciplines of the natural sciences, social sciences, and humanities. In addition to reflecting on the use of reason, there is another challenge for the sciences: the ability to self-reflect.

However, this requires the willingness to face one's possible misjudgments and an institutional framework in which self-reflection is methodically underpinned, accompanied, practiced, and monitored.

7.3 Forms of Knowledge

A whole range of categories has been proposed to deal with knowledge. These include factual, organizational, expert knowledge, or everyday expertise. Knowledge can further be differentiated into explicit and implicit knowledge. Explicit knowledge can be expressed in semantically correct sentences, mathematical formulas, or technical data. Due to its formalization and structuring, this knowledge can be duplicated, accessible to others, and communicated in various ways. Explicit knowledge is contrasted with tacit knowledge, which escapes formal linguistic expression. It is based on the experiences of the individual on their life path to date and concerns factors that are difficult to grasp, such as personal convictions, perspectives, value systems, thought patterns, and guiding principles.

Whereas in past decades, a journeyman could draw on the knowledge and skills imparted during his apprenticeship in his further professional life, "lifelong learning" is necessary today. There is no other way to find one's way in the abundance of new information and use it to realize one's interests. As a result of the increasing abundance of information, which enriches explicit knowledge with a vast amount of detailed aspects, a progressive specialization has occurred. We find this in all areas of life and, accordingly, also in food production. However, the various advantages of specialization are also accompanied by multiple disadvantages. These include, in particular, the danger of losing an overview of the relevant interrelationships and a decreasing ability to connect with other specialist areas.

Much of the previous information stock has been archived via electronic data processing. Search engines allow quick access to the already archived information. Knowledge about the characteristics of products, systems, and processes can be acquired differently than was possible for previous generations. New information has led to new technical developments and forms of organization, which, among other things, have made an enormous increase in productivity in food production possible and have made knowledge of the productive forces a dominant factor in production. However, knowledge is accompanied by varying degrees of certainty. At the same time, the real content of knowledge depends on the context in which it is acquired and in which it is brought to bear. Therefore, it is not by chance that knowledge is defined as a valid and justified opinion in **epistemology** (Mohr 1999). Knowledge only acquires a truth content when it can be justified to others. Otherwise, it is merely a personal opinion. However, in public discourse, it often remains unanswered to whom an opinion should or must be justified to be considered knowledge. Knowledge comes to us in three forms: Dispositional, orientational, and action-guiding knowledge. The following outlines the different forms of knowledge to take them up in the context of agricultural science and classify them better.

7.3.1 Dispositional Knowledge

Knowledge should be at your disposal if something is to be recognized, decided, or implemented. Disposable knowledge is applicable cognitive knowledge to which the individual can refer in theoretical argument and practical action. It has different dimensions that often cannot be sharply distinguished from each other. In this context, a reflection on experimental knowledge is particularly relevant. Experimental knowledge can be found in abundance in the respective professional journals (Wolff 1989). It requires an experiment's performance to determine whether the initial question or **hypothesis** that precedes a scientific experiment is confirmed or rejected. A particular requirement of an experiment is the repeatability of the results. The experimental design (material and methods) is to be understood as a set of instructions for the experimenter, with which the experiment can be repeated and should produce approximately the same results. Each experimental instruction refers to concrete framework conditions under which the experiment should be carried out (e.g., temperature conditions). Different framework conditions produce different or modified results.

Based on the knowledge of laws and regularities, which can be integrated into a theory depending on the known initial and boundary conditions, a **prognosis** can be formulated as a function of the action instructions and the specified boundary conditions. Under the premise that the action instructions are followed and the boundary conditions approximate the specifications, the hypothesis should be able to predict the experimental outcome with a certain probability. Consequently, experimental knowledge is given if the actual executions of an experimental set-up, with sufficient repetitions in which the framework conditions may also be modified to a limited extent, have consistently yielded the predicted results. An example of an experimental fact is the law of falling, which is considered to be a law of nature: for all objects x, for all times t, and all locations on Earth no higher than 1000 m above sea level, the following holds: if x is allowed to fall freely for t seconds in a highly air-dispersed space, the depth of fall is $T = 4.905\ t^2$ meters. Consequently, even this law of nature does not hold independently of specific boundary conditions.

Experimental facts do not express **causalities** but functional relations. Thus, the law of falling does not yield a causal relation between the depth and time of fall but an exact functional one. Experimental facts have a high degree of justified certainty of knowledge for the section of reality defined (delimited) by the framework conditions (i.e., initial and boundary conditions). These are not nature or reality per se, but they only concern a section, i.e., a delimited experimental reality. An experiment must be repeated according to instructions to obtain certainty about functional relationships under different initial and boundary conditions.

In many branches of science, replication of experiments is standard practice. However, various meta-studies have found that in recent years, many empirical studies related to biological processes have failed to replicate the results of other studies. This failure is no longer the exception to the rule. In several areas, the lack of replicability of scientific results has already developed into a "replication crisis" or "reproducibility crisis" that threatens to undermine public confidence in science (Fanelli 2018). Also for Drucker (2016) the crisis is evident:

> *"The available evidence from multiple fields indicates that we have a systemic reproducibility problem in basic science research."*

Weak **reproducibility** of experimental results is observed mainly in biology, pharmacy, medicine, economics, and social sciences. It has also been found to be relevant in farm animal sciences (Bello and Renter 2018). The crisis is causing great concern among scientists and the scientific community (Baker 2016). So far, the public has hardly noticed this crisis in the sciences (Mede et al. 2021). Only time will tell to what extent science will prove capable of drawing the correct conclusions from these setbacks and whether its self-correcting procedures will lead to overcoming the crisis.

Since there is an unlimited number of initial and boundary conditions, one helps oneself with the delimitation of different realities by defining standards (e.g., husbandry standards) or by using a **theory**, i.e., with a mental construct describing the context of the meaning of a theoretical section of the world. The connection between a theory and reality is made through an interpretation. There are theories about how reality can be interpreted and how the process of cognition can be theoretically penetrated. There has been a fundamental confrontation between empiricism and rationalism throughout the history of the philosophical discipline **of epistemology**. The lines of conflict run very simplistically as follows: The empiricist assumes everything we can know about the world comes from sensory experiences. The rationalist claims that we can make valid statements about the world independent of experience. The empiricists have difficulties encountering the true propositions of formal logic and mathematics. On the other hand, an empirical **induction**, i.e., an experiential generalization, must expect refutation at any time. In contrast, the propositions of mathematics and formal logic appear confident. However, the propositions do not directly apply to many agricultural practice areas.

Scientific and technical knowledge is naturally deficient. Scientists never tire of pointing out the existing gaps in knowledge in their scientific papers. In doing so, they usually refer to the state of scientific knowledge known from their perspective and to the methodological equipment available to them for dealing with gaps in knowledge. This self-referential system of revealing knowledge gaps is not free of self-interest. Ultimately, one serves a potential third-party funder with one's expertise or the discipline's expertise to use the available research funds to address the knowledge gaps. Further research activities then bring new insights and new knowledge deficits to light.

7.3.2 Orientation Knowledge

Orientation knowledge is defined as the **knowledge** that acts as a conduct guideline. The distinction between disposal and orientation knowledge, introduced by Mittelstraß, has become an integral part of describing and evaluating interdisciplinary scientific work.

> "Factual and dispositional knowledge is knowledge about causes, effects, and means; orientational knowledge is about justified purposes and goals. *Factual and dispositional knowledge is positive knowledge, and orientational knowledge is regulative knowledge. And it is precisely this regulative knowledge that is not at its best today. Science has lost sight of this knowledge—and so has society in many cases. The result is a lack of orientation (not a loss of orientation), self-doubt, and, more recently, a renewed susceptibility to all kinds of fundamentalism.*" (Mittelstraß 2001)

Orientation knowledge is reflective knowledge. It enables orientation about relevant states or processes and provides provisional answers to the questions: What should I do? What am I allowed to do? What should I not (or no longer) do? Cultural knowledge can restrain people from doing everything they can do. The context for grounds and reasons is provided by the context in which something is done and from which something emerges. Here, it is decided whether something is advanced to the advantage or disadvantage of, for example, the quality of a product, system, or process or to the advantage or disadvantage of the particular interests of individuals or the interests of the community. In contrast, **dispositional knowledge** results primarily from scientific knowledge and is brought into the public consciousness by experts, goals, and values. We can orient ourselves to this knowledge by ethical reflection. For Mohr "*the possession of orientations* ... becomes *orientation knowledge as soon as the orientations appear reasonably justified*" (Mohr 1999). Thus, orientation knowledge is not limited to ethical issues but extends to almost all topics and fields of action. Orientation is needed on both a small and a large scale, on the farm level, in agricultural and corporate policy, and among consumers. Because of the complexity associated with the production of food and the lack of clarity that has resulted, not least from specialization, the agricultural sciences also need orientation as to what level of knowledge is already available and what gaps in knowledge exist not only about self-referential but above all to public interests.

7.3.3 Action-Guiding Knowledge

The potential possibilities of access to dispositional knowledge and the orientation towards external or self-imposed standards do not automatically result in a level of knowledge that enables the actors in practice to make well-founded decisions and to implement effective and efficient measures. This is not only due to the exponentially increasing amount of detailed knowledge, which makes it difficult for decision-makers to identify which knowledge modules are relevant for the concrete situation, i.e., for the specific initial and boundary conditions and the solution of problems depending on the problem situation. Knowledge building blocks often represent invariant puzzle pieces, which usually lack connectivity to a system's context and objectives. To transfer the available knowledge via the orientation knowledge into goal-oriented actions, further knowledge is necessary, e.g., about the specific circumstances and the possibilities and limits of implementing the available knowledge. This action-guiding knowledge can be obtained neither from textbooks nor from search engines. Even at universities, one usually searches in vain for action-guiding knowledge.

The term "action-guiding knowledge" describes what someone needs to know to solve a task and behave competently in a specific situation. It is far more than practical experience (intrinsic knowledge) and more than procedural knowledge that characterizes any good craftsman, pianist, or surgeon. It can be described as a relationship between certain conditions and the resulting action. Action-guiding knowledge aims at the planning, decision, and implementation of concrete actions and includes the verifiability of the positive and negative consequences of the implementations. In addition to knowledge of details, which tends to be the domain of specialists, an overview is required to penetrate a given situation in its complexity mentally and to derive working hypotheses from it regarding the probability of changes occurring. Whether knowledge fragments are suitable for utilization is determined above all by whether they are conducive to the inherent function of a system and the achievement of the goal envisaged by the implementation of measures. First of all, however, it is necessary to be clear about the goal to be achieved by implementing measures. The more the goal of making knowledge usable is underpinned, the more precisely suitable knowledge fragments can be sought as a means to an end. Among other things, the following questions arise: Which pieces of the puzzle are still missing, which are more suitable than others, which may be misleading because they prove insufficiently resilient, not connectable, or not implementable? Knowledge for action must prove itself about the targeted achievement of objectives under the given internal structures, resource availabilities, and inter-company framework conditions. Far more than in connection with disposal and orientation knowledge, action-guiding knowledge does not remain without consequences for those who produce it.

If something is to be achieved and brought about, it is not just a matter of changing in the right direction. Because of generally scarce resources, the available means and resources (time, money, etc.) should be used as effectively and efficiently as possible. Whether this succeeds is determined by the availability of means and resources. These must also be checked to see whether they are adequate for the desired changes, i.e., whether they are tailored to the desired changes in quantity and composition (e.g., availability of feed in sufficient quantity and with an appropriate composition of ingredients). In addition, the expenditure of resources to achieve a goal is subject to a marginal utility function (see Sect. 5.2). The ratio of expenditure to benefit is generally significantly more favorable at a low output level than at a high output level. Accordingly, at a high output level, more expenditure must be invested to achieve the same increase in benefit.

Consequently, the cost-benefit ratio can only be estimated if the current level of the **target variable** concerning an external ranking or **benchmarking** is assessed. The classification of the initial level based on an external reference system leads to the question of whether the current level is already considered sufficient (remaining at the status quo), whether it is possibly associated with too much effort (downgrading the level), or whether the desire or even the necessity of an improvement is derived from the classification (increasing the level). The orientation to external standards can also be derived from the significant discrepancy between the actual situation and the target value. Only from the disparity between the actual and the target level, depending on the initial and boundary conditions, can estimates of the extent of the required resources and the specific characteristics be derived. It is in the nature of things that, to implement changes or improvements with effective means efficiently, it is not enough to know the means and their potential effectiveness. It also requires extensive knowledge of the context in which the means are used.

7.3.4 Values, Standards, and Goals in the Sciences

In addition to statements about specific interactions, scientific studies also include recommendations about what should be. For example, if factor A affects B in an undesirable way, the recommendation to avoid A follows. However, what is desirable or undesirable is subject to value setting. Values are what we value and prefer over other options. Therefore, value settings and the associated preferences are private matters for individuals or groups and are always controversial, given the different value settings made by individuals. What each individual considers relevant for themselves and their well-being or that of fellow creatures is subject to personal and, thus, subjective evaluations. Accordingly, the importance that an individual attaches to the welfare of farm animals or environmental protection is also based on their standard of values. Kant already formulated: "*What has a value also has a price.*" The results of surveys carried out in various studies on the willingness of consumers to pay when purchasing products with animal welfare-associated labels reflect the set range of values (Cicia and Colantuoni 2010). However, the survey values are not synonymous with purchasing behavior (see Sect. 6.5). The values that consumers expressed by their willingness to pay additional prices are, however, closely linked to what the individual associates with a label's advertising messages and how trustworthy and credible these are rated. Not only consumers but also livestock farmers have the right to attach a personal value to animal welfare concerns and to compare these with their interests in use, even if their values are not acceptable to the majority.

Recommendations based on values cannot be tested for truthfulness using scientific methods since freedom from values is a scientific-theoretical requirement for statements, according to which their truth should be independent of their normative content. However, many recommendations can be classified as conditional and thus made accessible to scientific scrutiny. Therefore, based on established evidence proving an effect relationship between A and B, the following scientific recommendation is legitimate: B should be considered a function of existing effect relationships if A is to be achieved. This conditional embedding frees scientists, at least superficially, from their value-setting and protects them from possible accusations of acting contrary to the premises of "value-free" science. In addition to well-founded statements on the interrelationships between specific factors, studies in the spectrum of applied sciences always include a classification of the effects within a value system. If, for example, high production performance is valued in an agricultural context, then the scientific discipline of breeding contributes to an increase in value with its methods. From the perspective of another discipline, such as veterinary medicine, a high-performance level of livestock is not classified as valuable but problematic because it can be associated with an increased risk of disease.

As a rule, the recommendations in human medicine prove to be scientifically unproblematic. Here, any studies suitable for reducing patients' pain are regarded as desirable and thus self-evident. The conditionality of such studies is indisputable, so this does not need to be explicitly mentioned without triggering a debate about a potential disregard for the value-free nature of science. However, when the question is raised as to what advantages and disadvantages the avoidance or reduction of pain may be associated with, one quickly comes to trade-offs based on different normative value settings. Accordingly, the premise of "value freedom" proves deceptive even in human medicine.

Unlike value systems, norms are about what is commanded, permitted, or forbidden. Every individual must obey the law regarding systems of norms such as the Basic Law or ordinances. Also, the protection of animals is anchored in legislation **normatively** formulated. The same applies to statutory minimum requirements for agricultural production processes; compliance with minimum requirements is mandatory, and non-compliance can be punished. However, no precise and binding guidelines for dealing with the environment or farm animals can be derived from generally formulated legal texts. Accordingly, there is a great deal of scope for interpretation, which, depending on preconceived values, leads to very different interpretations and, thus, to various actions. While one considers specific measures imperative to act according to the law, others see no reason to do so.

While human dignity is normatively set, i.e., non-negotiable and thus without price, the dignity of animals, although its protection is normatively formulated, is fundamentally negotiable. The state is not unconditionally obligated to **animal protection**, as there are conflicts of objectives with other goods and thus the necessity of weighing up between different public goods or gradations and restrictions. Animal welfare competes with many other goods, such as the pursuit of profit or the individual's freedom of choice. Accordingly, various groups of people assign animal welfare concerns a different ranking on a scale of other goods. The problem of protecting animals from excessive demands is divisible. It is divided into gradations on a scale ranging from a very high degree of realization of welfare to severe illness, which embodies the overstretching of adaptive capacity. The ultimate failure of efforts at self-preservation is sealed with the onset of death. Minimum requirements for husbandry conditions merely protect against an even greater degree of damage, suffering, and pain, which would be expected from non-compliance with the standardized minimum specifications. Consequently, animal welfare contains both normative and value-setting elements. The same applies to other social concerns such as environmental, nature, or climate protection. Consequently, the agricultural sciences operate within different value systems. The respective values cannot be established by the agricultural sciences themselves. However, within a previously defined framework of values, the sciences can make reliable statements about which measures are better suited than others to bring about an improvement (Fraser 2008). One of The key issues in the agricultural sciences is the lack of transparency in the value systems underpinning scientific investigations and the interpretation of research results. This lack of clarity can lead to many controversies and misunderstandings, emphasizing the need for a more transparent and open approach.

Normative elements, which extrinsically apply an evaluation standard to the structure and interpretation of the results of a study, are to be distinguished from those studies that are oriented toward the intrinsic characteristics of an abiotic or biotic system. Intrinsic value settings belong inseparably to the system and make it what it is. Thus, all objects and technological developments humans create have an underlying intention and purpose associated with the creation process. For example, bicycle technology aims to make it easier for a person to cover the distance from A to B and shorten the required time. What is significant about technology is how it works. Technical systems are strictly linked to internal processing rules and process flows: they either work or do not work, or only to a limited extent. If a bicycle can no longer be used for its intended purpose, it is no longer a bicycle but a pile of scrap. In the case of technical devices and developments, the purpose is determined by the technician. Functionality is amenable to examination, the result of which is independent of the persons who are examining it. If something has been proven to work, it requires no further explanation to outsiders as to why it works. One can operate a computer without having the slightest idea of which effective processes are caused by operating the keyboard. The mechanisms of action of technical processes are insignificant to users as long as they work. In addition, technology can usually be used and applied for various purposes. The use of technology is the responsibility of the user.

Living systems are also designed to function so that the goal of all living systems, to maintain themselves, can be ensured over as long a period as possible (**autopoiesis**) (Maturana and Varela 1987). The shape (morphology) of the living systems and the coordinated functions of the sub-systems are intrinsically oriented towards this goal. From a scientific point of view, it can be checked whether the interactions between the subsystems, for example, organs, tissues, and cell assemblies in an animal organism, function and are conducive to the purpose of self-preservation. While technology generally functions independently of the place of use (context-independent), the chances of survival of living systems are highly dependent on the living conditions and, thus, on the context. Living systems are open systems that absorb substances from the environment and release substances into it. Accordingly, survival depends not only on the functionality of the internal structures but also on the success of the interactions between the organism and the encountered living conditions. Furthermore, the survivability of living organisms is influenced by their potential to adapt to the respective changes in living conditions. By designing the living conditions under which livestock are kept in accordance with the specific and individual needs of the livestock, livestock keepers significantly influence the survivability of the animals or their ability to lead a largely undisturbed existence. In scientific studies, it is possible to examine the interactions between living organisms and their living conditions to determine whether they are beneficial or detrimental to the organisms' ability to survive. This verification is possible without the outcome being dependent on the subjective and normative premises of the scientists investigating the interactions. From the intrinsic value of living systems, in which the ability to survive is given the highest priority, it can be concluded that only the functional explanatory approaches of biology are suitable for gaining substantial knowledge about the value of self-preservation.

7.4 Agricultural Sciences as Applied Sciences

Agricultural sciences are the mother of all sciences (Mayer and Mayer 1974). In no other field was so early begun to generate knowledge and acquire **knowledge**. Driven by an intrinsic motivation, the aim was to understand what happens in nature and how it can be possible to make plants and animals usable to improve one's survival ability. The acquisition and availability of knowledge about the processes in nature were essential for survival for individuals and the group and, thus, the starting point of all that was considered worth knowing. Food was and still is "the daily bread" for which Christians pray throughout the centuries and to this day. This is true even if in the countries of the Western world, the universal availability of **aliments** has contributed decisively to the fact that the value of food has taken a back seat to other goods. However, oil, natural gas, or, more recently, renewable energy sources are not the fuel of life. For humans, it is and remains the negative **entropy** (Schrödinger 1992), which provides humans with the energy to live and the nutrients for their vital metabolism.

To gain knowledge in the context of food production, both empiricism and rationalism are needed. Which approach is adequate is not so much a question of principle as the result of a pragmatic search for an appropriate intersection between the two. Furthermore, natural science and socio-economic methods are essential, as modern agricultural research is concerned with elucidating and practically implementing scientific, economic, and ecological processes with the aim of qualitative and quantitative optimization of plant and animal production technology and socio-economic processes. Since the agricultural approach to nature, or to what remains of it, requires epistemological reflection about the values that are effective but not revealed, disciplines in the humanities should also be involved in the cognitive processes. At present, however, this is still a long way off.

7.4.1 Agricultural Sciences as an Isolated Branch of Science

The subject areas of agricultural sciences are highly diverse. They are located in the microcosm, mesocosm, and macrocosm and include both the abiotic and biotic spheres. Research activities are located in all four environmental spheres and use the spectrum of available scientific methods. The overarching goal is to generate valuable knowledge in the agricultural context. This mainly applies to more efficient and less costly food production. Furthermore, numerous additional benefits are produced that are of more or less interest to the various stakeholders involved. Besides the raw commodities produced for food being the positive externalities of agroecosystems, they also create various negative externalities. Animal science has historically accompanied the development of agricultural production by providing adequate knowledge to foster the intensification processes and to hedge the proliferation of adverse side effects; the first with outstanding, the latter only with limited success.

Since its beginnings, the agricultural sciences have formed their own *scientific community* and have remained so to this day (DFG 2005). They have created their scientific institutions and organizations, their target group, their own communication and publication channels, and, with the establishment of comprehensive advisory services, their public. At the same time, they operate within an independent political **system**. Politicians have their own federal and state ministries, advisory and administrative institutions, and departmental research, which is subordinate to the Federal Ministry of Agriculture and is endowed with extensive financial resources. On the one hand, the focus on the object of knowledge has led to a thoroughly strong science; the flip side is isolation and compartmentalization vis-à-vis other scientific disciplines.

The ambivalence resulting from this autonomy has already been described from an American perspective by Mayer and Mayer (1974) in a paper entitled "*Agriculture, the Island Empire*". They complain above all about the lack of communication and **cooperation** between the disciplines, which results in the relevant findings of one discipline being given too little attention in the other. According to the authors, the isolation from external criticism and a one-sided focus on performance characteristics have paved the way for many undesirable developments:

> *"There is a serious lack of scientific critics from outside looking at agriculture in an informed and constructive way. The lack of criticism of agricultural policy has been no less serious in the political and economic spheres. In Congress, only politicians identified with a farming interest have been willing to serve on the Agriculture Committees and sub-committees. This self-selection has tended to foster large-scale government programs designed to benefit narrow classes of producers without regard for consumers or even an overall production policy." (...) The development of intelligent and informed criticism will not come easily. In the past, criticism has come mainly from non-academic (...). Universities have largely failed at their task of encouraging dialogue among disciplines. The great universities with large agricultural schools have failed most strikingly."*

In describing the isolation of agricultural sciences from other disciplines as a tragedy, the authors leave no doubt about the extent of the consequential effects. As a result of this isolation, the agricultural sciences have not only lost touch with valuable models and systems approaches in other scientific disciplines, but they pose a tremendous potential threat to people and the world. Whether fundamental changes have occurred that make the criticisms made at the time obsolete today must be doubted. The lack of communication and cooperation in the agricultural sciences not only characterizes the relationship with other sciences. Also, communication and collaboration are poor among the multiple disciplines within the agricultural sciences. The general model for agricultural science is that disciplinary groupings set the logic and standards for research; the disciplinary sciences are set in a hierarchical arrangement, which should allow communication from the relevant basic sciences through applied research into technology development, use, and problem-solving. However, what seems reasonable has been subject to evaluations and criticisms about disciplinary narrowing in agricultural research (Kunkel 1988).

To communicate well with each other, a common language is needed, i.e., the same understanding of the meaning of central terms and a common objective of what is to be achieved by the exchange of knowledge and findings. The findings and results generated in the individual disciplines should also be compatible, i.e., they should have the same frame of reference. Simply belonging to the broad area of agricultural sciences is insufficient for this. The lack of clarity in using central terms can be observed in many subject areas of agricultural science. Uncertainties in the use of terms are not only because the agricultural sciences include both natural science and social and economic science disciplines, which adopt different perspectives. In addition, different interests overlap the uses and interpretations of terms. Ambiguous definitions of terms complicate communication between agricultural disciplines and discourse with other branches of science. This deprives interdisciplinary cooperation of its basis. Furthermore, the possibility of a valid assessment of facts is undermined if different groups interpret the same term differently. On the other hand, nebulous terms provide marketing with a wide range of possible uses. They offer an ideal projection surface for the self-referential associations of many interest groups. In this way, many people can come to terms despite differing and even contradicting interests without giving up previous positions (see also Sect. 6.4).

7.4.2 Agricultural Science Lost in Details and Over-Generalization

With the generation of knowledge, the various disciplines of agricultural science have substantially contributed to the increase in productivity within agricultural production. The enormous growth in scientific activities has inevitably led to a fragmentation of scientific knowledge. By answering key research questions within a discipline, new questions arise and generate a self-sustaining momentum, not unlike the **path dependency** in the farm business. The path of specialization in agricultural practice determined the various directions scientists felt compelled to follow in their search for topics to raise their profile. Differentiating disciplines and specialization of research areas are still considered the best recipe for scientific success.

Scientists must consider their career rewards when conducting research work. A scientific career primarily depends on publications in scientific journals. According to the guide for authors of many scientific journals, studies may be considered for publication only if the results are likely to be of international interest, i.e., it must be possible to generalize the findings using scientifically based approaches. This does not apply to all scientific journals, but generally, the journals that make such demands tend to have higher Impact Factors and thus are of higher attractiveness for scientists. As scientists compete for research funding and comparatively few permanent positions, promising scientists only have a chance to pursue a career when they carefully demonstrate their publishing skills. The general framework of research work affects the areas where scientists are looking for gaps of knowledge suited to generate generalizable publishable results.

The path of specialization promised not only economic but also publication-rich increases in productivity. Specialization also served the quest for visibility and distinction and made it possible to acquire a specific profile in the academic world. As in other branches of science, the incentive structures in agricultural sciences were also geared towards generating an individual reputation, which manifested itself in a general striving for "excellence." Proof of notable achievements is required to stand out and distinguish oneself from the competition. This will most likely be achieved through specialization and a one-sided methodological orientation. Once a path of knowledge acquisition has been taken, it is rarely abandoned. According to Isermeyer et al. (2002), the consequences are threatening:

> *"If, as a result, the principle of "have method, seek problem" becomes the guiding principle for researchers, interdisciplinary, problem-solving agricultural research is in danger of falling by the wayside. And if more and more management positions are filled by people who follow this principle, then agricultural research institutions will gradually decompose from within."*

Knowledge about tools, causes, and effects created under standardized conditions can be classified as dispositional knowledge, lacking orientation and action-guiding knowledge (see Sect. 7.3.1). It provides puzzle pieces in search of gaps within a large mosaic picture where it might fit. To be generalizable, the shape of dispositional knowledge should be rather unspecific. In contrast, envisaged effects can be expected when the knowledge input fits a high degree to the needs of a specific dysfunctional situation.

Generalizable results must satisfy the significant scientific criteria of internal validity. For example, improper randomization or missing data can undermine the validity of the results and conclusions of randomized controlled trials. Correspondingly, scientists are well recommended to control the experimental conditions and to exclude confounding factors wherever possible. Whereas study results might appear generalizable from a disciplinary perspective, they can appear highly context-specific concerning external relevance. Control of the environment is a cause of poor reproducibility of experimental outcomes. In addition, the variety of the interactions between influencing factors is subject to dynamic changes. In contrast, the combination of the respective factors affects the subjects of interest not separately but jointly. No less important is that highly standardized conditions that enable a high internal validity of the results can differ significantly from what is found on commercial farms, thus creating a low external and ecological validity. External validity depends on whether the study findings can be generalized to other contexts. Ecological validity refers to the degree to which research findings reflect real-life related functions and examine how closely an experiment aligns with real-world phenomena (Fahmie et al. 2023).

The focus on disciplinary research programs means that large parts of real-world complexity, particularly the complexity of living systems, are disregarded, associated with a vague expectation that other disciplines cover these. Differentiation and specialization result in perspectival knowledge asymmetries (Alrøe and Noe 2011). Scientists have their discipline-derived perspectives and see complex matters differently. Experts perceive knowledge gaps from within their professional paradigms and are widely incapable of forming judgments beyond their specific expertise (Millgram 2015). Each discipline follows its own goals and problem framing, encompassing the definition of system boundaries, the role of actors, and the selection of design tasks. They use different methodologies and refer to other reference systems, such as species, housing systems, or production methods. Thus, scientists are self-referential when choosing a specific perspective on agricultural issues or applying specific methodological tools they have taken ownership of in their specialist discipline. While this mono-disciplinary focus may provide a deeper insight into the interconnectedness of processes within sub-areas, it can readily become too narrow. This particularly applies to problems that develop beyond the boundaries of disciplinary specificities, which is the case for most of the problems in living systems, which are generally of multifactorial genesis and need a multi-disciplinary way of thinking. Excessive specialization can lead to the whole picture becoming increasingly lost out of sight, leading to a decline in the skills to solve complex problems (DFG 2005). As illustrated in Fig. 7.1, the general framework of research encourages scientists to look for gaps of knowledge, preferential in the micro- and macro-cosmos, suited to generate generalizable and, hence, publishable results.

Within living systems, the different parts and subsystems are not separate. They interact together towards the same goal: to keep the system alive. Each part needs the support and cooperation of nearly all other parts. A living system cannot be fully understood without considering the purpose behind the biological processes. This applies particularly

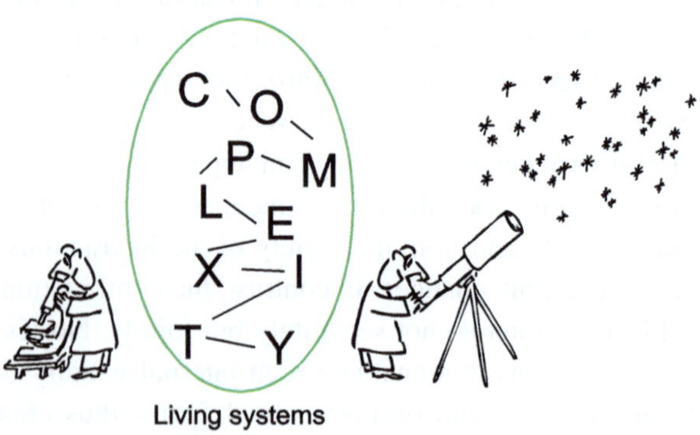

Fig. 7.1 Agricultural scientists focusing on questions in the micro- and macro-cosmos in striving for general and context-independent statements while widely disregarding the context-sensitive complexity of living systems (own illustration)

to adaptation as a functional and target-oriented process. Whether the whole living system responds adequately to challenges cannot be deduced from single parts or indicators. Whether adaptive measures and methods are beneficial depends also on the context in which they occur. The piecemeal approach of agricultural science blinds actors to the possible larger conflict constellation within farm systems. Controlled research trials require a focus on the experimental treatments while widely excluding possible confounding factors. Striving for generalizable and publishable results widely disregards the inherent goals of organisms and the inner logic of agroecosystems. Whereas in experimental studies, scientists look for the most possible narrowing of the variation, variation can be extended in farm practice. What is statistically valid for a group of farm animals under highly standardized conditions does not necessarily work for individual animals, deviating from what might be supportive when assessed regarding means values.

Solutions to dysfunctions and problems in farm practice do not follow a one-size-fits-all approach. Context-sensitive dysfunction and problems can only be solved when the underlying causes of the multifactorial genesis of problems are understood. Thus, contributing to their mitigation requires an integrative and, hence, inter- and transdisciplinary solution approach. However, inter- and transdisciplinary thinking are often not attractive for agricultural scientists. Furthermore, scientists are not well rewarded for translating their results for end-users. What is target-oriented when striving for a scientific career is not necessarily suited for solving problems that emerge from highly context-dependent dysfunctions. Thus, there is a fundamental trade-off between what scientists are striving for and what is required to grasp the complexity of processes within the farm systems.

As scientific findings always have some generality and specificity, it is essential to know more about the ambiguity and context dependence of methods and measures when implementing solution proposals in farm practice. Consequently, there is an urgent need to identify and assess the individual starting and boundary conditions represented by the current living conditions of plants and farm animals (see Fig. 7.2). The interconnectedness of starting and boundary conditions with the physiological laws and regularities results in individual circumstances responsible for a specific constellation of problems. To solve such problems, the respective constellation has to be analyzed for the most reasonable causes behind the problems within the farm-specific context. Identifying possible causes and the likely measures to mitigate them function as a hypothesis for a context-specific problem-solving approach. There is only a chance to externally validate the measures about their effectiveness when these are implemented and assessed for their success. In the face of the generally multifactorial genesis of problems, this procedure has to be repeated in an iterative approach until the desired outcome is achieved.

The scientific challenge lies in balancing the need to do great science with the need for impact delivery at scale. When agricultural scientists experience no personal benefits when focusing on context-variant problems, they cannot be expected to sacrifice their scientific careers (Sundrum 2024). In their current state, agricultural sciences appear more as auxiliary sciences of agribusiness than as service providers of public welfare

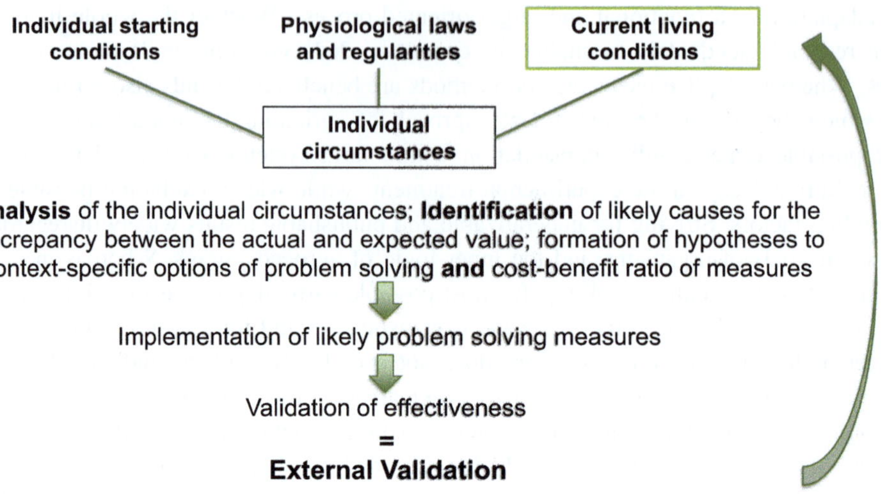

Fig. 7.2 Pathway for an iterative approach solving dysfunctions and problems that have emerged from individual circumstances (own illustration)

interests. Rather than continuing to hide behind the indefensible intellectual and class barriers that have been retained to protect their privilege and ego from contamination by undesired side effects, agricultural scientists have to contribute to public welfare interests in the first place. Finally, it is public money that enables agricultural scientists to follow their scientific interests and simultaneously make a living.

7.4.3 Agricultural Sciences in the Service of the Agricultural Economy

In the face of the food shortage after the Second World War and the persistent shortage in developing countries, agricultural science has been designed as a service provider for the agri-food sector, financed highly by agricultural policy as the primary contracting authority. Farms are generally too small to conduct their research. Further, agricultural growing conditions are highly diverse, so agricultural technologies must adapt to local needs. Public investment in agricultural research and development (R&D) has been a significant driver of productivity growth in high-income countries to ensure research performance in those areas of technology development where private incentives are inadequate to generate an economically or socially optimum level of research. The agricultural research community has taken considerable pride in its contribution to the remarkable economic performance of the agricultural sector over the last century. In livestock production, for instance, from 1957 to 2005, broiler growth rates increased by over 400%, with a concurrent 50% reduction in feed conversion ratio (Zuidhof et al. 2014). In dairy production, the annual milk yield per cow has increased over fourfold in the last 75 years, with no evidence of a plateau (Baumgard et al. 2017). The successful transition from resource-dependent to productivity-led agricultural growth during the latter half of the twentieth century largely stemmed from investments in agricultural R&D and the application of industrial inputs. Between 1961 and 2014, agricultural output in high-income countries increased by 98%, while total inputs declined by 14% (Heisey and Fuglie 2018). Aggregate total inputs grew slowly until the late 1970s and have declined ever since. The mix of inputs also changed, with capital and material inputs substituting labor and land. Overall, the growth in agricultural output and the decline in inputs imply that total factor productivity, the total productivity of the land, labor, capital, and material inputs employed in production, more than doubled over this period.

Parallel to other branches of the economy, the pressure of the market economy to increase labor productivity by introducing new techniques also increased in the agricultural economy. With a considerable expansion of financial resources, which flowed into agricultural research, agricultural science, technology, and exploitation grew into one complex **system**. For a long time, agricultural research was closely linked to government-commissioned research, which primarily focused on scientific and technical progress to increase productivity. In this way, agricultural research activities themselves became a decisive productive force. With their concepts and methods, the agricultural sciences have created a realm closely linked to biotic and abiotic resource utilization (exploitation) through the inherent logic of scientific and technical progress. The rules and characteristics of progress seem to produce the constraints that an agricultural policy obeying social needs must follow. However, agricultural policy has avoided openly stating which stakeholders are the beneficiaries of this procedure and at whose expense. As in other sectors of the economy, the productive forces brought to bear by scientific and technological developments have become the basis of legitimacy in the agricultural economy. According to the philosopher Habermas (1969), the scientific method provided both the concepts and the instruments for the ever-more-effective domination of man over nature. At the same time, it promotes subjugation to an industrial complex that increases the productivity of production processes and expands the convenience of human life.

In public proclamations, producing a sufficient quantity of **high-quality** food at low prices is claimed to be the overriding goal and the basis for legitimacy. However, the general and non-binding formulation of the objective leaves considerable room for interpretation. For instance, it leaves open, among other things, when the production quantities produced can be considered sufficient, how high quality differs from low quality, and whether low prices are a goal even if they lead to non-cost-covering prices for primary producers and high external costs. On the other hand, hardly any scientific publication dealing with the production of agricultural products believes it can get by without pointing out that the envisaged productivity increases will contribute to supplying the world's population. The global supply target has to serve as a justification for a scientific knowledge process that needs explanation.

Since the gap between what happens on individual farms regarding production outputs and their contribution to feeding the world's population is immeasurably large, the question arises as to what one has to do with the other. On the one hand, it is a matter of details that are pursued meticulously, and the results are backed up with various statistical methods. On the other hand, in classifying the results in the superordinate **context**, an Eldorado of speculative explanatory contexts opens up, which do not stand up to scientific scrutiny. While the results are generated as precisely as possible, the classification in the superordinate context is purely associative and speculative. Whether the individual results are suitable for contributing to feeding the world population or possibly even run counter to this goal, as suggested by the explanations in Sect. 5.2.3, is, however, as a rule not subject to any further reflection and certainly not to any examination concerning the validity of the justification. This is true even if land utilization for biomass and bioenergy production has created a further competitor from within its ranks that poses new problems for the primary production of food. In the meantime, the production of biomass and bioenergy has not only taken on the role of a high-income branch of farming on many farms but has also become very important in economic terms (Lanzerath et al. 2022).

Generally speaking, the agricultural sciences utilize biotic and abiotic resources to generate added value for the rural economy. They serve the rural economy with their methodological tools, with which the utilization of resources can be made more productive, efficient, and cost-effective. In this sense, scientific knowledge, analogous to technical, nutritional, or labor investments, is a means of production for increasing productivity. In contrast, only a few studies reflect on the meaningfulness of the prevailing forms of resource use in society and concerning the pursuit of public welfare interests.

The USA is the most crucial country in publication productivity, with 22.5% of the total publications and thirty-five of the 100 most productive research centers. However, the spectacular growth in research carried out in emerging countries, such as Brazil, China, and India, challenges this leadership. In Europe, production levels have experienced a moderate increase, with high levels of indicators of impact and collaboration standing out (Cañas-Guerrero et al. 2013). Agricultural sciences address the theme of ecosystem services via their traditional issues: yields/growth/soil/plants. Thus, agricultural sciences deal with ecosystem services biophysically and do not integrate the agro-socio-ecosystem as a whole. They pay little or no attention to the issue of the pollution (and restoration) of ecosystems, even though agriculture is one of the main activities responsible for this pollution. Agriculture is the world's single largest driver of global environmental change and, at the same time, is most affected by these changes (Field et al. 2014).

While emphasizing that food security is one of the critical global challenges, it is increasingly acknowledged that there is a pressing need for the 'sustainable intensification of global agriculture' (SIA), in which yields are increased without adverse environmental impact and more land is cultivated (Royal Society 2009):

> *"The governance of innovation for agriculture needs to maximise opportunities for increasing production, while at the same time protecting societies, economies and the environment*

from adverse side effects. Regulatory systems need to improve their assessment of benefits. Horizon scanning will ensure proactive consideration of technological options by governments. Assessment of benefits, risks and uncertainties should be seen broadly, and should include the wider impacts of new technologies and practices on economies and societies. Public and stakeholder dialogue—with NGOs, scientists, and farmers in particular—needs to be a part of all governance frameworks."

To seek to increase agricultural output while keeping the ecological footprint as small as possible is, no doubt, a practical and relatively important feature of sustainable agriculture. However, mainstream agriculture still concentrates on productivity and places limited focus on sustainability. It remains focused on avenues for resource efficiency, e.g., based on assumptions that efficiency in water and fertilizer use represents the avenues towards sustainable agriculture. The ideas of sustainable intensification are either not concrete enough or only partial. Particularly in agricultural development in poverty-stricken regions, this "productivity first" paradigm prevails while potentially reducing environmental impacts. Thus, it requires more than proclaiming new goals. SIA requires a radical refocusing of food production that encapsulates the twofold aims of increasing yields and the ecosystem services agriculture provides (Godfray and Garnett 2014). A paradigm shift towards sustainable intensification of agriculture has to integrate the dual and interdependent goals of using sustainable practices to meet rising human needs while contributing to the resilience and sustainability of landscapes, the biosphere, and the Earth system. Both of these, in turn, are required to sustain the future viability of agriculture. This shift demands a new framework for research and development to advance fresh, integrated whole-of-systems approaches for sustainable intensification (Rockström et al. 2017).

However, a quantitative and qualitative analysis of the scientific literature on ecosystem services revealed how little the ecological origins of the sustainability concept have been appropriated by agricultural sciences until now, despite the significant focus on the issue of agroecosystems in the literature (Tancoigne et al. 2014). Agricultural sciences tend to be more active in environmental services, defined as services rendered by humans to ecosystems. In contrast, less attention is paid to the antagonisms between different services. Generally accepted methods for linking research outcomes to sustainability impacts are missing. A scientific literature analysis of papers published between 2008 and 2016 revealed that most papers (56%) addressed economic impacts, such as the cost-effectiveness of research funding or macroeconomic effects (Weißhuhn et al. 2018). In total, 42% analyzed social impacts, like food security or aspects of equity. Very few papers (2%) examined environmental impacts, such as climate effects or ecosystem change. The authors found that most papers assess research impacts at the level of technologies, particularly for economic impacts. The most striking finding was the 'blind eye' towards environmental and sustainability implications in research impact assessments. The assessment of the environmental impacts of agricultural research or comprehensive sustainability assessments was exceptionally limited. They may have been systematically overlooked in the past for the reason of expected negative results, thought to be covered by other impact studies or methodological challenges.

7.4.4 Comparison of Natural, Technical and Agricultural Sciences

Students are trained to become agrarian engineers at the agricultural science faculties of universities and universities of applied sciences. Generally, the prevailing mindset is that engineering is a form of applying knowledge from the natural sciences in practice. The narrowing of natural sciences and engineering or technical sciences is—as Poser (2016) understandably deduced—historically and systematically misleading. The same applies to a reduction of agricultural sciences as engineering science. Public perception of the natural, engineering, and agricultural sciences is often misconceived and equated concerning the scientific approach, results, and utilization. Therefore, the scientific disciplines are compared and distinguished in the following.

The natural sciences are concerned with what takes place independently of human influence regarding cause-effect relationships between various influencing factors and are subject to specific natural laws. Natural scientists strive to grasp **reality** as it is; they want to understand it better and look for generally valid laws and cause-effect relationships. Even if there are assumptions about the outcome of scientific experiments, they are carried out theory-guided and with an open mind. The goal is to test a possible generalization of a scientific statement and to generate a law hypothesis. However, the natural sciences proceed **reductionistic** by explaining complex natural phenomena through the interaction of simple subsystems. This approach accounts for their success on the one hand and their limited access to the complexity of reality on the other.

While the **goals** of the natural sciences generally arise from questions and problems intrinsic to science, the technical sciences are asked questions to solve problems. The same applies to investigations in the agricultural sciences, at least to some degree. Technical and agricultural sciences presuppose a standardization or a setting of values about what is classified as a problem and in which direction a solution should be sought. Technicians and agricultural engineers change nature or what is left of it in the Anthropocene (Steffen et al. 2007). The central tasks include utilizing available resources and developing solution strategies for related problems. Above all, technical progress strives for increased efficiency in using and implementing resources and better solutions, primarily within locally limited contexts of action. In addition, artifacts are created, i.e., artificial modifications or alterations of natural conditions. Artifacts are not limited to technical objects but can also involve changes in living organisms or the chemical synthesis of building blocks (e.g., amino acids) or concern the production of artificial **meat**.

The agricultural sciences have a problem- or goal-oriented approach in common with the technical sciences. This distinguishes both from the natural sciences. In contrast to the technical sciences, the agricultural sciences are concerned not only with the inanimate matter but, above all, with processes in which living organisms are involved and in which living organisms interact with other living organisms and with inanimate matter. Therefore, many procedural rules in agricultural sciences refer to biological processes. Since, in addition to biology, related scientific fields such as medicine, biomedicine, pharmacy, molecular biology, biophysics, nutritional sciences, and food research up to bioeconomics also deal with living organisms, the term "life sciences" has become common in many scientific institutions. Correspondingly, various disciplines in the agricultural sciences are also classified as life sciences. On the other hand, the former clear distinction between basic research and application- and market-oriented research has partially dissolved, or there are fluid transitions. For the application-oriented areas of the agricultural sciences, the study of procedural rules describes functions to be classified primarily in connection with categories of ends and means. Purposes presuppose norms and values, which are pretty out of place in natural sciences.

Assessments of the knowledge gained from technical and agricultural sciences require a deeper insight into the fundamental processes of research activity. After all, it is only from an understanding of the processes that assessments of the informative value of the results can be derived. In this context, empirical research has contributed significantly to generating fundamental knowledge to initiate goal-oriented changes. The empirical sciences have used empirical **induction** as the basis of knowledge, i.e., the gain of knowledge results from the experience of specific facts, which are subjected to a certain generalization. Induction works particularly well for physical and technical processes that can be expressed in laws. In contrast, there are considerable doubts about whether induction works for biological and physiological processes, mainly when ambivalent underlying structures exist (Chalmers 1999).

The process of gaining knowledge is based on four sub-steps (Schlageter 2013):

1. Formulation of a **hypothesis**
2. Experimental testing of the hypothesis
3. Transfer of the hypothesis into practice under changed conditions
4. Renewed experimental verification (**external validation**

A hypothesis regarding a presumed causal relationship is developed based on concrete observations based on previous knowledge and experience. If an experiment confirms the effect relationship, it can be assumed that this relationship will also occur under modified conditions. If further investigations under similar initial and boundary conditions confirm the identified relationships, this leads to a generalizable statement. Statements or rules are accepted as valid if the system can correctly predict all future examples using this rule (Mohr 1999). However, experiments are not independent of time and place and of the interactions that occur under specific initial and boundary conditions. At the same time, if the experiment is to meet fundamental requirements, it must be protected from allowing ambiguous interpretations about the causes of the results obtained. Therefore, the ***ceteris paribus*** **assumptions** under largely standardized conditions are fundamental. This isolation principle of the *ceteris paribus* clause endeavors to capture those effects caused by changes in a single or very few factors by fixing all influencing factors as far as possible. The effects can then be causally attributed to the changed factors. Through further experiments, the boundary conditions of the experiment are changed and compared with the previous results. From the immeasurable variety of possible initial and boundary conditions, we can never represent the complete reality in experimental designs.

With each new experiment, we can continuously improve our understanding of reality. However, it is often overlooked that the results of an experiment are, strictly speaking, only valid for the conditions under which the experiment was conducted. This context dependency considerably limits the general validity of experiments. It is, therefore, all the more critical that the hypotheses tested in an experiment are tested again in practice under widely modified boundary conditions, i.e., validated externally, before a generalizable statement can be made. This statement is valid until it is altered or disproved by newer findings caused by further experiments. Experimental knowledge based on the first two steps corresponds to a **half-knowledge** that has not yet been substantiated but relies only on the internal validity of the study design and its realization. Whether the study findings can be generalized to other contexts remains an open question. The same applies to the degree to which research findings reflect real-life related functions and how closely the experiment aligns with real-world phenomena and possesses ecological validity.

In the technical sciences, it is less a matter of the validity of generalizable statements but rather a matter of the effectiveness of the **means** to be used in the sense of the effects aimed at with the technical possibilities. The findings should lead to improved solutions for specific problems, i.e., the technical means and procedural rules used should prove their worth in application (best practice). The methodological tools of modeling and testing are available to the technical sciences for verification purposes. Tests require initial and boundary conditions that are as realistic as possible. Unlike in the empirical sciences, they do not aim at hypotheses about regularities but are based on a functional expectation formulated as clearly and unambiguously as possible. It is determined whether the underlying rules of procedure and the means employed prove successful or not concerning the intended purpose (Schlageter 2013). The starting point of technical developments is situation A, understood and classified as unsatisfactory, and situation B understood and classified as more satisfactory or improved. Furthermore, an essential task of the technical sciences is to avoid undesirable side effects and late effects in their canon of rules and procedures from the outset. However, the problem is formulating criteria for the economic, social, psychological, and ecological conditions that technologies must fulfill in addition to their technical efficiency.

> *"Modern theories of technology must be multi-level theories; otherwise, technology assessment would be a farce. The technical sciences, based on their models extended to include the normative side, are precisely tasked with formulating tried-and-tested rules of procedure, issuing warnings, and demanding justifications where they fear dangers: Technical design is always future-oriented; therefore, the task of tackling existential problems arises for the technical sciences in interdisciplinary cooperation"* (Poser 2016).

What Poser formulates here as the maxim of the technical sciences has no equivalent in the agricultural sciences. Neither is the avoidance of undesirable side-effects and harmful effects, nor the associated long-term consequences, part of the original canon of research and teaching, nor is interdisciplinary cooperation with specialist disciplines within and outside the agricultural sciences sought and organized to deal with highly relevant social challenges and **problems**. The different perspectives and reference systems with which various disciplines view and analyze conditions in agriculture converge primarily to promote agricultural goals and improvements.

To implement improvements, a planned, systematic procedure for achieving the envisaged goal is indicated. At the beginning of developing improvements, solutions to problems, or partial solutions about a superordinate goal, there should be a well-founded problem analysis. In the technical sciences, problem analysis is primarily carried out in the immediate context of problem creation. It is based on the limited functionality of devices or processes. Thus, the proposed solutions are directly related to the results of the problem analysis. In addition, further tests are carried out to assess whether the proposed solutions are associated with a solution to the problem or an approximation to the set goal and whether the necessary use of resources is efficient. A technical solution must prove itself by restoring or improving the previously criticized functionality; if this is not the case, there is no solution and, thus, no progress in knowledge.

In agriculture, the processes are much more complex. The production outputs and the undesired side effects result from multifactorial interactions involving numerous abiotic and biotic components. In contrast to the technical sciences, the processes and their outcomes are not determined by unambiguous laws but are highly variable. While technology functions largely independently of the context, processes in agrobiology are influenced to a large extent by the initial and boundary conditions, which differ considerably from farm to farm. To make matters worse, it is seldom apparent which factors are involved in producing phenomena and to what extent. This is partly because many processes and interactions occur in areas that are not immediately visible, i.e., inside soils, plants, and animals. The dominant way of dealing with questions in agricultural science is through an experimental design. As far as possible, experiments are carried out under standardized experimental conditions and ***ceteris paribus*** **assumptions**. Suppose all potential influencing factors are essentially the same, and only one or two factors are designed differently. In this case, the result can be attributed to the changed factors compared to the control variant. Such experiments can only be carried out to a limited extent on farms. Still, they require specific conditions most likely present on experimental farms designed for this purpose. Since not all problems can be fitted into an experimental design, many practical problems remain scientifically unaddressed.

Due to the complexity of the agricultural impact structure, problem analysis rarely takes place in the concrete context of the specific problem development on a farm. Instead, a problem or question from agricultural practice is translated into a research question amenable to scientific treatment. Because of the efforts, the question should be relevant in the specific context of one farm and for many farms. If one were to scientifically accompany the problem analysis and the problem solution, including proof of effect in the company-specific context of the emergence of the problem, this would correspond to an individual case study. The advantage is that a concrete problem that has arisen in a specific context can be researched until a solution or partial solution is found. In addition to the immense effort that can be involved, the disadvantage from a scientific perspective is that the results of single-case studies can only be transferred to other operating systems to a limited extent, given the heterogeneity of the initial and boundary conditions. They, therefore, have a low degree of generalizability. Accordingly, the publication of single-case studies can hardly earn merits in the scientific community. In addition, only those case studies are published—if at all—that has a solution to the problem. In contrast, the failed attempts to solve a problem and the reasons for the failure usually remain hidden.

Consequently, agricultural research focuses not on solving farm-specific problems but on generating **dispositional knowledge** (see Sect. 7.3.1). This knowledge is made available to agricultural practice and advisory institutions in the search for solutions to individual farm problems and the pursuit of specific production goals. In this way, scientific findings are translated back into practice, where the questions came from. On the one hand, the results generated under standardized experimental conditions promise a higher degree of transferability to problem constellations under different initial and boundary conditions. Accordingly, the way to a potential problem solution is marked out by standardizing the experimental conditions. Thus, the scientific results are also attractive to other farm systems with similar problems. On the other hand, every investigated aspect is embedded in an overall system. The separation by experimenting under *ceteris paribus* conditions does not allow a back integration into the whole or integration into a system with different initial and boundary conditions without checking whether the partial aspect also unfolds the predicted effect. At this point, an elementary weakness of agricultural science is revealed: Without the implementation of **external validation**, scientific results are not scientifically sound.

Whether the transfer of scientific findings (e.g., via advisory recommendations) and the measures implemented based on the recommendations in agricultural practice also lead to success and efficient achievement of objectives or problem-solving is generally lacking scientifically substantiated proof. The extent to which a problem and weak point analysis are carried out in an operating system, which production goals are aimed for in the system, which external expertise and which level of knowledge is consulted, which measures are implemented, and in which way the effect of the implementations is checked, is left to the respective responsible persons at the farm level. The only exception is the quantitative recording of production performance, which is widely established in agricultural practice. In contrast to the practice in empirical and technical sciences, external validation of hypothetical assumptions (i.e., verification of effectiveness in agricultural practice) that meet scientific requirements is seldom conducted in agricultural sciences about the overall impacts of agricultural production processes. Thus, there is hardly any dispositional knowledge in the true sense regarding the positive and negative effects, let alone a balance between both. Consequently, experimental knowledge without external validation is merely "**half-knowledge of hypothetical character**": There is the hypothetical assumption that the implementation of specific measures experimentally tested under specific initial and boundary conditions in another operating system with a similarly positioned problem and with similar initial and boundary conditions leads with a certain probability to a similar result or a similar problem solution.

The complex interactions within an agricultural operating system pose various challenges for the sciences. Due to the high variability on the side of biotic and abiotic factors as well as the manifold interactions between them, which are furthermore subject to dynamic changes, it is not possible to conclude from partial aspects of an operating system inductively on superordinate system properties or the overall result of the interactions. At best, estimates of the probabilities of the occurrence of phenomena can be made. Whether or not experimentally tested measures prove successful when applied in a farm-specific context and associated interests would require individual farm testing using scientific methods. Since there is hardly any external validation in agricultural practice, the question of whether knowledge generated by agricultural science proves its worth when applied under specific initial and boundary conditions is left mainly to the self-referential assessment of the actors in practice.

The information that guides decisions and practical implementations is primarily driven by specific purposes, often leading to the disregard or ignorance of what is deemed useless. The success of companies in utilizing information and external expertise to ensure their economic viability initially remains a private matter. However, when pursuing private interests, these changes lead to undesired and harmful side effects, compromising the common good. Yet, as long as these effects don't impede the objectives of individual farms and society doesn't impose concrete demands or sanctions on primary producers, the polluters have little cause for concern. This is especially true if reducing harmful effects is associated with decreased performance or additional costs that counter business interests.

The fact that the agricultural industry has in the past paid so little attention to the undesirable and adverse effects that emanate from production processes is also because agricultural science has so far considered itself only rudimentarily responsible for assessing the consequential impacts of production processes, at least as long as the consequential effects do not jeopardize the original agricultural objectives. This means, among other things, that there is a lack of overview data with which the extent of the impairments could be classified and assigned to specific areas. The agricultural economic dictum that production processes must be primarily geared to increasing productivity to ensure the farm's financial viability under market economy conditions is seldom questioned in agricultural science. An impact assessment beyond the immediate on-farm effects, equivalent to an impact assessment in the technical sciences, is not to be found in agricultural sciences. If this were the case, the agricultural sciences would have to deal with the conflicting goals and characteristic antagonisms that inevitably arise when particular and public welfare interests clash. Such a reflection on the advantages and disadvantages of existing structures for society would have no less a consequence than that the prevailing market economy conditions and the contribution of agricultural economics to their creation would have to be subjected to critical appraisal and questions.

Derivations

Agricultural sciences are embedded in many different ways and in the various forms of scientific work that have emerged over the past centuries to gain knowledge. Since our ancestors developed settled communities, the aim has been to gain insights into how food production can be made more efficient and productive and thus increase yields. It was only based on an adequate supply of food that cultural freedom and, subsequently, other branches of science could develop. The agricultural sciences have remained closely linked to their original goals, even though the spectrum of methods has expanded immensely in parallel with developments in other branches of science. Consequently, various intersections exist with the natural sciences, social sciences, humanities, and technology. Suppose one uses the metaphor of the tree of knowledge, the efforts to secure one's livelihood and survival through mastering nature. In that case, agricultural sciences represent the trunk from which the diverse branches of the sciences take their starting point.

The productivity of modern agriculture is a result of a remarkable fusion of technology and science. However, it becomes problematic if research in agricultural sciences is reduced to a natural science or engineering approach. Even though agricultural sciences are often classified as empirical or engineering sciences, profound differences exist between them and these branches of science. Compared to the empirical sciences, many disciplines in the agricultural sciences lack one central premise, namely **external validation**. In the empirical sciences, only external validation entitles the researcher to make scientifically reliable, i.e., generalizable, statements. From the perspective of empirical science, the approach in agricultural sciences often gets stuck with the experimental testing of hypotheses without a comprehensive examination of the overall effects in the context of farming enterprises' heterogeneous initial and boundary conditions. The problem is that the associated provisional nature of the findings as "half-knowledge with hypothesis character" is not explicitly communicated to agricultural practice. This raises expectations of the general validity of agricultural science statements, which cannot be fulfilled with this approach and inevitably result in misjudgments and disappointments.

Examining whether knowledge and problem-solving strategies prove their worth is mainly limited to the effects on production performance and single aspects of productivity. This one-sided focus underscores the productivist orientation of agricultural science. Agricultural scientists do not deal with nature properly and only to a limited extent with natural processes. The research field of agricultural sciences is embedded in a framework that is thoroughly shaped by initial and boundary conditions determined by humans and economic principles. The rural economy is undoubtedly oriented towards norms and values and securing economic existence by maintaining competitiveness. The elementary contradiction between supposedly value-free agricultural research in a research field characterized by diverse and diverging interests and conflicting goals like no other points to an irresolvable incoherence in the agricultural sciences. While **coherence** can only be regained by establishing a systems science, adherence to a natural science approach requires much mental effort to suppress cognitive dissonance. This displacement effort may also explain the strong urge to specialize in ever more differentiated sub-disciplines and in the world of the microcosm, where one can avoid the contradictions in real agricultural life. However, the agricultural sciences are paying a very high price for this. In the process, they lose nothing less than the competence and the associated reputation to solve problems in agricultural practice.

The specialists may hardly notice the loss of problem-solving competence. They may be satisfied with the experimental testing of specific hypotheses and chalk this up as a contribution to basic research. When focusing on specialist disciplinary issues, the central problems and undesirable developments in food production are quickly lost from view. The loss of problem-solving competence is of little concern to the specialists as long as a research activity predominantly secured by government funding does not demand a reorientation from them. While the past successes of agricultural sciences are overwhelming in terms of the productivity gains achieved, the same is increasingly valid for the associated extent of undesirable and harmful side effects caused by the intensification of production processes. So far, agricultural science has not adequately considered the Janus face and ambivalence of the envisaged successes. In the one-sided focus on productivity advances, dealing with the diminishing marginal utility to which all biological processes are subject as a natural law in the course of growth and intensification processes has been neglected.

Research has a central role in society's endeavor for sustainable development. This is particularly true for agricultural research since agriculture is a nexus between numerous sustainable development goals. While generally conceding that the investment in agricultural research has paid high social dividends in the past, there is an increasing concern that the agricultural system is losing its capacity to make comparable contributions in the future. In this respect, it has to be considered that agricultural scientists are neither free from pursuing primary self-interests nor independent from third-party donors, which are the fundamentals for qualified research. Agricultural sciences are, to a high degree, financed by public money. Correspondingly, mutual interdependencies exist between agricultural policy and agricultural scientists. The latter benefit from the "belief in progress" highlighted by agricultural politicians and other stakeholders. These do not tire of maintaining that scientific studies are a good investment because they will be rewarded by problem-solving strategies in the future. Although agricultural sciences have largely failed to provide solutions, they are essential for agricultural policy, maintaining faith in technical progress and optimistic future projections. This prevents agricultural policy from the confession that fundamental changes in the economic framework conditions are essential to solving societal problems and that agricultural policy is not able to implement these changes because they are not able to withstand the pressure of the market environment and the powerful lobbying from the representatives of various stakeholder groups. Thus, the illusion of the problem-solving capacity of agricultural sciences is a win–win situation for the one receiving even more money and the other being protected from a political oath of disclosure.

From the perspective of public welfare interests, agricultural science can be regarded as a powerful instrument of technical and social change captured by organized agribusiness and misdirected its energies against the people and the institutions it was designed to serve. The emerging skepticism about the role of science has led to questioning the benefits of technical change in agriculture. There is a rising demand for more effective social control over the development and use of agricultural technology. It is in society's interest to insist that agricultural science embrace an agenda that includes a concern for the undesired side effects of agricultural production processes on the health and safety of farm animals, a concern for the nutrition and health of consumers, a concern for the quality of life in rural communities; a problem for the pollution of the environment and last but not least for the climatic change.

To find a solution to the ever-increasing problems of the agricultural economy, it is not enough to accumulate more **dispositional knowledge** (more of the same). It is also illusory to rely primarily on technical developments in the search for solutions to problems, as these do not do justice to the complexity of agricultural production processes involving many living systems. To face the challenges, it is necessary to generate **orientation knowledge** and thus a purpose and value orientation that must emancipate itself from the pure doctrine of a value-free scientific approach. No less important is the generation of **action-guiding knowledge** and, thus, the possibilities of subjecting various options to scientific scrutiny concerning their problem-solving ability in the specific context of an agroecosystem.

Due to the manifold interactions between biotic and abiotic factors and the very heterogeneous, dynamically changing interactions, the research subject of agricultural sciences cannot be surpassed in complexity compared to other branches of science. Dealing with complexity in living systems requires a strategy of complexity reduction that is fundamentally different from that in other branches of science:

i. The complexity suggests an approach in which the facts as the object of consideration are approached from different perspectives. Each partial perspective involves a "partial participation" in reality. The various perspectives must be integrated into an overall picture. They must not be considered separately as a "partial reality" that can be extrapolated or transferred into a different context without checking for validity.

ii. It also follows that the individual agroecosystem is the reference system in which dysfunctions and associated problems emerge and in which they have to be solved. Negative externalities emerge from dysfunctional interactions within and between subsystems embedded in the individual agroecosystem, representing the subsystems' environment. Thus, the internal validity of scientific results is insufficient to solve problems, but it must be supplemented with ecological validity tested against the farm-specific context. The challenge lies in generating an evidence base that can rigorously address whether and how approaches lead to efficiency and effectiveness gains on the agroecosystem level compared to other methods. Thus, the challenges are context and location-specific and require, among other things, the conduction of single case studies and their meta-analysis.
iii. Correspondingly, the significance of the respective partial sections concerning the whole system's functionality can only be understood from a superordinate overall view or a comprehensive scientific theory of agricultural sciences, understood as a systems research. However, this approach in the agricultural sciences has been torpedoed by the particular interests of the single scientific disciplines.

The lack of competence to solve socially relevant agricultural problems calls for a reorientation and the implementation of agricultural research programs that will consider the socio-agroecosystem as a whole and broaden the traditional issues addressed by agricultural sciences, particularly by increasing interdisciplinary collaborations. Becoming part of the solution would require nothing less than a **paradigm shift** from an input to an output orientation that focuses on the system performance of agricultural enterprises. Since no overarching perspective can be developed from individual partial views alone, the agricultural sciences cannot take this step of their own accord. The much-vaunted ability of science to self-correction remains limited in the agricultural sciences to the respective disciplines. A cross-disciplinary reorientation is not to be expected because of the dominance of agricultural lobby interests. The impetus must come from outside, i.e., from society's responsibility. It must come up with clear targets for what agroecosystems must achieve in the future, namely, to produce high-quality products under the respective initial and boundary conditions of agricultural ecosystems without excessively impairing farm animals, nature, the environment, the climate, and consumers. Only under clear objectives could the various disciplines of agricultural sciences unfold their potential for solving problems within systems. At the same time, a superordinate and independent institution is needed to assess the evidence of effectiveness and efficiency in the use of resources and the achievement of objectives in agricultural practice.

Suppose the agricultural policy would change the goals of agricultural research work by reorienting the funding policy. In that case, agricultural scientists are expected to follow this paradigm shift, provided they are appropriately financed. Suppose the agricultural policy prevails against the particular interests of the stakeholders and implements a tool by which the negative externalities of farm systems are assessed and benchmarked, as is partly the case in the Netherlands (Brouwer et al. 2015; Daatselaar et al. 2015). This would provide orientation not only for farm management but also for agricultural science. With support from agricultural science, farmers would be encouraged to develop system-based farming practices that integrate land, water, nutrients, livestock, and crop management within the boundaries of a farm system or a combined system of several single farms to reduce negative externalities. An essential prerequisite would be to honor positive outcomes and impose sanctions to reduce negative externalities and prevent free riders from participating without providing appropriate services for the public welfare interests.

References

Alrøe HF, Noe E (2011) The paradox of scientific expertise: a perspectivist approach to knowledge asymmetries. Int J Spec Commun 24:152–167

Baker M (2016) 1,500 scientists lift the lid on reproducibility. Nature News 533:452. https://doi.org/10.1038/533452a

Baumgard LH, Collier RJ, Bauman DE (2017) A 100-year review: regulation of nutrient partitioning to support lactation. J Dairy Sci 100:10353–10366. https://doi.org/10.3168/jds.2017-13242

Bello NM, Renter DG (2018) Invited review: Reproducible research from noisy data: revisiting key statistical principles for the animal sciences. J Dairy Sci 101:5679–5701. https://doi.org/10.3168/jds.2017-13978

Brouwer H, Stegeman JA, Straatsma JW, Hooijer GA, van Schaik G (2015) The validity of a monitoring system based on routinely collected dairy cattle health data relative to a standardized herd check. Prev Vet Med 122:76–82. https://doi.org/10.1016/j.prevetmed.2015.09.009

Cañas-Guerrero I, Mazarrón FR, Pou-Merina A, Calleja-Perucho C, Díaz-Rubio G (2013) Bibliometric analysis of research activity in the "Agronomy" category from the Web of Science, 1997–2011. Eur J Agron 50:19–28. https://doi.org/10.1016/j.eja.2013.05.002

Carrier M (2011) Wissenschaft. In: Jordan S, Nimtz C (eds) Lexikon Philosophie: Hundert Grundbegriffe. Reclam, Stuttgart, pp 312–315

Cassirer E (1974) Das Erkenntnisproblem in der Philosophie und Wissenschaft der neueren Zeit, vol 2. Wissenschaftliche Buchgesellschaft, Darmstadt

Chalmers AF (1999) Wege der Wissenschaft: Einführung in die Wissenschaftstheorie, 4th edn. Springer-Verlag, Berlin

Cicia G, Colantuoni F (2010) Willingness to pay for traceable meat attributes: a meta-analysis. Int J Food Syst Dyn 1:252–263

Daatselaar CH, Reijs JR, Oenema J, Doornewaard GJ, Aarts HFM (2015) Variation in nitrogen use efficiencies on Dutch dairy farms. J Sci Food Agric 95:3055–3058. https://doi.org/10.1002/jsfa.7250

DFG (2005) Future perspectives of agricultural science research. Wiley-VCH, Weinheim

Drucker DJ (2016) Never waste a good crisis: confronting reproducibility in translational research. Cell Metab 24:348–360. https://doi.org/10.1016/j.cmet.2016.08.006

Fahmie TA, Rodriguez NM, Luczynski KC, Rahaman JA, Charles BM, Zangrillo AN (2023) Toward an explicit technology of ecological validity. J Appl Behav Anal 56:302–322. https://doi.org/10.1002/jaba.972

Fanelli D (2018) Opinion: is science really facing a reproducibility crisis, and do we need it to? Proc Natl Acad Sci 115:2628–2631. https://doi.org/10.1073/pnas.1708272114

Field CB, Barros VR, Dokken DJ, Mach KJ, Mastrandrea MD (2014) Climate change 2014: impacts, adaptation, and vulnerability. Cambridge University Press

Fraser D (2008) Understanding animal welfare. Acta Vet Scand 50:1–7. https://doi.org/10.1186/1751-0147-50-S1-S1

Godfray HCJ, Garnett T (2014) Food security and sustainable intensification. Philos Trans R Soc Lond B Biol Sci 369:20120273. https://doi.org/10.1098/rstb.2012.0273

Habermas J (1969) Technik und Wissenschaft als "Ideologie". Edition Suhrkamp. Suhrkamp, Frankfurt am Main

Heisey P, Fuglie K (2018) Agricultural Research Investment and Policy Reform in High-Income Countries: ERReport-249. U.S. Department of Agriculture, Economic

Isermeyer F, Breitschuh G, Hensche HU, Kalm E, Petersen B, Schön H (2002) Agrar- und Ernährungsforschung in Deutschland: Probleme und Lösungsvorschläge. Agrarspectrum, vol 35. DLG-Verl; BLV-Verlagsges; Landwirtschaftsverl, Frankfurt a M, München, Basel, Wien, Münster

Kunkel HO (1988) Issues of academic disciplines in agricultural research. Agric Hum Values 5:16–25. https://doi.org/10.1007/BF02217644

Ladenthin V (2020) Kann man Wissenschaft evaluieren? Forschung & Lehre 27:828–829

Lanzerath D, Schurr U, Pinsdorf C, Stake M (eds) (2022) Bioeconomy and sustainability: perspectives from natural and social sciences, economics and ethics. Springer, Cham

Maturana HR, Varela FJ (1987) Der Baum der Erkenntnis: Die biologischen Wurzeln menschlichen Erkennens, 3rd edn. Scherz, Bern

Mayer A, Mayer J (1974) Agriculture, the Island empire. Daedalus 103:83–95

Mede NG, Schäfer MS, Ziegler R, Weißkopf M (2021) The "replication crisis" in the public eye: Germans' awareness and perceptions of the (ir)reproducibility of scientific research. Public Underst Sci 30:91–102. https://doi.org/10.1177/0963662520954370

Millgram E (2015) The great endarkenment: philosophy for an age of hyperspecialization. Oxford University Press, Oxford, New York, Auckland

Mittelstraß J (2001) Für und wider eine Wissenschaftsethik. In: Mittelstraß J (ed) Wissen und Grenzen: Philosophische Studien, 1st edn. Suhrkamp, Frankfurt am Main, pp 68–88

Mohr H (1999) Wissen – Prinzip und Ressource, 1st edn. Springer, Berlin Heidelberg, Berlin, Heidelberg

Nurse P (ed) (2020) What is life?: Five great ideas in biology. David Fickling Books, Oxford

Poser H (2016) Homo Creator: Technik als philosophische Herausforderung, 1st edn. Anthropologie – Technikphilosophie – Gesellschaft. Springer; Springer Fachmedien Wiesbaden; Springer International Publishing AG, Wiesbaden, Cham

Rockström J, Williams J, Daily G, Noble A, Matthews N, Gordon L, Wetterstrand H, DeClerck F, Shah M, Steduto P, de Fraiture C, Hatibu N, Unver O, Bird J, Sibanda L, Smith J (2017) Sustainable intensification of agriculture for human prosperity and global sustainability. Ambio 46:4–17. https://doi.org/10.1007/s13280-016-0793-6

Royal Society (2009) Reaping the benefits: Science and the sustainable intensification of global agriculture, 11/09. Royal Society, London

Schlageter W (2013) Wissen im Sinne der Wissenschaften: Exaktes Wissen, empirisches Wissen, Grenzen des Wissens. August-von-Goethe-Literaturverl., Frankfurt, M.

Schrödinger E (1992) What is life?: The physical aspect of the living cell. Cambridge Univ. Press, Cambridge

Steffen W, Crutzen PJ, McNeill JR (2007) The anthropocene: are humans now overwhelming the great forces of nature? Ambio 36:614–621. https://doi.org/10.1579/0044-7447(2007)36[614:TAAHNO]2.0.CO;2
Sundrum A (2024) Why has animal science not led to improved farm animal health and welfare? Front Anim Sci 4. https://doi.org/10.3389/fanim.2023.1214889
Tancoigne E, Barbier M, Cointet J-P, Richard G (2014) The place of agricultural sciences in the literature on ecosystem services. Ecosyst Serv 10:35–48. https://doi.org/10.1016/j.ecoser.2014.07.004
Weißhuhn P, Helming K, Ferretti J (2018) Research impact assessment in agriculture—a review of approaches and impact areas. Res Eval 27:36–42. https://doi.org/10.1093/reseval/rvx034
Wolff M (1989) Naturwissenschaftliche Erkenntnis – ihr Status und ihre Rolle bei rational-ethischen Entscheidungen. In: Gatzemeier M (ed) Verantwortung in Wissenschaft und Technik. BI-Wiss.-Verl., Mannheim, Wien, Zürich, pp 102–113
Zuidhof MJ, Schneider BL, Carney VL, Korver DR, Robinson FE (2014) Growth, efficiency, and yield of commercial broilers from 1957, 1978, and 2005. Poult Sci 93:2970–2982. https://doi.org/10.3382/ps.2014-04291

Economic and Ecological (Thinking) Approaches

8

Abstract

Even though economics and ecology have the same prefix and are both dedicated to household management, the two scientific fields are based on very different approaches and objectives. Thus, the frequently cited need to reconcile economics and ecology has little common ground. To understand both approaches in their diversity, they will first be reflected concerning their relation to agricultural economics. It can be concluded that the dominance of the economy over ecology caused the consequences and maneuvered the agrarian economy into the impasse in which it currently finds itself. Contrary to what the theory-laden instructions of economics suggest, decision-making processes at the farm level follow a logic that can only be understood on a farm-by-farm basis. Thus, agricultural economics has, at best, a macroeconomic but not a microeconomic approach to reconcile economic and ecological requirements. Instead of providing orientation, the dominance of agricultural economic models is more likely to cause disorientation. This will only change if the methodological tools of economics are made available to realize ecological goals in the first place. How this can be achieved and how a transformation to an eco-social market economy can be prepared are discussed below.

The Greek word "oikos" is the origin of the syllable "eco," which precedes the terms "economy" and **"ecology."** "Oikos" means household. Accordingly, ecology is the study of the natural household, and economy is the study of the household in human communities. The households of human communities have taken possession of the natural household and have subordinated nature to their economic interests. The original meaning of ecology was formulated in 1866 by the physician and zoologist Ernst Haeckel formulated:

A. Sundrum, *Public Welfare-Oriented Production of Food*,
https://doi.org/10.1007/978-3-662-69040-6_8

> *"By ecology, we understand the entire science of the relations of the organism to the surrounding external world, to which we can include in a broader sense all "conditions of existence." These are partly of an organic, partly of an inorganic nature; both these and those are, as we have shown before, of the greatest importance for the form of organisms because they force them to adapt themselves to them."* (Haeckel 1866)

The term ecology, however, did not become established in biology until the late nineteenth century. During its historical development, the definition of the term underwent various modifications, sometimes more restrictive and sometimes very broad. In southern Germany, the term economy was often used instead of ecology when referring to the relationships between biotic and abiotic factors in an agricultural enterprise. In Anglo-Saxon, the two terms are sometimes used overlappingly today, as can be seen among others in the book title: "Ecology: The Economy of Nature" by Ricklefs and Relyea (2014). The linguistic origin points to the close interconnection between biological and economic processes that have existed since the beginnings of food production and which also persists in the term "agriculture."

8.1 Economic Access to the Agricultural Economy

The economy is embedded in and thus part of the natural balance, which comprises the totality of the interactions between all components of the environment and nature. Resources have been extracted from the natural balance for millennia to ensure human subsistence. As the world's population grew, the interactions between humans and nature became more intense. However, it was not until after the Second World War that the exploitation of natural resources increased at an almost exponential rate. What is no longer needed and what accumulates as waste materials is returned to the environment. To the extent that the waste materials are not (or cannot be) degraded, they accumulate in the environment and nature.

Economics as a field of knowledge in its own right began in 1776 when Adam Smith published his famous work: '*The Wealth of Nations.*' In it, he substantiated his assessment that the free market, in the form of the much-cited metaphor of the "invisible hand," can better organize the exchange of goods than can be performed by state intervention. Smith's work provided groundbreaking impetus for the theory of classical economics. Economic activities revolved primarily around how matter and energy were exchanged with the environment. In this context, property relations and the rules of exchange between people play a central role. An increase in efficiency in the utilization of resources is propagated above all through forms of division of labor, intensification, and specialization. The definitive object of economics is, first of all, **aliments**. They are offered in markets and are demanded by consumers. Commercial goods possess a use-value by consumers and a monetary value, which is realized in a variable price. Goods compete with each other in terms of cost, utility, and value in favor of buyers. The logic of the market is based on competition between different suppliers. When there is a high degree of interchangeability and an oversupply of raw materials, primary producers, in particular, come under competitive and, thus, cost pressure.

From the Middle Ages to the eighteenth century, the European economy was primarily a static agrarian society (Niemann 2009). By 1800, approximately 70–95% of the working population in the various European countries was engaged in agriculture. During this period, the rural economy was primarily characterized by the problem of scarcity and, to a large extent, by a **subsistence economy** with low growth (see Sect. 2.2). Economists of the time held rather pessimistic expectations for future development. In particular, they saw the possibilities for raising living standards as limited. This worldview is primarily associated with the economist Thomas Malthus (1766–1834). In his theory, Malthus concluded from the limitations of the agricultural land available for growing food that, with the concurrent trend of world population growth, a condition limiting prosperity to subsistence levels was to be expected. Retrospectively, we know that these and many other economic theories have been proven wrong. The main reason why Malthus was wrong with his theory is that he and his contemporaries did not foresee the potential of technological development. By analogy with Malthus' misjudgments, it is reasonable to conclude that the utility theorem of contemporary economics is also off the mark partly because it significantly overestimates the potential of technological developments and significantly underestimates their undesirable side effects. These misconceptions will be discussed in more detail in the following sections.

In the course of the Industrial Revolution, conditions changed fundamentally. The hitherto predominant agricultural economy was pushed to the margins and assigned its still enduring role as a food procurer for a steadily growing population. The instruments of industrialization, which achieved considerable increases in productivity through the standardization of goods production, progressive specialization and division of labor, accelerated mechanization, and increased use of external resources, were also applied in agriculture, albeit with considerable delay in some cases. After the Second World War, the agro-industrial revolution in England and the USA spread to other countries. This development continues to shape agriculture and its appearance to this day. Unlike in the eighteenth century, the belief in unlimited technological progress made it conceivable that the standard of living of all people would rise even if the world's population continued to grow. The pursuit of economic growth became an overriding and self-perpetuating **objective**. The idea of continuous and unlimited growth quickly found many supporters, as it promised the prospect of alleviating poverty and hunger and an efficient distribution of goods by the "invisible hand" of the free market economy.

From its very beginnings, economics, as the study of the economy, has claimed to be committed to satisfying people's needs and desires. However, already in the past and until today, it remains unclear and invisible for the majority of the population who benefits most, who less, who hardly, and who does not at all from the economic processes. Similarly, the economy does not explain at whose expense the prevailing business models are practiced. The theoretical superstructure of economics is the theory of rationality, according to which the optimization of one's advantage is the rational motive for action (Nida-Rümelin 2020). This theory finds its personification in the form of *homo* economicus, a virtual economic actor who, based on comprehensive information, directs his actions in a purposive and rational manner, i.e., towards maximizing utility. One of the strengths of this theoretical conception is that the practical **rationality** is neutral in content, i.e., the concrete motivations of the respective actors can be disregarded. In this way, rationality theory does not interfere with the respective individual evaluations of the actors. Thus, it can maintain the appearance of **objectivity**. Rationality is understood primarily in instrumental terms as choosing appropriate means to achieve one's goals. In the form of the utility theorem (Neumann and Morgenstern 1947), economics has had a concept of economic rationality since the middle of the last century. Provided that a person's preferences fulfill certain coherence conditions, the qualitative idea of preference can be transformed into a quantitative concept of utility. Preferences are interpreted as decisions between alternatives: A person prefers x over y if, when faced with the decision to choose x or y, she chooses x. The rationality criterion requires maximizing expected utility. By analogy with individuals who determine what is in their interest and act accordingly, social groups also have an interest and are motivated to act according to their interests.

The theoretical construct of *homo economicus* has long been criticized for a variety of reasons. There are hardly any economists in the literature defending it against this criticism, but also no theoretical concept that is unanimously accepted by the scientific community as a counter-model and thus as a substitute. In his Theory of Practical Reason, Nida-Rümelin (2020) has extensively reflected economic models of thought. From his perspective, the economic utility theorem conceals the conflict of goals between different interests. The relationship between individual and social interests is rather complex. It is by no means a foregone conclusion that what is in the individual's interest is also in the interest of society. What significance is to be attached to the benefit theorem if it turns out that those who behave particularly ruthlessly towards the interests of society and pursue their advantage at the expense of the public welfare interests have the most significant benefit?

Since the beginning of the post-war period, agricultural economics has focused on rationalization, mechanization, specialization, and agro-industrial mass production of raw goods. The overriding production goal is to increase productivity. Only a high level of productivity promises a competitive advantage with which the players can hold their own against competitors at home and abroad and realize an operating profit sufficient to secure economic viability. The standardization of production processes and the raw goods produced through procedural optimization in breeding, nutrition, and further processing has been accompanied by considerable increases in yields and productivity. These far exceeded original expectations and strengthened the belief in unstoppable progress. Largely ignored, however, was the fact that the focus on productivity gains also produced many losers who failed to secure economic viability. It is true that, from an economic perspective, the aim was never to ensure the economic viability of all farms. Instead, structural change aimed to induce unprofitable farms to give up. They should make room for an enlargement of the production units of the remaining farms to bring to bear the economies of scale that are considered particularly relevant from an economic point of view. The structural change, which the agricultural associations also welcomed, manifested itself in a drastic number of farm closures. This was accompanied by reduced production costs per product unit, which initially confirmed the expectations placed on economies of scale. Contrary to the proclaimed expectations, however, this strategy did not ensure the economic survival of the remaining companies that were still in production mode. It is foreseeable that this will not be achieved even with further productivity increases, further expansions of operations, and further reductions in production costs. What is the point of sticking to the current strategies if they rob those supposed to implement them of their livelihoods? What is the point of participating in the global "race to the bottom" when no bottom is in sight? Agricultural economics has so far failed to provide a coherent answer to this question and a sustainable solution to the crisis of "farm death." Unfortunately, this also applies to other crises.

Already in the 80s of the last century, the produced food goods exceeded the demand of the domestic population and caused a first **saturation crisis**. This could only be halfway defused at first with the help of enormous funds from the public and thus a visible hand by shifting the surplus goods to export. This strategy was not without severe effects on the agricultural markets in the countries where the products were exported with considerable subsidies. The increased export orientation did not bring lasting relief to many export-oriented nations either. In the meantime, global agricultural markets have also become saturated to a certain extent. This does not concern the absolute demand for food by the world's population but the demand of those who can afford food at world market prices. The consequence of the excessive global supply relative to solvent demand is that, on the one hand, the comparatively low world market prices do not provide domestic primary producers with market prices with which they can fully cover their production costs. On the other hand, world market prices for food are too high for many population groups in developing countries, so needy people cannot afford enough appropriate food to cover their individual needs.

In the agricultural economy, a productivity crisis has joined the saturation **crisis**. This term refers to the crisis of classical industrial capitalism, in which profitability growth was inhibited by the exhaustion of the technical and organizational possibilities of highly standardized mass production (Reckwitz 2019). An analogous development is also emerging in the agricultural economy. This does not mean that further increases in plant and animal production output are reaching their absolute limits. The growth limits are not brought about by reaching maximum performance limits but by marginal utility effects in which the ratio of additional effort concerning the benefit of further growth processes becomes negative (see Sect. 5.2). In this context, utility gains are not limited to productivity growth. Still, they are increasingly linked to undesired and harmful side effects associated with intensified production processes and general attempts to increase productivity further. In many cases, an economy of increased production of the same goods has long since ceased to promise additional operating profits and is accompanied by negative marginal profits and externalities. There is much to suggest that the economic concept of the "economy of scale" has had its day on many farms.

Due to a lack of reliable data, the importance of the marginal utility function of production processes is greatly underestimated in agricultural economics. Even if the law of diminishing marginal utility is part of the general knowledge in economic theory and agricultural practice, the vast majority of farmers are unaware of the concrete course of the function in the respective farm situation due to a lack of specific data. Therefore, expenditures on further increases in output can result in a reduction in the efficiency of the use of resources (including working time and investment resources) because a corresponding increase does not match the total additional expenditure in yield. Furthermore, negative marginal returns and marginal gains result if the increases in output cause undesirable side effects (e.g., in the form of increased mortality and morbidity rates of farm animals), so an increase in the production can cause higher failure costs and, thus, more harm than good. However, no theoretical models or generalized statements about these relationships can be developed. Whether the use of specific means of production is income-generating or not is not decided in theory but exclusively in the farm-specific context, revealed only on the base of a proper accounting that does not rely on average or rule-of-thumb figures. As explained in Sect. 5.2, on many farms, there is no verification of whether the use of resources is adequate regarding the effectiveness of measures for achieving a goal and the cost-benefit ratio. This is true despite the importance of the degree of efficiency in the use of means of production for securing economic viability.

Many problems in securing the livelihood of farms, as well as the generation of negative externalities, are related to inefficient use of inputs. On the other hand, an unbroken **belief in progress** contributes to the expectation that many of the problems will be solved by future technical developments. At present, hopes are being projected, in particular, onto digital technologies. It is undisputed that further developments in digital technology offer new possibilities for information processing that were not available before. The hopes are based, above all, on new options for dealing with large volumes of data (**Big Data**) and the identification of regularities via algorithms, which could not be dealt with using previous methods due to the complexity of processes. However, there are good reasons for doubting whether these findings will lead to relevant solutions to most problems. Regardless of new technical developments, whose potential can hardly be predicted in principle and therefore cannot be negated, potential developments are already overlaid by other core questions: How are investments in new technologies to be refinanced if the use of means of production is already not sufficiently cost-covering in many cases? How much more do the latest means of production have to increase productivity to be refinanced under the given market prices?

Furthermore, it is unclear how technical developments can solve problems that have been awaiting a solution for decades. On the other hand, a wide range of solution strategies are already at hand. These include nitrate and ammonia release, the knock-on effects of pesticide and antibiotic use, the high level of production diseases, tail biting in pigs, and cannibalism in poultry, to name but a few of the relevant problems. In many cases, these problems can be solved or significantly reduced and eliminated with the existing *know-how*. However, this is impossible under the current economic framework conditions, where the additional expenditures required to solve problems do not adequately pay off. This does not only concern additional financial and working time expenditures but is also to be located on the mental level when it is a matter of racking one's brain over the pros and cons of changes (Kahneman 2011). The example of tail biting in pig farming, particularly relevant to animal welfare, has shown that the problem cannot be solved on many farms under the current framework conditions because the expenditure is greater than the financial benefit expected from curbing it (Sundrum 2020). What applies to the problem of tail biting also applies to many other issues. Any investment and expenditure in measures to improve consumer, animal, environmental, nature, or climate protection is subject to cost. They only have a chance of being implemented if not accompanied by increased production costs and, thus, an impairment of competitiveness. Since competition is primarily fought for cost leadership, many measures that can be expected to protect **public welfare interests** are not applied for cost reasons alone. Suppose hopes are placed in technical developments to solve problems. In that case, the potential solutions to the problem are expected to be more cost-effective than previous approaches. However, if the solution of socially relevant problems costs farmers more than maintaining the status quo, they will refrain from using those measures to solve problems.

In the past, low market prices, which often did not fully cover the production costs, combined with rising costs for production inputs, have driven many farms to ruin. In this constellation, state subsidies such as the EU's direct payments to farmers, which are paid per hectare of agricultural land, provide only limited relief. After all, many **subsidies** are directly or indirectly handed down to the landowners. For instance, Germany's share of leased land in 2016 was about 60% (German Farmers' Federation 2020). Farmers' lease expenditure, which was €2.8 billion in 2016, has doubled since introducing direct payments in 1992. At the same time, there are enormous local and regional differences in rental prices, primarily geared to yield expectations. Thus, area-based direct payments' intended initial income-stabilizing effect is considerably relativized.

Public subsidies that the recipients use to finance lease expenditures are lost for further investments that could increase productivity, which would be the most apparent goal from the perspective of public welfare interests to decrease negative externalities. Those who rely on leased land to a relevant degree must exploit even more available resources to compete with other privileged farms that do not depend on leased land. The Common Agricultural Policy is heavily criticized, partly for its lack of contribution towards achieving sustainable agriculture and the large payments received by wealthier landowners due to its flawed criteria (Antonopoulos et al. 2022). The emerging approach of "**public money for public goods**" means that the award of subsidies will depend on the contribution to specific goals, such as environmental protection and sustainable food production. This new form of subsidies should be examined through a human rights lens in determining whether human rights law could guide the interpretation of "property ownership" and "control of the use of land." The significance of "public money for public goods" in the context of property rights indicates a historical clash between the enjoyment of property and state control of the use of property for environmental purposes.

Even if the agricultural industry is already confronted with various crises, the problem areas outlined so far do not fully describe the crisis. The fact that the agro-industrial mode of production is increasingly losing acceptance in society is likely to have a very problematic effect on the future development of the agricultural economy. Criticism of the prevailing production processes is being voiced increasingly offensively and with more significant public impact by various groups (see Sect. 6.6). It is associated with a drastic loss of image for the agricultural sector. Critical reports on the agricultural and food sectors have increased significantly recently (Anonym 2020). In 2020, 1103 critical publications were recorded, 19% more than in 2019, with the meat industry being the most frequently thematized sector. The persistent criticism also undermines the willingness of citizens to promote a mode of production through taxpayers' money that causes harm to society. An ever-increasing number of domestic consumers are reducing their consumption of products of animal origin or are saying goodbye to livestock farming and its products as vegetarians or vegans. The decreasing consumption of products of animal origin is accompanied by the growth of **competition** by industrially manufactured substitute products (see Sect. 6.3), which is still in its infancy but must be taken very seriously. The manufacturers of these products have set out to compete massively with products of animal origin and to wrest substantial market shares from them. This competition is of a different caliber than the previous competitors, as they will not shy away from addressing the various weaknesses of conventional production (including animal suffering, pollutant emissions into the environment, energy, and water consumption) and denouncing them with advertising and publicity effects.

Disputes about the future course of agriculture occur in many European countries and Switzerland, which is otherwise famous for its prudence. In June 2021, the population had to vote on two agricultural policy amendments to the constitution: the pesticide initiative and the drinking water initiative. The initiators wanted to ban synthetic pesticides nationwide and prohibit importing food produced with pesticides. The drinking water initiative was about ensuring that only farmers who use antibiotics sparingly, produce without pesticides, and feed their animals with the feed they grow on their farms would receive government subsidies. Some farmers fight for a better future, others for their economic existence. In the end, the initiatives have not (yet) been able to prevail. It is remarkable, however, that they won a share of the vote of about 40%. Even if, unlike in Switzerland, citizens in Germany and other countries are not allowed to vote on such issues in a referendum, these referendums show the potential for conflict between the interests of primary producers, the interests of large groups of the population, and the direction that future developments will inevitably take.

The agro-industrial processes have made it possible to produce large quantities of food at comparably low prices. At the same time, the negative consequences of the production processes have reached a hitherto unknown extent and have cost the economic existence of the majority of the originally existing agricultural enterprises. Given the damaging effects outlined in Sect. 5.3, a "business as usual" approach seems unacceptable from the point of view of society as a whole. Nonetheless, the agricultural economy continues to adhere to its basic premise of competition for **cost leadership** to sell the surpluses produced on the global markets at world market prices. The incompatibility between undercutting competition worldwide and agricultural production that considers domestic social demands more than has been the case to date is being addressed primarily by raising production standards (e.g., husbandry standards). These will be implemented voluntarily and with financial support from the state and consumers to meet rising domestic demand. All other primary producers are allowed to continue operating as before at the expense of the public welfare interests and fair competition in the agricultural sector.

With these proposals, agricultural economics reveals that it cannot provide convincing answers to the crises and problems of the rural economy. The essentially untouched utility theorem, based on maximizing particular interests and mainly ignoring public welfare interests, is neither suitable for providing the remaining primary producers with a sufficient income nor solving the societal problems that have accumulated for decades. While other sectors of the economy have already agreed in good time on qualitative differentiation to prevent market prices from falling, agricultural economics clings to the proclamation of increasing productivity and minimizing costs in producing interchangeable raw goods. It seems trapped in its theoretical explanations. Apart from the normative forces of fact, there are currently no forces in sight in the "outside world" of agricultural economics to which agricultural economics would feel compelled to adapt. The scientific community left agricultural science and economics to their devices decades ago (see Sect. 7.4), so there is no corrective in sight from this side either. The "outside world" to which farms must adapt is represented by world market prices for saleable commodities and purchased inputs. Market prices have a normative character for the management of conventional farms. To survive economically, they must produce saleable products at market prices that enable them to cover their on-farm production costs. Many farms find themselves increasingly forced to **externalize** to maintain economic viability. The solid and radical reactions of many farmers to the attempts of the legislator to curb externalization underline the importance that many farmers attach to the possibilities of externalization for securing their economic existence. At the same time, due to the oversupply of commodities in global markets, primary producers are responsible for the market prices in world markets being mainly stagnant. With the oversupply of commodities, they are undermining their economic viability. On the other hand, due to the existing path dependencies, they have few options to deviate from the existing practice. It is indeed no tiny dilemma that the agricultural economy has maneuvered itself into here.

8.2 Eco-Systemic Approach to Agriculture

In the early days of agriculture and the **subsistence economy**, household management in rural communities was geared to the prevailing conditions. With the Industrial Revolution and the delayed development of the agro-industrial sector, drastic changes in the earth's natural balance took place due to human intervention. From year to year, the impact of human activity on the natural balance became ever more significant. These changes justify the proposal to name the current geochronological epoch of the Anthropocene, namely an age in which humans have become the most critical factor influencing Earth's biological, geological, and atmospheric processes (Crutzen and Stoermer 2000). Agriculture is a significant player in the Anthropocene. It has not only changed the landscape but has also involved many other changes taking place at local, regional, and global levels, encompassing ecological and economic dimensions.

Many changes in the natural balance are closely related to the drastic increase in the intensification of production processes. The responsible farm managers initiated and promoted the use of the available resources of the soil and the organisms living in the soil and on the surface for their own interests. Using on-farm and purchased resources (including energy, fertilizer, and feed) can be achieved differently. Accordingly, they are associated with a wide range of different internal and external consequences. At one end of the scale is a form of overexploitation in which resources are exploited until "nothing more can be extracted," and the resources have lost mainly their regenerative capacity (desertification). At the other end of the scale is a form of economy that comes close to a natural **ecosystem**. Here, human intervention is comparatively low. This also applies to the yields of foodstuffs that can be taken from the operating system. There is a great diversity between the two extreme forms of management, which means that the effects can sometimes be assigned more or less towards one or the other corner of the spectrum.

The forms of operation differ not only in terms of the degree to which resources are utilized but also in the degree of complexity. Access can be gained through various thought models to grasp the complexity of on-farm processes, at least to some extent, and identify and understand the decisive forces. One widespread model sees the agricultural enterprise as a factory farm (State Historical Society of Iowa 2004). A counter-model sees the farm as an agroecosystem characterized by distinct interrelationships between subsystems. To approach the complexity of an agricultural operating system through a change of perspective and comparison, the two perspectives are first explained separately before being compared.

A factory is an industrial building where workers manufacture uniform products or machines from various raw materials. In an "**agricultural factory**," plant and animal production uses industrial production processes. Crops are cultivated, and livestock is kept to produce saleable raw goods as cheaply as possible. The process sequences can show high conformity with industrial production processes. This includes realizing economic economies of scale and standardizing the production processes as much as possible so that a product that is as standardized as possible emerges. This has to do, among other things, with raw goods being paid for according to specific criteria, such as classification according to trade classes. Accordingly, the profile of requirements set externally by trade practices dictates the procedures and biological processes to be followed. If carcasses with a high lean meat content are reimbursed at a higher rate, the production processes are aligned to achieve a higher lean meat content. The same applies to plant production, e.g., the raw protein content of baking wheat. A high degree of mechanization and specialization helps to keep costs per product unit low through economies of scale. The individual production processes are continuously examined for opportunities to reduce costs and increase efficiency when using cost-intensive production factors, including labor. It often seems more cost-effective to replace existing technical and biological production factors (machinery, livestock) with new ones than to repair or treat them. Also, the reuse of means of production, such as the **recycling** of nutrients within an inner farm cycle, is only an issue of concern if production costs can be saved in this way. After all, even saleable raw goods are interchangeable products. Only those who can produce them cost-effectively will be ahead of the competition for cost leadership in the long term.

Unlike in other industrial sectors, farms do not have to pay waste charges for releasing pollutants into the environment. Since no costs are incurred for the "disposal" of the waste materials into the environment, the external effects caused by this do not play any role in business management considerations. The **rationality** of individual decisions and actions results from the rationality of the overriding objective of generating operating profits. This requires managers to adapt internal structures and processes to the conditions of the economic framework. Biological disturbance factors (e.g., diseases) should affect the production processes as little as possible. Production aids such as antibiotics, which can keep the germ load in the stables in check, are indispensable on many farms. This is true even if the agents used are so-called reserve antibiotics, which should be reserved for use in human medicine to be able to fall back on them in the event of resistance developing. Here, too, the interests of society must take second place to the interests of farm managers. After all, it is their economic existence that is at stake, and social concerns cannot be given much consideration.

The model of a farm as an **"agroecosystem"** is far less standard than an agricultural factory's. It is most likely to be associated with organic farming. However, the biological processes of agriculture management have evolved in natural ecosystems. Even if the biological processes are modified to a greater or lesser extent by human intervention, significant mechanisms of action have remained essentially the same. Microorganisms, plants, and farm animals are subject to the same regulatory and functional principles as their conspecifics in the so-called wild. Crop plants and animals are not new creations but modifications of species that evolution has produced in adapting to changing conditions. I.e., a pig remains a pig, and a cow remains a cow, even if the performance potentials have increased many times compared to the wild forms.

In natural ecosystems, plants capable of photosynthesis are at the top of the food chain. Plants serve herbivores as suppliers of nutrients and energy they use for survival. However, many herbivores are not immune to being killed and are used as food for carnivores. In an ecosystem, the food chain comes full circle in that the substances contained in the organisms after their death and the excretions during the life processes are broken down again by destruents and recycled in the material cycle. In this way, a loss of the substances in an ecosystem and organic bonds is counteracted. The intra-ecosystem

metabolism is the basis of life for various living organisms, depending on the initial and boundary conditions (soil conditions, climatic conditions, etc.). An ecosystem is designed for diversity and sustainability and characterized by high efficiency in the use of available nutrients in the overall system and by low nutrient losses from the **system** (Reichholf 2004). Self-regulating processes limit the consequences of excessive growth of one species at the expense of others. Viewed from the outside, we assume that natural ecosystems are based on a high degree of self-organization. Ultimately, however, the manifestations of an ecosystem are nothing other than a forced adaptation of the organisms living in the system to the availability of resources in the surroundings to secure their ability to survive, as has already been pointed out by Haeckel (1866) as a decisive principle of **ecology**.

The basis of adaptability in natural ecosystems is biological diversity. Not conformity, but variation in the expression of traits within species is the prerequisite for a high degree of adaptability. This can be found in microorganisms no less than in plants and animals. A high richness of trait expression gives an advantage to the species and those organisms within the species that have traits advantageous in the face of changes in the "outside world." Given the limited availability of resources, bacteria that are insensitive (resistant) to substances such as antibiotics have a much-improved reproductive potential if the other members of the species are restricted in reproduction by appropriate substances. This applies analogously to plants confronted with heat or water shortages or animals exposed to food shortages. An ecosystem is a community of life designed to maintain the system as such and the species (not the individuals) that reside within it. All subsystems are relevant to the functioning of the higher-level whole, while the whole provides the subsystems with the conditions they need for their functioning and survival.

Regarding adaptive capacity, natural ecosystems have some advantages over agricultural factories. However, ecosystems cannot produce biomass, which is taken from the system in large quantities to make it available to other systems (e.g., human communities) as food. This brings us to farms as **agroecosystems** (DFG 2005). All farms share features with ecosystems and agricultural factories, albeit to different degrees. One of the things they share with the latter is the need to produce and sell products. In return, they receive a fee, which they, in turn, need to maintain on-farm production processes and secure their economic livelihoods. In the case of agroecosystems, we are dealing with self-sustaining (**autopoietic**) **systems** of different order, which are embedded in a hierarchical order, between which there is a high degree of interaction and mutual dependence (see Fig. 8.1).

Living cells (autopoietic systems of the first order) are, as a component of the organism, dependent on the supply and removal of nutrients (matter) to contribute through their functional capacity to the functional capacity of the tissue or organ and thus to the survivability of the organism as a whole. The organization of an autopoietic system, such as a cell and an organism, is oriented towards preserving the system and, concerning the active processes, is only accessible via a functional and teleological approach to explanation. The subsystems contribute to the functioning and survivability of the superordinate system and, thus, simultaneously to their self-preservation (Maturana and Varela 1980).

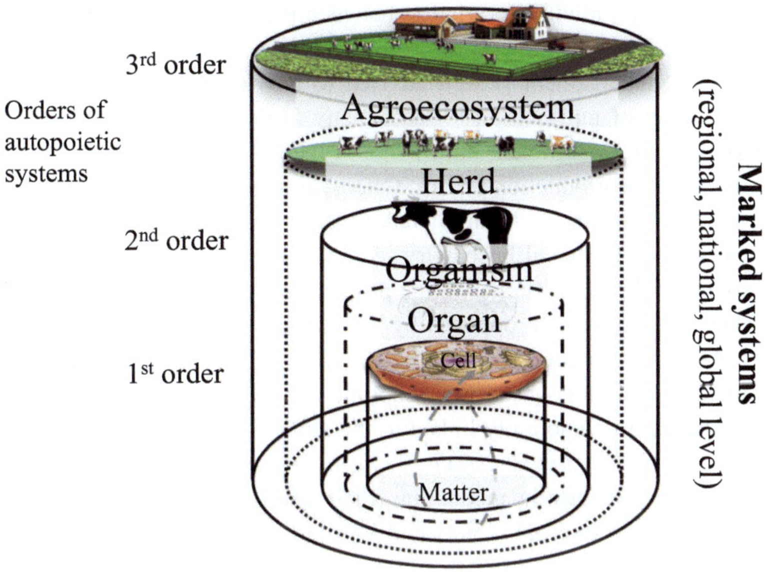

Fig. 8.1 Farm as agroecosystem using the example of a dairy farm (own representation)

Maintaining the ability to function is, therefore, the highest priority. This task provides the measure for the intake of nutrients in the required quantity and quality. These must then be distributed within the organism so that, in the most efficient way possible, all organs and cells in the organs get what they need. If the availability of nutrients is limited, as is inevitably the case with high-yielding animals in certain phases of life, a preferential supply of organs or cells at the expense of the supply of other cells can result in a disturbance of the whole organism, which threatens the survival of the animal. In livestock farming, there is no lack of examples of one-sided breeding orientation towards specific production characteristics (e.g., milk yield), which influence the distribution of nutrients within the organism (partitioning) so that the survivability of dairy cows is endangered (Habel and Sundrum 2020).

What is coordinated at the animal level by the nervous system to maintain the functioning of physiological processes has to be done at the farm level by the people responsible for the farm. They decide on the availability of resources, among other things, through on-farm production or purchasing fertilizer and feed. This also applies to allocating nutrients to individual animals through organizational and technical measures. In the case of lactating animals such as dairy cows and sows with enormous nutrient requirements, there are often significant gaps (imbalances) between the varying requirements of the individual animals and the respective supply. The individual animals must compensate for these gaps through physiological adaptation processes. While some animals succeed better than others, excessive demands on their ability to adapt to existing nutrient imbalances manifest themselves in subclinical and clinical diseases, thereby showing a large intra- and interindividual variation between the animals on a farm and between farms.

Although farm management is well advised to provide individual animals with the quantity and quality of nutrients they need, many farms fail to meet these challenges. This is not only due to insufficient knowledge of the requirements of the individual animals in their respective production phases. Often, farms also lack the resources needed to counteract nutrient imbalances. A supply tailored to the animal's needs is associated with additional work time, technical husbandry requirements, and investments in purchasing external resources. The same applies to the protection of animals from biotic and abiotic stressors. These necessities are opposed by the fact that milk prices have been low for many years. As a result, many farms are neither able to cover their running costs nor to provide individual farm animals with appropriate resources, although the economic sustainability of the farm is inherently linked to the ability of cows to cope with their environment (Hoischen-Taubner et al. 2021).

Farms differ considerably in reconciling their interest in securing economic viability by avoiding adverse internal effects and limiting externalities in the form of excessive discharges of waste and pollutants into the environment. Ongoing efforts to increase efficiency and minimize costs by applying agro-industrial strategies come up against the limits of adaptability to which biological systems of different autopoietic orders are subject. In an agroecosystem, the focus on sub-sectors is usually at the expense of other sectors and subsystems. The practices of externalizing the waste materials produced and ignoring the undesirable consequential and harmful effects are also increasingly reaching their limits and bringing the conflicts between particular and common interests to light.

8.3 "What Matters is on the Court"

Agroecosystems, as autopoietic systems, are forced to adapt to the economic framework conditions to be economically viable and to have a chance of passing on the business to the next generation. At the same time, the operators of agroecosystems compete with each other and must try to produce the products for sale more cheaply than their competitors. This is sought mainly through growth (economies of scale), intensification (input of external resources, including capital), and specialization and mechanization (labor productivity). The various instruments exploit the potential of the biological growth processes of living organisms in the operating system as far as possible (see Sect. 4.1).

The transitions from beneficial use for living organisms to overuse and the associated harmful effects are fluid. Overuse is synonymous with an excessive demand for adaptability, which increases susceptibility to disease in crops and livestock. It is often overlooked that the means of production used to "create value" are subject to the law of diminishing marginal utility. As the level of productivity increases, the gain in output that can be realized with increased input of resources decreases. At the same time, as output levels rise, the expenditure required to achieve higher output increases (see Sect. 5.2).

This can go so far that the additional spending is higher than the gain, resulting in a negative marginal profit. This corresponds to an excessive demand for management to adapt to the economic framework conditions. If regeneration does not occur, the farm system loses its financial ability to survive, analogous to livestock.

The law of marginal utility also explains why increasing quantities of nutrients absorbed into the farm systems during intensification are only partially used. The following impacts are an increased discharge of nutrients and pollutants into the environment and an overload of the natural ecosystems outside the agricultural farm systems, overtaxing their ability to regenerate and adapt. The consequences of an overtaxed adaptive capacity are diseases and the death of living organisms, resulting in considerable **biodiversity loss**. Consequently, it is a matter of nothing less than existential challenges, which the intensification processes in agriculture conjure up and affect both the remaining agroecosystems and the natural ecosystems that still exist.

Agroecosystems differ considerably in their structures, their resource availability, and thus in their adaptability. Changes usually occur gradually. Hence, transgressions of the limits of adaptive capacity often become visible only after a considerable delay. They also only become visible to those with an eye for them and possess the necessary powers of judgment. For those who prefer to overlook the signs of dysfunctions and disorders or to ignore them because they are incompatible with the prevailing patterns of thought, the overstretching of adaptability is often only perceived and considered to be true when the consequences have assumed drastic and incalculable proportions. To safeguard livelihoods and avoid excessive demands on the adaptive capacity of agroecosystems, it is, therefore, necessary to have reliable assessments that allow actors to grasp the extent of disturbances at an early stage and to classify them adequately. However, these assessments prove difficult, as each agroecosystem and the management responsible for it have their particular strengths and weaknesses and thus represent an individual unit. Therefore, general assessments and generalized advisory recommendations on how to deal appropriately are not very useful. Before measures that might be suitable for securing the livelihood of the individual agroecosystem are considered, a system analysis is first needed.

"*All theory is grey—it's on the pitch that counts*" (Potofski 2014). This wisdom, attributed to the commentator Alfred Preißler, not only applies to the football pitches or other sports fields of this world but is also justified in connection with agriculture. Agricultural sciences have accumulated a large arsenal of knowledge about the mechanisms of individual processes within a multifaceted process chain. However, this only proves to be "**half-knowledge with hypothesis character**" (see Sect. 7.4.4) as long as it is not validated in the respective context. The manifold interactions within an agroecosystem produce an emergent overall performance. As in team sports, the partial performances and the interactions between the different areas determine the final success. In sports, an independent authority watches over the observance of the rules. It certifies the final result that the teams achieve in the interaction of the individual players and the confrontation with the opposition of the opposing team. Within a season, there are continuous replays in which the team play has to prove itself. The points scored per match are added so that at the end of a season, the ranking position on a table reflects the annual performance of the teams in the respective competition.

Unlike sporting events, what goes on in the fields and stables of farms is hidden from public view. Often, even the farm managers are unaware of how much their activities harm the public interest. Consequently, this also applies to the relationship between a farm system's positive and negative externalities. Thus, neither those responsible for the farm nor the public have sufficient knowledge of the extent to which farms pursue their particular interests at the expense of the common good. Nor do the competent authorities in the agricultural sector know whether the subsidies paid to farmers positively contribute to the common good or are used to maintain or even force production processes accompanied by drastic adverse effects on public interests. Without knowledge of the harmful effects caused by the individual farms, neither the responsible decision-makers nor the supra-farm authorities can intervene in a regulating and controlling manner. Indeed, farm managers are generally aware of the level of production output generated in the respective farm sectors. Rarely, however, do they have a concrete data basis for assessing how the farm-specific relationships between costs and benefits can be classified.

For farm management, the central question is whether further increases in expenditure and investment are accompanied by increased benefits that exceed the additional spending or whether the optimum ratio between expenditure and benefits has already been exceeded. This applies to expenses and investments for more animal welfare and environmental protection in the same way as those for increasing production output. The decreasing **marginal utility**, which occurs with increasing output or benefit, not only limits the potential benefit but also reverses the benefit into a loss or a negative marginal gain in the case of excessive expenditure. In the long run, farms cannot afford negative marginal profits without undermining their economic existence. Whether investments or additional expenditures still make sense, i.e., produce a positive benefit, and whether it is counterproductive cannot be answered in general terms based on model calculations but is decided in the farm-specific context. No state regulation can prevent entrepreneurs from leading their companies into ruin. The question, however, is whether it can be left to the decision of individual farmers to act against public interests while at the same time receiving a considerable amount of public money.

A reliable assessment of the cost-benefit ratio requires a sufficiently solid data basis. The various data types from a farm, e.g., economic and biological data, commonly lack connectivity and compatibility at the animal level. In livestock farming, individual animals' cost-benefit ratio and monetary gain vary greatly, as demonstrated by a comprehensive data collection and analysis on selected dairy farms (Habel et al. 2021). The focus was on the cost-benefit ratios of those dairy cows that "involuntarily" left the production process to compare the monetary gains achieved up to the time of departure with the financial expenditure during the animals' lifetime. The involuntary departures during the year, ranging between farms from 18.6 to 41.8%, were due to death, emergency culling, or slaughter. As a rule, culling was preceded by production diseases, which led to mortality or emergency culls or gave the farmers a reason to have the diseased dairy cows slaughtered to achieve at least a slaughter price. The link between biological and economic performance revealed a considerable variation in the single-animal contributions to the farm profit. For the 4,962 animals evaluated, the range was between a loss of 11,818 and a gain of 13,680 Euros. The proportion of dairy cows that left the farm and earned at least their rearing and husbandry costs varied between 0 and 74%. On average, the involuntarily departed dairy cows generated a negative profit return. Only 41% of the farms investigated generated a positive marginal profit. Various factors could be identified that influence the individual animal and farm results. However, the interaction of these factors was different for each cow and on each farm (Hoischen-Taubner et al. 2021). Farms with a low proportion of dairy cows in the profit phase do not succeed in providing the individual animals with the resources and protection they need for self-preservation. Involuntary departures express the failure of dairy cows to adapt to their living conditions. At the same time, the departures document the failure of the management in the attempt to adjust the living conditions to the individual animal needs of the farm animals concerning the performance requirements. Thus, most of the farms investigated succeed neither in meeting their economic interests nor the interests of the farm animals entrusted to them and thus not in meeting the requirements of animal welfare and the interests of the common good.

The proportion of dairy cows in the profit phase constitutes a normative target for the economic viability of the farm in question. In addition, if the proportion is too low to secure the financial existence of the farm, the "reasonable cause," which may justify that livestock farmers impose restrictions on farm animals that might be associated with pain and suffering, is forfeited. The proportion of animals in the profit phase thus proves to be an objective, reproducible target variable with a high explanatory power, which provides farm managers with an economic and ecological perspective on the complex interactions in an agroecosystem and offers them orientation for individual farm decisions and the need for action.

8.4 Lack of Orientation and Orientation Knowledge

Because of the continued ruinous competition in the global markets, many farmers feel compelled to continue gearing their production processes toward productivity increases. Much suggests that most farm managers are not sufficiently aware of how much the optimum cost-benefit ratio has already been exceeded in the various subsystems, and economic losses are incurred. In addition, there are also external threat scenarios that are no less powerful. The negative consequences of the intensified production processes provoke criticism from various public sectors, which is being voiced ever more loudly and whose resonance has increased through media amplification. Although there is no lack of attempts to ward off criticism across the board, these defensive efforts are increasingly coming to nothing due to a lack of validity and evidential value.

The food retail sector is responding to criticism of the existing conditions by successively raising the **minimum requirements** on production processes. This puts even more pressure on primary producers. After all, the increased requirements are usually accompanied by additional expenditure, but without the retail trade rewarding the additional expenditure appropriately, i.e., in a way that covers the extra costs. Many farm managers have or see little room for maneuvering to both further minimize production costs and, at the same time, cope with higher expenses. While the situation is becoming increasingly acute, the complexity of the on-farm processes in the interdependencies of an agro-ecosystem stands in the way of simple solutions and action strategies. Nevertheless, the belief of many actors in the continuous progress of knowledge and technical developments is unbroken. The outlined examples of negative marginal gains due to excessive intensification show that many problems of agricultural enterprises will not be solved by general progress of knowledge about further details. Technical innovations are also subject to the economic law of diminishing marginal utility. The law does not necessarily describe the end of growth, but first of all, only the loss of sense in growth processes that run beyond the optimum ratio of effort and benefit. Depending on the initial and boundary conditions, individual farms differ considerably regarding their challenges. This also applies to their options to defuse the conflicts of interest. This makes the need for the reorientation of management all the more significant.

However, more information and detailed knowledge do not offer a new orientation. At a time when the flood of information continues to increase, the difficulties of identifying the factual expertise relevant to the solution of context-dependent problems from the existing mass of data are growing. Therefore, increasing **dispositional knowledge** should not be equated with increasing solution approaches. In addition to searching for suitable solution approaches, it must also be assessed whether utilizing specific available knowledge is worthwhile. The ability to distinguish between the important and the less important, the unimportant and the valuable, and the wasteful is not exactly strengthened by an increasing flood of data and information. Whereas dispositional knowledge is concerned with technically possible actions, why we should act in one way or another is paramount for orientation knowledge. It is knowledge about justified purposes and goals when using (production) means (see Sect. 7.3.2). In **action-guiding knowledge,** dispositional and orientational knowledge come together by linking what is technically feasible with what leads to the goal, depending on the available resources, and by justifying a reasonable use of means and resources (Poser 2016).

Orientation knowledge is not available off the shelf; it is also rarely found in textbooks. Orientation means first knowing where one is in relation to an external reference point. Without reliable orientation, there is a risk of disorientation or confusion. Whether it is a question of spatial localization in an unclear terrain, whether the view is obstructed by fogging or darkening, or whether it is a question of the localization of internal efficiency or the harmful effects emanating from the operation, the localization of one's position always takes place in relation to other parameters. The sun's rising in the Orient, the setting of the sun in the West, and fixed stars or lighthouses at night are vivid examples of spatial orientation. They enable us to find our way in the outside world. The ability to orient oneself is based on the cognitive ability to establish a mental triangular relationship between the orientation point, one's point of view, and the intended target point. Those who know which target they are heading for and the relationship between the target point and their current position can deduce the direction they must move to get closer to the target point. On the other hand, it is difficult for those who do not know exactly where they want or should go. The Roman philosopher Seneca (65 AD) put this in the memorable formula:

"If you don't know the port you're sailing into, no wind is right for you."

It is of central importance for agricultural enterprises to orient themselves towards future developments and to secure their economic existence. The question is whether the partial objectives pursued so far are still justified concerning the overriding objective or whether, sooner or later, there is a risk of ending up at a dead end. One can continue to (unthinkingly) trust the signposts set up in the past, especially by agricultural economics, because one has also trusted them and done well with them so far. You can try to convince the outside world (including politicians, citizens, trade, and consumers) that your business must maintain the previous orientation because you are not prepared or willing to adapt. However, one can also try to clarify whether one is still well positioned and equipped in one's operational situation when the wind changes, i.e., when societal demands are placed on agricultural production goals and thus on production processes. When the framework conditions change, the orientation points must also change. The old orientation points may not be able to meet the new challenges.

The conventional goal of management is primarily to design production processes in such a way that productivity can be increased. The close link between production performance and monetary income suggests that farmers are more interested in information about their crop and livestock production performance than other farms. In this way, farms can orient themselves as to whether their level of performance is at the upper, middle, or lower end of the spectrum. The comparison provokes the question of whether the management can be satisfied with its performance level or whether there is reason or even the necessity to increase performance to position itself better in competition. Consequently, various structures and measurement systems have been established in the agricultural industry to record production performance. This includes weighing the products for sale with calibrated scales and dividing them into trade classes based on specific criteria. In the past, performance testing institutes have been set up to test the performance potential of animals of different breeds and genetic origins or feed rations of different compositions under standardized conditions. These findings were then passed on to the practice as recommendations. Even though the recommendations have significantly contributed to raising the performance level on farms, it has also become apparent that the results generated under standardized conditions can only be transferred to conditions in farming practice to a minimal extent. In the meantime, many testing stations have been closed again. On many dairy farms, performance tests of individual animals are still carried out today (usually monthly). The test results provide farmers with information on the individual performance of their animals, which they can use primarily for breeding and feeding measures. In addition, they give some selected information about disorders present in dairy cows.

However, the information content inherent in monthly milk records is rarely exploited. For simplicity, many farmers are content with aggregated averages to which the performances of individual animals are added up. Average values are not only of little significance, but they do not provide any information about the efforts required and, thus, the efficiency to achieve them. Aggregating individual values to mean values masks and levels out the variation between individual animals. The ignorance of individual animal variation, often observed in agricultural practice and science, can be classified as the starting point of severe misjudgments and disorientations. The cost-benefit ratio at the individual animal level determines the economic success of livestock production and the proportion of animals reaching the profit phase. An inadequate reference level can hardly be underestimated regarding its effects on the farms' economy and the consequences for public welfare interests.

What can be observed about sales products of plant and animal origin, namely that farmers primarily orientate themselves towards aggregated average values, is even more true for recording groups of substances not sold but used on the farm. This applies, for example, to the crop yields of fodder and the growth performance of rearing animals used on the farm. One deficit in particular seems to be responsible for the low level of data and information. During the agro-industrial intensification processes, many instruments were adopted that could be used to increase productivity in agriculture. One instrument, however, was not one of them: detailed **accounting** of all inputs and outputs occurring in the various subsystems and stages of production. What is taken for granted in other sectors of the economy and has become an indispensable source of information for management decisions presents many farm managers with more than methodological challenges.

Methodological challenges for detailed accounting are far more demanding in living systems than in uniformly running technical production processes. Even more critical for many farm managers is that data collection and evaluation involve considerable time. In addition, they require a particular affinity for numbers and presuppose a willingness to work at a desk. What many farm managers in agricultural practice lack, above all, is time. In hardly any other economic sector has working time become such a scarce commodity as in agriculture. The increase in the size of farms that has accompanied structural change has rarely been accompanied by increased labor capacity. In various areas, it has been possible to save on routine work through the increased use of technology and, in this way, to manage more extensive stocks of crops and livestock with less labor input. While mechanization and automation have often led to significantly reduced workload per production unit, production units have increased, and new tasks have been added elsewhere. As a result, many farm managers are operating at the edge of what is still reasonable regarding working hours. They only keep their heads above water through a high degree of self-exploitation.

Because of the considerable overload on their working time, it is no coincidence that farmers and their representatives publicly complain at every opportunity about the heavy bureaucratic burden imposed on them by various parties. From the point of view of many farmers, the bureaucratic effort is entirely unnecessary and a waste of time. After all, subsidies could be paid without unnecessarily complicating the process with various application forms. However, the resistance to bureaucratic expenditure cannot be explained solely by an aversion to spending time on measures that are considered dispensable. After all, from the farmers' point of view, there are more urgent things to do. In addition, there is also the concern that the data could be used to gain external insight into on-farm processes and thus exercise a degree of control. After all, according to a frequently encountered essential attitude in agriculture, it is none of the outsiders' business what is happening within the farms.

To outsiders, a farm is a "**black box**". When agricultural associations repeatedly point out that farmers know best what is happening in their farms, this statement cannot be contradicted in principle. How could anyone outside know more than those in the farm system? However, this statement belies that even most farmers have limited insight into the processes at work, such as the on-farm nutrient flows and their effects on livestock or the discharge of pollutants into the environment. Since most biological processes and their effects are beyond direct observation and thus hidden from the eye of the beholder, comprehensive data and robust data evaluation are of central importance. Management can estimate what input variables enter the farm and what sales products leave the company again. However, the farm manager remains largely unaware of the **efficiency** with which the methods are implemented in the individual plots and farm animals and what consequential effects result from the storage and use of production inputs in the form of releases of pollutants into the environment. If output and input variables are only aggregated at the farm level, e.g., in the case of a farm gate balance sheet, and are not also recorded in the subsystems, no valid statements about the **environmental compatibility** of the production processes can be derived (Sundrum 2019). In the various subsystems of the farm, the extent to which nutrients are organically bound or escape into the environment is determined depending on various influencing variables. It is only in the specific context that it becomes apparent whether targeted measures can be taken to prevent substances from being emitted from the farm and whether associated partial effects contribute to a positive or negative overall impact on the agroecosystem or the environment. Accordingly, comprehensive balancing at the different system levels with data-based parameters and estimates is required to gain insight and a meaningful data basis. The knowledge gained in this way not only provides an overview but is an elementary prerequisite for controlling the internal nutrient flows and improving the efficiency of their use in the farm by improving their **allocation**. At the same time, organic sequestration and more efficient use of nutrients can reduce substance discharges into the environment. This shows how economic and ecological requirements can be brought into synergistic balance and how a classic win–win constellation exists. However, the potential cannot only be exploited by average figures but requires sufficiently detailed knowledge of substance conversion at the relevant process levels.

While many farmers save themselves the effort of comprehensive data acquisition and evaluation, they pay a high price. Due to a lack of well-founded data on the on-farm material conversions within and between the subsystems, as well as on the course of the growth processes in the plant and animal populations and insufficient knowledge of the resource availability in the manure and fodder storage areas, they do not have sufficient insights into the relationship between effort and benefit. Therefore, they do not recognize if and when effort exceeds benefit at the relevant process levels, and negative marginal returns and gains occur. At the same time, there needs to be more preconditions for checking the effects of targeted measures intended to bring about improvements to see whether they prove effective in the specific context. A lack of or inadequate monitoring of the success of the implementation of measures runs counter to the effective and efficient use of funds.

On the other hand, the lack of data or the restriction to highly aggregated data offers a broad scope for interpretation. These can be used to counter critical objections and to maintain one's patterns of interpretation without any in-depth clarification of the facts under discussion. As part of being human, we constantly strive to form a picture of what surrounds us in our immediate and broader environment. We need the world view to orient ourselves, to classify events and developments with a specific pattern of interpretation, to be able to deal with the complexity of the processes, and to bring a particular order into the chaos of the sensory impressions. Our worldview expresses how we classify partial aspects into a superordinate whole, establish mental (associative) connections, and what we consider authentic and accurate when we look at a partial object. The philosopher Liessmann summed this up as follows:

> *"Our worldview forms the framework for how we think of the world, according to which regularities, according to which aspects it appears to be ordered for us, according to which maxims we orient our cognition and action in this world."* (Liessmann 2021)

The fact that many farms have to give up every year because they can no longer cope with the prevailing competition for cost leadership would be reason enough, alone out of self-interest, to fundamentally reflect on the current practice of inadequate accounting in agricultural operations. When considering the degree of impairments that originate from production processes compared to the amount of public money invested in agriculture, society has every reason to demand an adequate accounting of substance conversion processes in farm systems.

8.5 Forms of Complexity Reduction

Agroecosystems are complex adaptive systems with interactions between diverse biotic and abiotic factors. Just as the user of technical devices such as a car or a computer does not need to understand in detail how the individual components interact with each other to use them for his interests, the farm manager does not necessarily need to understand the complex interactions within his farm in any great detail. The primary concern for the farm manager is that the web of interactions functions following its intended purpose and enables value creation by adapting on-farm production processes to the challenges of the outside world. These are shaped, above all, by market conditions in the form of the prices that can be gained for the products sold and that have to be paid for the means of production. Long-term sustainability can only be achieved through corresponding operating profits. However, more than an economic approach is required to do business sustainably. Analogous to the evolutionary processes of adaptation, the sustainability of an agroecosystem cannot be achieved without organizing a high efficiency in the use of available resources. Unlike natural ecosystems, an agricultural operating system does not organize itself. The farm manager must bring the production processes at the various levels into balance with each other and steer them in the desired direction. Negative marginal utility effects must be avoided as far as possible. At the same time, care must be taken to prevent overtaxing the adaptability of living systems because the resulting disruptions run counter to securing economic survival.

Within the framework outlined, the central challenge of farm management is to deal with the **complexity** that characterizes the interplay of biological processes in the respective agroecosystem. **Management** is essentially based on a mental reduction of complexity (Baecker 2014). Without a reduction of complexity, there is inevitably an information overload, which means that the abundance of possible information can no longer be processed meaningfully. On the other hand, complexity reduction is fundamentally associated with a considerable loss of information. All the more, the question of the right degree of reduction arises. Which data is necessary, which is dispensable, and what is the appropriate ratio of effort and benefit of information acquisition and processing? To find the right balance between too little and too much, one cannot fall back on general rules of procedure. The degree of complexity in the respective structure of effects and the capacity for intellectual penetration are so heterogeneous that general recommendations are inevitably doomed to failure.

Here, too, purpose orientation provides the necessary assistance. With the question of what is to be achieved with the mental penetration of complexity, complexity reduction is classified as a means to achieve a goal. The more the interactions can be understood mentally, the more effectively and purposefully they can be influenced. The spectrum of possibilities for reducing the complexity of agroecosystems is shaped not least by the primary objectives and the respective technical perspective that should be used to deal with dysfunctions and solve problems. However, repairing a technical device suggests a different approach than dealing with dysfunctions within and between living systems. And yet, farm management is required to balance the diverse challenges and perspectives. In the following, we distinguish between an agroeconomic and an agroecological approach in dealing with complex adaptive systems and discuss the resulting implications.

8.5.1 Agroeconomic Approach

The proclamation of the free market economy grants agricultural enterprises economic viability only as long as they can hold their own against their competitors in an undercutting competition. If the number of farms that have closed down in recent decades alone is taken as a basis, the chances of the remaining farms remaining viable long-term are slim. It appears to be less a question of whether but rather a question of when the enterprises will (have to) give up individual branches of their operations or even farming as a whole. Unless the land has to be sold because of over-indebtedness, farmers generally remain in possession and receive rent from those who continue to struggle to profit from the farmland. In many cases, the return on land ownership generates higher income than can be achieved by cultivating farmland.

The farm-specific production goals determine where management's attention is directed internally. The internal orientation focuses on the relevant processes and probes whether and to what extent the actual state deviates from the expected parameters. Whether and which corrective interventions are made is determined by the attention and importance that management attaches to the degree of deviation observed and the resources still available internally. In addition, management cannot avoid keeping an eye on developments outside the farm system. This applies above all to changes in the market about the development of prices and demand for raw goods and means of production. The more accurately internal and external developments can be anticipated, the sooner and, more specifically, measures can be initiated to adapt processes to the changes.

Given the limited resources available, management is constantly confronted with the question of in which areas and to what extent these can be used, targeted, and activities bundled or saved. Management on farms is primarily concerned with managing scarcities. Consequently, decisions must be made continuously regarding which requirements must be prioritized or subordinated with limited resources. Orientation for the prioritization of urgencies is provided by one's values and the established thought patterns in assessing circumstances. Accordingly, management decisions are made self-referential according to its standards. In contrast, norms applied to the farm from outside (e.g., minimum legal requirements and defined standards) are of secondary importance for day-to-day business practice. They concern only a few partial aspects (e.g., husbandry conditions, fertilization level), which must be complied with, at least if unforeseeable controls and sanctions are to be expected in the event of rule transgressions. Otherwise, farm management is largely left to itself and its self-referential judgment when it comes to reconciling on-farm processes with the requirements of the competition.

To hold one's own in competition, the rationality of decisions from an economic point of view is to increase production performance further and to reduce production costs where possible, i.e., to minimize the use of resources in the form of time, money, and other expenditure per unit of product to a minimum. Measures include, for example, using plants and livestock with the highest genetic performance potential and feeding them with a highly concentrated supply of nutrients. What applies to other sectors of the economy also applies to agribusiness. Modern action theory is dealing with selective points:

> *"It takes the individual act at a given moment as the result of an individual calculation of advantage as the paradigm of human action."* (Nida-Rümelin 2020)

However, the selective optimization of partial aspects in the sense of maximizing the expected value of the utility function by no means implies that selective measures combine themselves to form a reasonable overall practice. After all, it is necessary to consider not only production performance and cost minimization but also the functioning of the subsystems involved, the relationship between effort and benefit, and undesirable side effects. Punctual acts are confronted with the problem of structurally embedding the individual act in the larger practice context. In this context, punctual optimizing activities do not necessarily produce a coherent (cohesive) picture of a structurally rational practice. On the contrary, selective rationality often does not add to a reasonable overall practice (Hoischen-Taubner et al. 2021).

If only highly aggregated data on monetary expenditures and benefits are available, it remains unclear how efficiently the various resources and means of production are used at the relevant process level in the respective subareas of the farm and in relation to the individual production units (e.g., individual animals or arable fields). As a rule, the relationship between costs and benefits varies significantly between the individual production units. The average obscures the view of the part of the production units that deviate considerably from this average, so that—as discussed above—average values can be misleading. In addition, there are usually no estimates of the course of the marginal utility function of the production processes in the operational **context**. Thus, it cannot be assessed whether the additional input of resources possibly already exceeds the optimum. The same applies to determining whether reducing production costs is successful or leaves gaps in other areas, the effects of which have to be paid for dearly. If, for example, savings are made in the rearing of young animals, this can have far-reaching negative consequences in the later stages of the livestock's lives in the form of additional costs or reduced performance that far exceed the original savings potential (Blume et al. 2021). Focusing on selective measures and their effects in a limited area does not prevent deficits and grievances from being caused in other places when there is a complex interplay of effects.

With an economically motivated focus on increasing performance and minimizing costs, it can be easy to lose sight of the fact that agricultural production processes are confronted with dysfunctions. From a rural economics perspective, the focus is on technical approaches to solutions (Bundesministerium für Ernährung und Landwirtschaft 2019). These reduce the complexity of the facts and the cause-effect chains to technical process sequences. In this way, other cause-effect relationships are lost from view and remain primarily unconsidered. At the same time, the processes are mentally detached from the context with which they are, in reality, closely linked by manifold interactions. Even if it is undisputed that technical developments can lead to relevant improvements, past developments have shown that many problem areas cannot be solved by technology and selective measures. There is a great danger that a one-sided focus will obscure the view of other problem areas and the system as a whole.

Despite the considerable productivity gains, the primary production sector has been gloomy for years. Criticism of various grievances from the agricultural industry has become much harsher in recent years, thus damaging its image. From the perspective of rural economics, regulatory law primarily aims to ensure a minimum of protective measures that take account of the interests of the common good (Scientific Advisory Board of BMEL 2018). Since legal requirements at the national or European level apply to all primary producers, regulatory law does not seriously influence market processes. However, this changes if, in the case of a predominant export orientation to third countries, an increase in **minimum requirements** impairs the competitiveness vis-à-vis the competition abroad. From the point of view of agricultural economics, individual farm measures that go beyond the statutory minimum requirements justify financial subsidies by the state. However, for the statutory minimum requirements and requirements beyond them, positive effects are expected without individual farm **proof** of the services rendered for the public welfare interests.

To take account of the increased social demands at home, **brand programs** or seals of quality are repeatedly brought into play by the agricultural economy. These are characterized by the fact that producers comply with a higher standard than the legally prescribed minimum requirements in individual sub-areas and thus set new production standards. Whereas in the past, many attempts have failed to increase consumer confidence in the quality and safety of the products via branded products and quality marks (Spiller 2001), two strategies, in particular, have been successful in bucking the general trend of cost minimization and realizing higher product prices with increased production standards. On the one hand, these are organic farming and, on the other hand, the production of eggs in floor and free-range systems. In both cases, the requirements for production conditions are increased compared to the legal minimum level. Both strategies are not limited to regional or national levels but are established at the European level. Through corresponding trade agreements with third countries, they also develop their validity beyond the borders of the EU. The prerequisite for this was that the production method was defined by law and that independent institutions regularly monitored compliance with the higher production standards. This ensures that the raw goods processed into higher-priced products come from farms with higher production standards and that free riders do not mislead consumers. Finally, the aim is to ensure consumers retain confidence in the higher standards.

Production standards are a specific form of complexity reduction that is very much in line with the thinking and approach of economics. It is therefore not surprising that standardization is also very popular in the agricultural sector, both on the part of economists and policy-makers, and that they intensively promote it (Bundesministerium für Ernährung und Landwirtschaft 2020). In technology and economics, standardization corresponds to the unification of components, production methods, units of measurement, processes, structures, types, goods, or services. With standardization, both cost and efficiency advantages can be achieved. The cost benefits of standardizing production processes have been discussed in detail in Sect. 6.4. Because of saturated markets, diminishing marginal benefits from further productivity increases, and the damaged image of the agricultural industry, the potential effects of standardization in marketing are now coming to the fore in agricultural economic strategies. From a rural marketing perspective, it is essential,

> *"that the core of positioning is not aimed at the real, objectively measurable side of the offer (quality tests) and also not at the self-perception of the companies, but at the* ***evaluation by the demanders."*** (Spiller 2001)

The effectiveness of standardization is based primarily on the transmission effects generated by advertising (spill-over effects). This is the term used to describe the transfer of the image of a product or a company to another product or company. For example, the positive image of a **brand** can be transferred to a new product of the same brand. Transfer effects are particularly interesting for agricultural marketing if an associative equation of increased production standards with improved quality of products and processes is booming, and the equation of production standards with **"Quality standard"** catches on with the addressees. If one follows this associative transfer, all products in the production of higher demands placed in selected sub-areas are attributed to a higher level of quality. Since the highest legal requirements are placed on the production method in organic farming, these products are consequently ascribed the highest "quality standard." In this way, products whose production standards are situated between the lowest and the highest requirements also find a classification. In terms of transfer effects, these can also claim an additional price compared to the conventional goods, which is located between the highest and lowest prices. This also serves consumers who do not want to or cannot pay high prices for organic products and do not want to disregard a higher quality standard completely. The classification of products from pig farming according to the production standards of previously defined farming systems into categories from 1 (minimum standard) to 4 (analogous to the organic standard) is a prominent example of transfer effects through standardization. Transfer effects also take place in many other areas of consumer, animal, or environmental protection. They always occur when selective emphasis and individual measures induce an associative transfer to other qualitative characteristics. Although there is no quality assessment per se, the impression is given to the addressees that an increased level in one sub-area also includes all other areas.

With the help of standardization, the complexity of the interactions in an agroecosystem and the effects of individual measures and their consequences can be reduced to simple schemata. This facilitates communication between the supply and demand side. **Standards** form the bridge of understanding; they are the common denominator without further explanation of the complex background in connection with qualitative statements. Different standards serve the purpose of manageable product differentiation for trade and consumers alike and can be easily coupled with corresponding price differences. Standardization is synonymous with generalization. The fact that standardization does not automatically serve the purpose of improving the living conditions of farm animals or providing benefits for the environment or other common goods is of secondary importance for marketing as long as consumer confidence is not disappointed. What matters for marketing is not what improvements have been realized but what consumers think might have been realized with increased standards.

Unlike the standardization of technical products such as computers, cars, or technical processes, biological processes always result in many manifestations and variations in their effects. Consequently, no two natural products or foods based on biological processes are alike. Even if farm animals are raised in identical husbandry systems, the physiological and pathological manifestations of farm animals are very heterogeneous. The same applies to the behavior of farm animals. Standardization highlights individual aspects, such as the movement area of farm animals, in terms of their importance for protecting the animals. At the same time, this form of complexity reduction allows all other aspects to fade into the background. After all—according to the apparent basic idea—an improvement in one area is better than no improvement at all. This is where the connection, extensively researched by the cognitive sciences, comes into play: the human brain always focuses on one aspect by actively blocking the surrounding areas (Lega et al. 2019). A brain circuit filters out unwanted background noise or other distracting sensory stimuli. Even if there were concerns and an effort to understand and correctly classify the background, this would initially raise more questions than provide answers for the observer. This is all the more true if one does not have sufficient expertise to make a reliable assessment. Who has the time to approach the complexity of interactions when the offer of complexity reduction via standards already provides a means of differentiating products based on first impressions conveyed by advertising?

Classification of products by standards corresponds to categorization. **Categories** are needed to order the manifold impressions fed to the brain in the context of perception. In this sense, category formation is fundamental in interpreting and evaluating perceptual content. Categories are, therefore, basic concepts of thinking. According to Hofstadter and Sander (2018), a category is defined as follows:

> *"We understand a category to be a mental structure that emerges over time, containing information in an organized form that can be accessed given the appropriate circumstances. The act of categorization is the groping, graded, gradual association of an entity or situation with a category given in one's consciousness."*

Consequently, the categorization of facts takes place in the observer's mind by assigning information about the properties of an entity or situation to already existing mentally created categories. While categorization rather denotes the unconscious, intuitive, or traditional process of class formation for arbitrary objects of everyday perception, standardizations are the consciously planned ordering of knowledge in a concrete observation according to objectifiable, uniform criteria. A standard is, therefore, a comparatively uniform or unified way of describing something, which serves as a guideline for others, for example, when making a purchasing decision. Accordingly, marketing is very interested in using standardization in advertising messages to increase the sales of higher-priced products. In this context, it is exciting to find out which qualitative cues consumers respond to and how advertising messages affect the willingness to buy or pay additional prices. With the help of various publicly funded research projects, agricultural economics has supported the trade in better-aligning marketing strategies with consumers' responsiveness and willingness to pay when specific standards are offered and matching them with their associations (Grethe 2017). These findings have proven very helpful for food retailers, helping them increase sales of specific product lines and improve their image.

Instead of informing about the relationships between standards, the variation of quality attributes, and the background of advertising strategies, publicly funded agricultural marketing studies function as an auxiliary agribusiness science. This becomes particularly clear in connection with the **animal welfare** concerns anchored in wide circles of the population; however, analogies also exist in dealing with environmental services. As a comprehensive discourse analysis reveals, the "species-appropriate husbandry" of farm animals is used as an advertising claim that points to a higher "quality of products" and, therefore, justifies a higher price (Sauerberg and Wierzbitza 2013). In the communication with the consumer, the supposed higher quality of the products due to higher husbandry standards is not presented as an advantage for the quality of life of the farm animals, but in the sense of a higher quality product refined by a higher husbandry standard. In this way, the consumer can imagine acquiring a supposedly "healthier food" for himself and simultaneously relieve his conscience about the animal welfare issue. According to the authors above, the agricultural industry propagates to the outside world that "happy animals" provide higher-quality human food. On the other hand, the actual relevance of husbandry standards for the welfare of farm animals is completely overshadowed.

Husbandry standards are, on the one hand, a means of production for primary production and, on the other hand, a means of agricultural marketing for complexity reduction, categorization, and communication with consumers. In both cases, there is no verification that the effects attributed to the husbandry systems are produced in the farm animals. Extended movement areas offer the animals more opportunities to exercise their natural behavior. However, whether higher husbandry standards are associated with increased protection of farm animals from pain, suffering, and harm remains to be tested. Various reviews show that organic livestock with the highest standards of husbandry have, on average, the same degree of incidence of production diseases in dairy cattle, cattle for fattening, pigs, and poultry as conventionally kept animals (Åkerfeldt et al. 2021). Thus, the association between husbandry systems and reduced disease rates as an essential prerequisite for animal welfare proves to be an illusion. Sickness and **welfare** are mutually exclusive, although freedom from disease is only a necessary and not a sufficient prerequisite for welfare.

Even though representatives of agricultural economics have repeatedly been made aware of the extensive body of evidence which, from a scientific perspective, rejects an equation of husbandry standards with "animal welfare standards" as not robust, the representatives refuse to revise their positions. The example reveals an interest-led procedure in agricultural science and that the often-invoked self-correction of science rarely occurs across disciplines. The numerous review studies published on the relationships between husbandry conditions and the absence of disease underline the complexity of the interrelationships. **Animal protection service** is an overall performance of the farm management that cannot be reduced to individual factors without committing an inductive fallacy. The **"Quality standard"** used by agricultural marketing to advertise a higher level of quality and justify higher market prices is primarily a virtual idea of quality intended to evoke a response in the minds of consumers without having a sufficient connection to reality.

The renunciation of **external validation,** characteristic of the agricultural sciences (see Sect. 7.4.2), has paved the way for the belief that it is possible to dispense by checking the effectiveness of production standards in the operational context. Although agricultural scientists would be most capable with their methodological toolbox, they show limited interest in external validation of their findings. As outlined in Chap. 6, no fundamental interest in external validation can be expected from other stakeholders involved either. Ultimately, this might result in the positions previously held in public being subjected to revision.

While food retailers can generate different market prices with products allocated to various standards, of which they only pass on a fraction to primary producers, competition for cost leadership remains. Raised minimum standards act as a norm in the same way that statutory minimum standards do. All farms that want to sell their products under a defined label must adhere to the enhanced standard level. At the same time, they are forced to implement the improved minimum standard as cost-effectively as possible to be competitive with competitors in the higher segment. As a result, any measures that could improve the protection of common property will continue to be subject to cost. Standardization utilizing increased minimum requirements consequently does not solve the problems farmers face in securing economic existence.

Consumers who expect higher quality performance from higher production standards and underline this with a higher willingness to pay are being deceived about their expectations. Those who, by propagating higher production standards, claim to be contributing to the fulfillment of public welfare interests not only fail to provide evidence of a relevant effect in the context of the respective farms. Pretending that there is a fundamental change in the conditions of agricultural practice when some farms voluntarily commit to higher production standards takes the pressure for action off the remaining farms. At the same time, the undesirable side-effects and harmful effects of the production processes caused by these farms are lost from view. The same applies to the polluter pays principle. Instead, the weight of responsibility is passed on to consumers. After all, it is now supposedly in their hands and their purchase behavior whether there will be any change in the proportion of farms with higher production standards. Until then, farms can continue their production processes demonstrably damaging to public welfare interests. They can expand them further if the undercutting competition inextricably linked to export orientation requires it. In this way, the protagonists of complexity-reducing standardization perpetuate the suffering of farm animals, the damage to natural ecosystems, and the release of pollutants into the environment indefinitely.

8.5.2 Agroecological Approach

Autopoietic systems are functionally designed to survive as long as possible. To achieve this goal, complex interactions have developed in their evolutionary development. These not only promote the functionality of the processes within the system but also the ability to adapt to changes with which living systems are continuously confronted. Even the processes in a living cell as a first-order autopoietic system are highly complex. They can only be understood if they are seen as a purposeful contribution to the self-preservation of the cell. The same applies to the living organism (second-order autopoietic system), whose cell aggregations in the tissues and organs and their network of interrelationships have evolved to maintain and reproduce themselves and to adapt as efficiently as possible to the given habitats. Adaptation occurs via morphological and physiological modifications in phylogenesis, ontogenesis, and the adult organism, enabling the organism to react to changes in the internal and external environment in the long, medium, and short term.

By analogy to the adaptation processes of 1st and 2nd order autopoietic systems, it makes sense to conceive of an **ecosystem** as a 3rd order autopoietic system. As in an organism, the individual subsystems are part of a close network of effects in which the coordinated interaction of the parts determines the manifestations of the whole. The interactions between the subsystems follow well-rehearsed paths that have proven themselves in evolution. Efficiency in the conversion of substances acts as a decisive selection criterion within and between the species competing for the limited resources available in an ecosystem. Unlike in natural ecosystems, the adaptations of an agroecosystem to changes within and outside the system boundaries do not occur via evolutionarily optimized self-organization processes. In agroecosystems, people are part of the system. Farm managers and others involved in decision-making and implementation processes have a significant steering role, helping to determine the success or failure of adaptation efforts. This challenge cannot be mastered without various forms of complexity reduction.

From an agroecological perspective, the first step is to broaden the field of vision for all actors by including additional perspectives. In addition to the predominant agroeconomic perspective with its focus on monetary parameters, this also includes the perspective of the technical sciences (see Sect. 7.4.4) and the utilization of technical developments for agricultural production processes. Technical devices can appear quite complicated if one needs to learn the processes underlying the device's function. In contrast, experts usually understand technical processes easily. Technical devices are, therefore, not complex. Subjectively, something appears complicated if one does not see through it. Consequently, the extent to which complex procedures can be understood and difficult tasks can be solved depends on the competence and the available resources of time and money. In the case of living systems, complexity arises from the interaction of subsystems with additional influencing variables. Effects can rarely be traced back to single causes and do not change linearly. Instead, unforeseen developments are among the characteristic elements of complex systems. Accordingly, complex issues are only accessible to simple solutions to a limited extent.

Technical processes have the advantage over complex facts in that they are easy to interpret in terms of functionality. Even if the inner workings of a technical device are not accessible to most users, the user can still judge whether the device is functioning. Malfunctions are an expression of an impairment of the functionality of a system. If a technical device such as a computer or a tractor malfunctions, its usability for the intended purpose is limited or even impossible. The causes of malfunctions are usually first attributed to the technology itself and not to an operating error, which would be the user's responsibility. If the user himself cannot remedy the malfunction, someone who is more familiar with it is called in. Depending on the extent of the malfunction, a repair or a replacement of the technical device with a new acquisition is considered. The order for repair is usually only given if the repair costs appear to be more favorable than the purchase of a new device. Compared to an old device, a new device also allows a gain in technical progress to be expected. Since technical equipment is generally indispensable for maintaining agricultural production processes, the restoration of the functionality of technical equipment in the event of malfunctions is rarely in question.

This fundamentally differs for biological systems, i.e., complex autopoietic and adaptive systems. The assessment of impairments to the functioning of living systems does not follow a binary classification: functioning or not functioning. Impairments of biological systems rank on a scale from very minor to major. Even recognizing biological malfunctions requires a certain amount of effort and competence; there are no control or warning lights here to indicate malfunctions. Unlike technical devices, living systems have a certain degree of regenerative capacity. This is part of the adaptability of living systems. While the organism tries to avoid the occurrence of severe disturbances by adaptation processes, organisms can fall back on a repertoire of self-healing processes to deal with disturbances that have already occurred and maintain the organism's survivability as far as possible. Depending on the severity of the disturbances, the original functional state can even be restored entirely (*restitutio ad integrum*). However, a variety of prerequisites are necessary for this. These include the elimination of the causes of the disturbance. Since this is often not the case to the required extent on farms, self-healing processes are frequently not successful. Usually, they are associated with more or less pronounced subclinical or clinical diseases. In this sense, disorders in farm animals express an excessive demand on the animals' ability to adapt and regenerate to the given living conditions (Sundrum 2015). This limits the vital forces, but not necessarily to the extent that the farm animals can no longer produce.

Managing an agroecosystem does not necessarily require detailed knowledge and understanding of the physiological and pathological processes in living organisms such as crops and livestock. However, it would be helpful if symptoms indicating disorders were identified at an early stage and used as an opportunity to get to the bottom of the causes of the disorders with the help of specialist expertise. As soon as these have been identified in the course of diagnostic investigations, it would be expedient to eliminate the causes of the disturbances and to create living conditions that support the organisms in their ability to regenerate. However, farms have considerable differences in dealing with disturbances in living organisms. The backgrounds are manifold and complex and thus seldom directly comprehensible to outsiders. In connection with disturbances of farm animals, it is likely to be essential whether the farm animals are regarded primarily as means to the end of monetary value creation or also as living beings that have the end of securing their ability to survive. There are fluid transitions between these cornerstones of the approach and a wide range of variations concerning the associated outcomes.

In an agroeconomic view of livestock as a means of production, the focus is that functional disorders in livestock are associated with economic losses in the form of productivity losses due to reduced output and additional expenditure. The extent of the disturbances is assessed from the perspective of management, i.e., from an anthropocentric point of view, in terms of economic relevance for farm profits. This is contrasted with an agroecological viewpoint, which sees the disturbances that occur as a threat to the organisms' ability to function and survive, resulting from disturbed interactions with the living conditions for which the farm management is responsible. Whereas malfunctions in technical equipment generally cause prompt rectification or replacement of defective equipment, malfunctions that occur in farm animals are only sometimes rectified and often only after a time lag. The typically high prevalence of production diseases in farm animals (see Sect. 4.1.) justifies the statement that health problems in farm animals often do not have absolute priority in the range of management tasks. This is probably also related to farm animals still producing despite diseases. For example, health disorders such as pneumonia in fattening animals or limb diseases in dairy cows do not immediately affect meat or milk production or only reduce it slightly. From an economic perspective, the question arises whether the effort required to restore the animal to an unimpaired state of health is worth the effort involved. However, the efforts that would have to be expended to improve the living conditions of the farm animals to prevent health disorders and production diseases is a major unknown unless appropriate measures are implemented in the farm-specific context up to the point of success. Instead, the inherent self-healing powers of farm animals may be considered the more cost-effective alternative to enlisting veterinary expertise and extensive treatment measures. As an alternative, supposedly more cost-effective attempts at self-medication are undertaken. If these are unsuccessful, in the end, there is still the option of slaughtering diseased farm animals and replacing them with animals with a higher genetic performance potential.

In theory, costs and benefits can be easily reconciled mathematically. The difficulty, however, lies in precisely quantifying the costs and benefits and correctly assessing or predicting the functional relationship between the two parameters while considering the diminishing marginal utility in the respective operational context. The difficulties are due not least to the complexity of the processes in agroecosystems. Unlike technical equipment, the interactions within and between livestock and the respective living conditions can only be penetrated mentally to a limited extent. Accordingly, management decisions in dealing with disturbances are challenging. This is all the more true if those responsible want to save themselves the trouble of weighing up the costs and benefits by utilizing comprehensive diagnostic measures, weak point analyses, and economic calculations on a sound data basis. From a psychological point of view, the situation is further aggravated by the fact that those responsible would be required to undergo an ongoing self-critical analysis about their share of responsibility for the health problems of farm animals. After all, the primary causes of production diseases are home-made. They arise when management fails to provide farm animals with the resources and protection they need to adapt successfully. For many livestock farmers, it is more obvious to declare the livestock themselves, with their specific genome, as the weak point and to rely on the hope principle that breeding measures could succeed in reducing the extent of health disorders in livestock, at least in the long term, even though since several decades these attempts have failed to succeed (Brito et al. 2021).

The limitations in the availability of time and monetary resources in dealing with disturbances that repeatedly occur in agroecosystems at various points on the different process levels reveal the dilemma in which many responsible parties find themselves. If they were to follow the requirements of the Animal Welfare Act consistently, then pain, suffering, and damage to farm animals could be reduced to a minimum. After all, every single animal is a living being capable of suffering. Without "reasonable cause," they must not be stressed or harmed. The EU regulation on organic farming even stipulates that every sick animal must be treated as soon as possible and, if necessary, separated from the healthy animals. However, this legal requirement is not consistently monitored and is rarely implemented appropriately in farming practice (Keller et al. 2019). On the other hand, consistent implementation of the guiding principle of the Animal Welfare Act would go beyond the economic framework set by market prices. The idea that farms competing for cost leadership could organize their operating systems to prevent increased incidence of production diseases in the first place is way off the mark.

Since disorders and production diseases are context-specific, they cannot be addressed by increased minimum requirements in partial aspects but must be investigated and resolved in the context of the overall system of an agroecosystem. Health disorders result from the interactions between individual animals and their living conditions. The individual animals are, therefore, the decisive **reference system**. To reduce disturbances and dysfunctions, the individual animals would have to be accommodated in their respective individual performance, need, and requirement situation to a significantly greater extent than has been the case to date. This would not only run counter to a general trend of simplification and reduction in the complexity of management measures but would also require additional resources that are hardly available in farm structures that have been trimmed to reduce costs. Furthermore, additional expenditure would be necessary to review the success of targeted measures in the farm-specific context to derive further conclusions and necessities for action.

Based on the results of numerous scientific studies, there is no fundamental doubt that production diseases are not the problem of individual farm animals and farms but are the result of the intensification processes in livestock farming. Nevertheless, the disorders are still not perceived and acknowledged by interested parties as a problem inherent in the agricultural system. As was comprehensively reflected in Chap. 6, neither the processing industry (including slaughterhouses and dairies) nor food retailers or various NGOs have taken an interest in the fact that under the given conditions on many farms, management is not up to the challenges concerning the organization of a largely disturbance-free system. The high prevalence of mortality and production diseases shows not only the overstrain of the adaptability of farm animals to suboptimal living conditions but also the insufficient protection offered by farm animal owners. At the same time, the overstrain of the farms in successfully adapting to the market conditions becomes visible. Thus, the economic system of the "free" market, oriented towards cost leadership, prevents farms from trying to avoid disturbances relevant to animal health and welfare.

What was discussed in the example of production diseases of livestock about the areas of conflict between particular and common interests in the agricultural economy also applies to other problem areas. These include contaminating plant and animal products with pathological and antibiotic-resistant germs affecting consumer protection. In plant cultivation, plant protection measures and the use and, thus, the discharge of pesticides are particularly relevant. Since agroecosystems are open systems, disturbances are not limited to on-farm areas. With their discharges of waste and pollutants into the environment, the managers of agroecosystems are responsible for considerable harmful effects (see Sect. 5.3) without being prosecuted for them to date. The external impacts of agriculture could best be countered by limiting the discharge of waste and pollutants caused by production processes to a minimum in relation to the quantity of raw goods sold. Instead of enforcing the polluter pays principle, the state merely complies with minimum legal requirements. This is happening despite the scientific evidence that the many billions spent by the European Union's Common Agricultural Policy on promoting sustainability have failed to achieve their objective (Scown et al. 2020). The European Court of Auditors also leaves no room for doubt in its assessment that the agricultural subsidies paid by the EU between 2014 and 2020 have not affected the protection of the environment; instead, emissions have been more or less stagnant for many years. Although minimum requirements have proven ineffective, agricultural policy continues to adhere to them.

Farm management that wants to meet the societal demands for reducing pollutant emissions from farms is confronted with considerable conflicts between economic and ecological requirements without any support from agricultural policy. An agroeconomic approach focuses on maximizing the utilization of production potentials in the individual farm sectors. External effects are neglected, as are the intra-farm interactions between subsystems. For example, the fertilization measures for particular crops are geared to the maximum possible yields. Because of the site-related differences between the individual arable and grassland fields, considerable overfertilization inevitably occurs (see also Sect. 5.1.2). These can only be limited if fertilization measures are not geared solely to maximizing yield but also to the pre and post-cropping effects and reducing external effects. The same applies to a diet for livestock that is geared to the individual needs of the animals. Adaptation measures are associated with additional expenditure. Many practitioners assume that these extra costs do not pay off economically, so they are seldom considered, let alone implemented.

From an agroecological perspective, applying means of production tailored to the subsystems, i.e., adapted to the specific demand situations, is one of the priority strategies for reducing the discharges of residues and pollutants from the farm systems. The more the supply corresponds to the demand, the more efficient the use of the means of production. However, this requires evaluating the output-to-input quantities ratio in the various subsystems to calculate the input's **efficiency**. Only in the respective **context** of the subsystems is the assessment of the optimal relationship between input and output in the use of means of production meaningful. An agroecological approach, therefore, requires, first of all, a comprehensive and profound data basis and evaluation. This includes, among other things, knowledge about the extent of the material flows between the individual subsystems of an agricultural operating system. In Fig. 8.2, the possibilities of a hierarchical structuring of an operating system into subsystems at different process levels are shown, with which the control of material flows and targeted distribution of resources can be organized according to the respective demand situation.

Subsystems are characterized by inputs and outputs of specific substances. With the aid of available parameters and estimating equations, the inputs and outputs can be quantified sufficiently reliably, and the efficiency of substance conversion in the individual subsystems can be recorded and compared from the output ratio to the input variables. The better the data basis and the more differentiated the evaluation, the more targeted measures can be taken, and the more efficiently the resources available in the overall

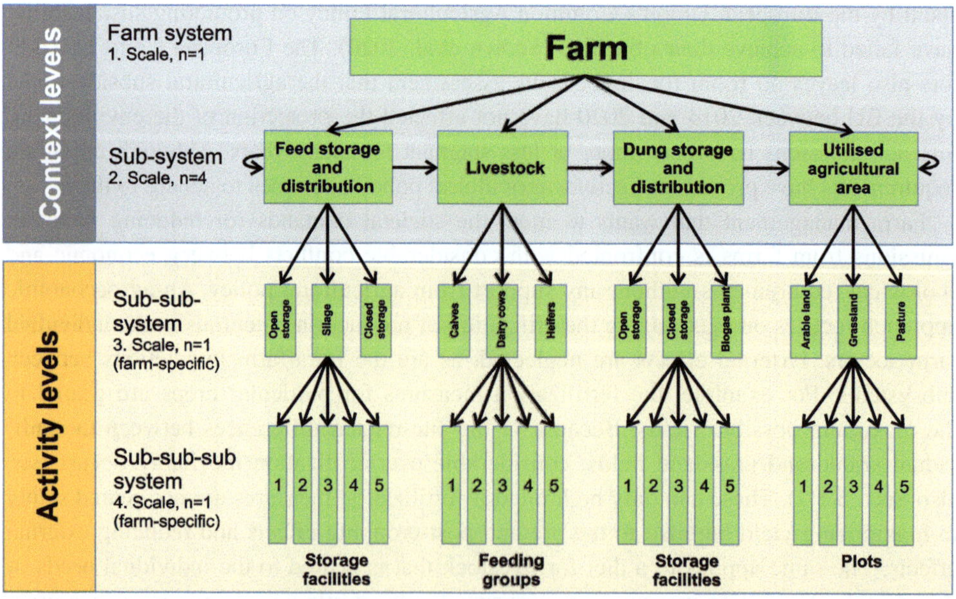

Fig. 8.2 Organization of a nutrient cycle between subsystems at different process levels and hierarchical structuring of subsystems to control material flows and allocate resources at different process levels (Sundrum 2019)

system can be used to produce sales products (Sundrum 2019). Other strategies include avoiding inefficient purchases of fertilizers and feedstuffs and increasing the organic binding capacity of nutrients to prevent them from escaping into the environment. By creating suitable conditions, the recycling of nutrients in the **material cycle** of an agroecosystem and a considerable potential for increasing efficiency can be realized. No less important is the fact that knowledge of the specific requirements of a given plot or farm animal can facilitate a more targeted **allocation** of the limited **resources** available. In this way, resource use efficiency can be increased, material discharges into the environment can be reduced, as well as the occurrence of disturbances.

It is not economic economies of scale but ecological synergy effects representing the value creation potential to be exploited in a win–win-win situation. This requires structural **rationality**. The coordination between the subsystems and the sub-goals and the decisions and implementations aimed at efficiency follow the path from a superordinate orientation to the sub-actions, i.e., from the whole to the parts and from the parts to the whole in an iterative approach. The manager does not act selectively rationally but structurally rationally when he refrains from maximizing in the sub-areas if this is incompatible with the structural optimization of the overall operational performance (Nida-Rümelin 2020). **Selective rationality**, as it is anchored in the utility theorem of economics and is also pursued by agricultural economics, is incompatible with structural rationality, which forms the basis of an agroecological approach.

An agroecological approach is oriented towards **cooperation**. On the one hand, this extends to the on-farm interactions between the subsystems of an agroecosystem. On the other hand, cooperation can also be essential beyond the farm level. This applies, for example, to producer groups, which, in addition to agricultural enterprises, can also involve processing and trading organizations and consumers. The latter may be involved in a producer group to a greater or lesser extent. Still, they contribute to the maintenance of the producer group primarily through their willingness to pay additional prices. For Nida-Rümelin (2020), cooperation is the paradigmatic case of structural rationality. Rationality theory fails when it comes to the phenomenon of cooperation. Cooperative motives for action require a break with an exclusive orientation towards punctual utility optimization. The actors' intentions guiding their actions form the bridge from a punctual to a structural perspective. The will to cooperate is an expression of belonging to a community. The individual wants to contribute his share to the possible joint success, expecting others to contribute also. To do this, the individual actors must assess what functions the individual actions have in a larger structural fabric to judge whether the selective actions are reasonable. Otherwise, they are still left to trust that they are performing a task relevant to maintaining the overall system in the positions assigned to them by others. Nida-Rümelin (2020) expresses this as follows:

> *"Control is not from punctual to structural, but from structural to punctual action. (...) The rationality of punctual action results from the rationality of overall intentions. (...) Accordingly, we act structurally rationally insofar as we refrain from optimization by individual acts if this optimization is incompatible with structural optimization."*

Derivations

Farms are complex adaptive agroecosystems. The ability of the **management** to design appropriate internal structures and to organize processes decisively determines the overall performance of the operating system. This ability also determines whether adaptation to changes in the outside world is successful and whether the economic existence of the farm can be secured for a foreseeable period. The mental access of the management to the complexity of the operating system, the strategies of complexity reduction, as well as the handling of goal conflicts and disturbances have a decisive influence on whether and to what extent the value creation potentials inherent in the operating system can be realized and whether a satisfactory balance can be achieved between the particular interests and the **public welfare interests**.

There are severe differences between an agroeconomic and an agroecological approach, both in the prioritization of objectives and in dealing with the complexity of biological interactions. Due to the internal limitations in the availability of resources and the associated conflicts of objectives in their allocation between the different areas, a one-sided focus on increasing performance and minimizing costs is always at the expense of other system-relevant areas. It is, therefore, all the more critical to identify those measures in the farm-specific context whose implementation considers both the farmer's economic interests and the ecologically oriented public welfare interests. Instead of exacerbating conflicting goals by focusing on specific areas, conditions that promote synergistic effects and benefit ecological and economic requirements should be created.

A considerable loss of information accompanies the complexity reduction strategies resulting from an agroeconomic approach. With an under-complex approach, production processes cannot be oriented towards synergistic effects. The primary reference points of an agroeconomic approach are the yield quantities and the sales value of the raw goods. The prices that primary producers receive for saleable raw goods are often not correlated with the monetary costs or the extent of undesirable side effects of the production processes and the product characteristics of the raw goods. As long as the adverse side effects of the production processes are not assessed and monetarized, they are irrelevant to the agricultural economic approach. The focus on market success shows the core problem of an economic way of thinking: it largely ignores the undesirable side effects and harmful effects. This means that it does not have the necessary problem-solving competence to integrate processes in the sub-sectors into an overarching strategy for the good of the whole.

Although various new orientations are propagated in strategy papers, the economic premises that cause the problems remain largely untouched. This also applies to the access of the agricultural economy to quality production, which is pursued exclusively via **standardization**. In this way, the intra- and inter-farm variations in the quality services produced are not made accessible to differentiation but concealed. Without access to the significant variation in the characteristics and the desired and undesired results of processes, the agricultural economic explanatory approaches are denied access to the ambivalence inherent in all agricultural production processes. Thus, the prerequisites for assessing production processes and their marginal utility trajectories in the respective contexts and the significance of disturbances for the functioning and efficiency of subsystems are also missing. **Selective rationality** inherent in the agricultural economic way of thinking thus proves counterproductive when trying to solve complex problems.

References

Åkerfeldt MP, Gunnarsson S, Bernes G, Blanco-Penedo I (2021) Health and welfare in organic livestock production systems—a systematic mapping of current knowledge. Org Agric 11:105–132. https://doi.org/10.1007/s13165-020-00334-y

Anonym (2020) Studie des AFC Risk & Crisis Consult. Allgemeine Fleischerzeitung 138:24

Antonopoulos I, Bell MJ, Čavoški A, Petetin L (eds) (2022) The governance of agriculture in post-Brexit UK. Earthscan food and agriculture series. Routledge, New York

Baecker D (2014) Organisation und Störung: Aufsätze, 1st edn. Suhrkamp-Taschenbuch Wissenschaft, vol 2012. Suhrkamp, Berlin

Blume L, Hoischen-Taubner S, Möller D, Sundrum A (2021) Status quo der nutritiven und ökonomischen Situation sowie Potentiale des Einsatzes heimischer Proteinträger auf ökologisch wirtschaftenden Geflügel-und Schweinebetrieben: Teil 1: Bedarfsgerechte Nährstoffversorgung, Tierverluste, Ressourceneffizienz und Wirtschaftlichkeit. Ber Landwirtsch 99. https://doi.org/10.12767/buel.v99i2.349

Brito LF, Bedere N, Douhard F, Oliveira HR, Arnal M, Peñagaricano F, Schinckel AP, Baes CF, Miglior F (2021) Review: genetic selection of high-yielding dairy cattle toward sustainable farming systems in a rapidly changing world. Animal 15(Suppl 1):100292. https://doi.org/10.1016/j.animal.2021.100292

Bundesministerium für Ernährung und Landwirtschaft (2020) Empfehlungen des Kompetenznetzwerks Nutztierhaltung. https://www.bmel.de/SharedDocs/Downloads/DE/_Tiere/Nutztiere/200211-empfehlung-kompetenznetzwerk-nutztierhaltung.html

Crutzen PJ, Stoermer EF (2000) The "Anthropocene": Global Change Newsletter 41:17–18

DFG (2005) Future perspectives of agricultural science research. Wiley-VCH, Weinheim

German Farmers' Federation (ed) (2020) Situationsbericht 2020/21: Trends und Fakten zur Landwirtschaft. Situationsbericht, 2020/21. Deutscher Bauernverband e. V., Berlin

Grethe H (2017) The economics of farm animal welfare. Ann Rev Resour Econ 9:75–94. https://doi.org/10.1146/annurev-resource-100516-053419

Habel J, Sundrum A (2020) Mismatch of glucose allocation between different life functions in the transition period of dairy cows. Animals 10:1028. https://doi.org/10.3390/ani10061028

Habel J, Uhlig V, Hoischen-Tauber S, Schwabenbauer E-M, Rumphorst T, Ebert L, Möller D, Sundrum A (2021) Income over service life cost—estimation of individual profitability of dairy cows at time of death reveals farm-specific economic trade-offs. Livest Sci 254:104765. https://doi.org/10.1016/j.livsci.2021.104765

Haeckel E (1866) Generelle Morphologie der Organismen: Bd. 2 Allgemeine Entwickelungsgeschichte der Organismen, vol 2. G. Reimer, Berlin

Hofstadter DR, Sander E (2018) Die Analogie: Das Herz des Denkens. Klett-Cotta, Stuttgart

Hoischen-Taubner S, Habel J, Uhlig V, Schwabenbauer E-M, Rumphorst T, Ebert L, Möller D, Sundrum A (2021) The whole and the parts—a new perspective on production diseases and economic sustainability in dairy farming. Sustainability 13:9044. https://doi.org/10.3390/su13169044

Kahneman D (2011) Schnelles Denken, langsames Denken, 11th edn. Siedler, München

Keller D, Blanco-Penedo I, de Joybert M, Sundrum A (2019) How target-orientated is the use of homeopathy in dairy farming?—A survey in France. Germany and Spain. Acta Veterinaria Scandinavica 61:30. https://doi.org/10.1186/s13028-019-0463-3

Lega C, Ferrante O, Marini F, Santandrea E, Cattaneo L, Chelazzi L (2019) Probing the neural mechanisms for distractor filtering and their history-contingent modulation by means of TMS. J Neurosci 39:7591–7603. https://doi.org/10.1523/JNEUROSCI.2740-18.2019

Liessmann KP (2021) Alle Lust will Ewigkeit: Mitternächtliche Versuchungen. Paul Zsolnay Verlag, Wien

Maturana HR, Varela FJ (1980) Autopoiesis and cognition: the realization of the living. Boston Studies in the Philosophy of Science, vol 42. Springer Netherlands, Dodrecht

von Neumann J, Morgenstern O (1947) Theory of games and economic behavior, 2nd edn. Princeton Univ. Press, Princeton

Nida-Rümelin J (2020) Eine Theorie praktischer Vernunft, 1st edn. de Gruyter, Berlin/Boston

Niemann H-W (2009) Europäische Wirtschaftsgeschichte: Vom Mittelalter bis heute. Geschichte kompakt. Wiss. Buchges, Darmstadt

Poser H (2016) Homo Creator: Technik als philosophische Herausforderung, 1st edn. Anthropologie – Technikphilosophie – Gesellschaft. Springer; Springer Fachmedien Wiesbaden; Springer International Publishing AG, Wiesbaden, Cham

Potofski U (2014) Entscheidend ist auf'm Platz: Die verrückte Welt des Fußballs und seiner Kommentatoren. Gütersloher Verlagshaus, S.l.

Reckwitz A (2019) Das Ende der Illusionen: Politik, Ökonomie und Kultur in der Spätmoderne. Edition Suhrkamp, vol 2735. Suhrkamp, Frankfurt

Ricklefs RE, Relyea R (2014) Ecology: the economy of nature. W. H. Freeman and Company a Macmillan Higher Education Company, New York, NY

Sauerberg A, Wierzbitza S (2013) Das Tierbild der Agrarökonomie. Eine Diskursanalyse zum Mensch-Tier-Verhältnis. In: Pfau-Effinger B, Buschka S (eds) Gesellschaft und Tiere: Soziologische Analysen zu einem ambivalenten Verhältnis, 1st edn. Springer; Springer Fachmedien Wiesbaden; Springer International Publishing AG, Wiesbaden, Cham, pp 73–96

Scientific Advisory Board of BMEL (2018) Für eine gemeinwohlorientierte Gemeinsame Agrarpolitik der EU nach 2020; Grundsatzfragen und Empfehlungen. https://www.bmel.de/SharedDocs/Downloads/DE/_Ministerium/Beiraete/agrarpolitik/GAP-GrundsatzfragenEmpfehlungen.html

Scown MW, Brady MV, Nicholas KA (2020) Billions in misspent EU agricultural subsidies could support the sustainable development goals. One Earth 3:237–250. https://doi.org/10.1016/j.oneear.2020.07.011

Spiller A (2001) Gütezeichen oder Markenartikel? Fleischwirtschaft 81:47–50

State Historical Society of Iowa (2004) Every farm a factory: the industrial ideal in american agriculture. Ann Iowa 63:86–88. https://doi.org/10.17077/0003-4827.10780

Sundrum A (2015) Metabolic disorders in the transition period indicate that the dairy cows' ability to adapt is overstressed. Animals 5:978–1020. https://doi.org/10.3390/ani5040395

Sundrum A (2019) Real-farming emissions of reactive nitrogen—necessities and challenges. J Environ Manage 240:9–18. https://doi.org/10.1016/j.jenvman.2019.03.080

Sundrum A (2020) Schwanzbeißen – ein systeminhärentes Problem. Der Praktische Tierarzt 101:1213–1227. https://doi.org/10.2376/0032-681X-2046

9 Quality Production—The Most Promising Approach for the Realization of Public Welfare Interests

Abstract

Public welfare interests must not only be defended against the dominance of particular interests. The latter must, at the same time, be integrated into an overall strategy. This requires an approach that can hold its own in theory and in the context of very heterogeneous initial and boundary conditions. Evidence-based quality production proves to be the key to accessing the realization of common good interests. Process qualities such as animal and environmental protection services correspond directly to improving common goods. At the same time, they offer options for increased value creation through synergy effects. In connection with increased product quality, the farms' gradually graduated animal and environmental protection services can create the preconditions to compensate for the indispensable additional labor and financial costs. A well-balanced mix of political and market instruments will be needed to reward farms' quality and public welfare services with appropriate compensation. This raises various questions concerning a valid assessment of quality services, the definition of system boundaries, and how to deal with the actors necessary to achieve a breakthrough in quality production.

It should be apparent from the preceding that, in addition to the undesirable developments based on selective rationalization, it is, above all, the economic dictum of cost minimization that stands in the way of realizing **public welfare interests**. Clear objectives and quality assurance measures for implementation and access to the necessary resources are required. Two sources come into consideration: On the one hand, these are additional financial resources from the public purse and the redistribution of previous subsidy payments to those who deliver common goods above the average while removing subsidies from those over the average, producing negative externalities. Secondly,

A. Sundrum, *Public Welfare-Oriented Production of Food*,
https://doi.org/10.1007/978-3-662-69040-6_9

quality production will not be possible without higher prices for food products, which consumers have to pay at the point of sale, and from which it must be ensured that a large proportion reaches those who realize public welfare interests. The prerequisite for increased welfare benefits is an increase in counter-benefits, which are directly linked to the concept of "quality production." Quality production is the key to generating additional income for increased benefits. However, preconditions are a turning from the previous focus on **cost leadership** to **quality leadership**. Making this shift requires more than setting new production targets. As outlined in the last chapter, it means nothing less than turning away from traditional decision-making, action processes, and thought patterns with a change from selective to structural rationalization. Such a change cannot be accomplished overnight. It takes time, even if many believe they do not have it. Nor will it be possible without the support of a cooperative collaboration of diverse actors from different groups. In addition, various forms of state regulation are needed to support the reorientation processes and, with the help of suitable guard rails, to protect against a relapse into old behavior patterns. In return, a win–win-win situation beckons, in which the interests of the **primary producers** and citizens about the common good are protected. Moreover, the interests of consumers looking for an excellent cost-benefit ratio must be brought out to mutual advantage. Various prerequisites must be fulfilled to establish quality production as an alternative to the current orientation of agricultural (mass) production. The potential of evidence-based quality production is explained below.

Moving away from the cost-minimization strategy

The agricultural sector can look back on a great success story. What is most impressive is the considerable growth in the amount of food produced per year. Still, productivity increases have been achieved through various strategies for intensifying production processes. These successes are the basis for the considerable increase in the world's population, which could not have occurred without the achievements of the agricultural economy. Primary producers and members of other interest groups involved in these developments could be proud of this success if it did not also come at a very high price. The pioneers of productivity increases could not yet have imagined this price. However, the adverse knock-on effects are evident today, although not all those in charge are prepared to see them. The knock-on effects tarnish the successes achieved so far but are cause to fundamentally question the continuation of previous agricultural practices and the dominance of agroeconomic patterns of thought.

The one-sided orientation of the agricultural economy towards productivity advances is still justified by the need to feed the growing number of world citizens. The extent to which this justification is misleading can be seen, among other things, in the fact that despite many years of sustained surplus production, it has not been possible to substantially reduce the number of hungry people on the globe (Food and Agriculture Organization 2018). At the same time, future production potentials are being reduced on a large scale by, among other things, soil erosion, soil compaction, humus, and nutrient losses caused by current production processes. In addition, a significant proportion of arable land is used not for food production to feed insufficiently nourished people but for producing energy-based biomass. In the production chain from harvest to household, a considerable proportion of the food produced at great expense ends up in the garbage, even though the food is still fit for consumption. The extent of internal and external disturbances has long since exceeded the threshold of what is still reasonable. As if the damaging effects were not dramatic enough, the economic structures rob most primary producers of their livelihood sooner or later. A negative marginal utility is increasingly eating up the successes achieved in the past and requires a fundamental reflection on previous food production strategies.

The cost-minimizing undercutting competition to which primary producers are subjected in expanding participation in world trade undermines the availability of resources farm management needs to contain the undesired and harmful side effects. Undercutting competition represents a global exploitation machine for the benefit of the few and at the expense of the many. Suppose leading agricultural economists have their way, in addition to the billions in subsidies already pumped into agriculture for many decades. In that case, billions more will be wrested from taxpayers to enable domestic farms to continue participating in this undercutting competition. The justification is that the additional money will be used to contain the harmful effects on public welfare interests. However, the newly launched environmental or animal protection programs appear to be a strategy to conceal a "business as usual." After all, the initial constellations that are primarily responsible for the harmful effects and the driving forces of the market economy are to remain largely untouched. This means that the true extent of the negative consequences of the economic system and the reasons behind them continue to be ignored.

Psychology, as the science of human behavior, has produced many findings that can be used to explain what tempts people to bury their heads in the sand figuratively. Uncomfortable truths always provide a solid motive to doubt them. On the other hand, "business as usual" is not only problematic because it offers no solutions to increasingly pressing problems. By pretending to have solutions at hand but, in reality, only offering pseudo-solutions, much time is wasted, and many billions continue to be invested in an ailing economic system, which is urgently needed for a reorientation towards an eco-social market economy. For decades, farms have been forced to adapt their production processes to falling market prices. In the majority of cases, they received too little in return to be able to secure their economic existence. It is time to end the ruinous dependency ratios and the doomed attempts to surrender to global undercutting competition.

Alternatives are necessary if only to enable a change of perspective to the prevailing explanatory approaches as an intellectual counter-model and to break through the narrowness of thought. Further explanations should be preceded by the fact that not all agricultural enterprises can follow the path of quality production outlined below. Nor can it be the solution to all the problems previously addressed. Just as the nomads, in their search for new sources of food, have to leave behind those who are not able to contribute to the survival of the community and endanger the survival chances of others, the development of new value-added strategies cannot be about taking all farms along this path.

The slogan issued by representatives of the farmers' associations that all farms should be taken along on the path of change in the upcoming innovations is not only an expression of a denial of reality. Still, it ignores the reasons for the decline of farms that have been going on for decades. The notion of farms as part of a large community of like-minded people that this invokes serves primarily as a defensive bulwark against any attempts to differentiate farms in terms of their contributions to the public welfare interests. Such an approach undermines the position of farmers' unions. Accordingly, those who, through their belief in progress and their blindness to the negative consequences, have led the agricultural economy to the sidelines and very many farms into economic misery are trying to save their position by invoking the spirit of a community of solidarity that has never existed in this way.

In contrast to generalized and thus misleading proposals for solutions, the following remarks take as their starting point the preconditions that must be in place for an agroecosystem to survive in the long term. First, the question of how to escape the constraints of undercutting competition and generate financial resources to gain new room for maneuver is at the forefront. The prerequisite is that internal resources must no longer be drained in cost minimization strategies but can be used for increased value creation. There is no alternative to producing higher-quality foodstuffs, as this is the only way to achieve a higher price on the markets and, thus, a greater availability of resources. The higher value must not be limited to the assessments of primary producers and trade but must also be compatible with the public welfare interests. It must also be a question of keeping cost and price increases within reasonable limits by exploiting synergy effects. The way out of a strategy of specialization and minimization of production costs, which has long since passed its zenith and is therefore caught in a trap of progress, requires additional resources, new production goals, changed approaches, and adapted methods. It is, thus, a question of producing services for public interests, which must be offered on the markets and for which a demand must be sought and found. Suppose one is still convinced—as most primary producers are—that the market will only demand cheap goods in the long run. In that case, one can say goodbye to the considerations of alternative strategies and does not have to read the book to the end if one has made it this far.

Organic farming and many other examples of branded programs are unquestionable evidence that higher-priced products also find sales on an appreciable scale. In further discussions, these and other examples serve to learn from experience with higher-priced products. In doing so, the author can look back on decades of experience in organic farming. However, the conclusion of the experiences is not to tout organic farming as a "gold standard" for solving many problems. Instead, the aim is to show which mistakes have been made in organic farming and thus should be avoided as far as possible if one wants to come closer to the abovementioned goal.

Despite various undesirable developments, the representatives of organic agriculture have succeeded in bundling the private initiatives that began about 100 years ago and developing them into a separate, globally operating market. Organic agriculture has been defined by production standards since 1954. The starting point was trademark law, which required clear criteria for labeling organically produced products. Due to the diversity of locations and resulting product characteristics, it was impossible to tie labeling to a product quality that could be precisely described and analytically traced. Therefore, the production method itself was elevated to a labeling criterion (Sundrum 2014). This fundamental principle has been retained and adopted in the EC regulation on organic production. In the EU, the developing market was protected from unfair distortions of competition and free riders by the EU Commission through legal rules in the 1990s. Since then, this branch has flourished with continuously increasing turnover rates. Organically managed farms' performance consists of the fact that they comply with the statutory organic standards. In many areas, these go far beyond the minimum legal requirements of conventional agriculture. Independent inspection bodies monitor compliance. Products that comply with the higher production standards can be labeled with the organic label and thus clearly distinguished from other products on the market. This prevents competitors from being lured in, offering supposedly equivalent products at a lower price. The following premises, in particular, are decisive: a clear definition of the production method, clear labeling of the products, and a higher demand compared to the supply of accordingly produced goods, which enable higher prices for partly significantly higher production costs.

As with conventional products, the relationship between supply and demand determines the price of organic products. Unlike in traditional agriculture, where an oversupply of raw goods floods the markets and pushes prices down to a low level that often does not cover production costs, the development of the supply of organically produced products is lower than the development of demand. The decisive factor for the scarcity of organic products is, on the one hand, an increased demand caused by the image of organic products and, on the other hand, the limited possibilities to significantly increase the supply. There are significant obstacles to conversion from conventional to organic farming methods. Among other things, sufficient acreage is required, mineral nitrogen fertilizers and pesticides are banned, and farm animals must be kept under high husbandry conditions. During a two-year conversion period, the various restrictions on the production method take effect, significantly reducing productivity and production output. However, the products produced during this period, such as "conversion goods," do not achieve the organic price level. If it were comparatively easy to convert to organic farming and receive higher selling prices for the products, many more farmers would take this step. On the other hand, if the increased supply of organically produced products were not accompanied by growing demand, the high price level would be jeopardized and the basis of the business model undermined.

A shortage of product supply to increase the sales price is a business model practiced in many sectors of the economy. Except for organic farming, this strategy could only be implemented in the agricultural industry for a few specially designated products, such as ham from Iberico pigs. On the other hand, the oversupply of conventional raw goods will hardly be contained in the future, as global production capacities for agricultural products are still far from being exhausted. If the market prices for conventional products were to rise, this would result in an expansion of production capacities in various countries of the world and, with a time lag, a renewed flattening of the price curves. In addition, a further threat scenario will provide conventional producers with further and very potent competitors in the future. The synthetic production of food, which is already being planned and launched on a large industrial scale, will further increase the food supply in the coming years. Given the potential attributed to the synthetic production of substitute products, some prophets even see the end of agriculture looming on the horizon (Stengel 2021). Whatever the scale of the increase in the production of substitute products, one does not have to be a prophet to predict that they will exert additional price pressure on conventional products due to the overhang of production capacity. Traditional production is threatened with the loss of relevant market shares. Added to this is Sect. 6.5 outlines the loss of image, stating that discrediting conventional products with substitute products will harm the agricultural industry. Because of the new competition, the golden age of agribusiness seems to have receded into the distance. The vision that has now faded was based on the expectation of a shortage in food availability, particularly of animal origin, which was expected to result from a drastic increase in demand from an affluent population in emerging countries.

For people who have a sufficient income, food is more or less easily accessible. What is limited is, at most, the willingness to pay an excessive price for it, i.e., a higher price than necessary. Only from foodstuffs that are available in limited quantities and at the same time enjoy sufficient demand among consumers can higher prices be realized that exceed production costs and thus are associated with an operating profit. Since higher-priced products are expected to have slow and limited market penetration, they incur significantly higher transformation costs, making marketing more difficult. The demand must, therefore, exceed a critical size so that the additional logistical expenses incurred from the farm to the consumer's fork can be covered in monetary terms.

To establish a successful and sustainable business model that builds on the higher price of the products, another prerequisite is elementary: The product must be of higher quality or at least have a high reputation so that customers are willing to pay a higher price. In the long run, it is not sufficient to advertise products with various attributes without them being associated with proof of superiority. In the case of organic farming, the wide range of variation between farms prevents a clear distinction between organically and conventionally produced products. Under the current framework conditions, those farms that pursue a minimalist implementation of the standards—i.e., behave conventionally—gain a competitive advantage. In contrast, farms that strive for and implement a high level of product and process quality must accept higher production costs in terms of working time and investments without the additional expenditures being rewarded separately. Standardizing higher minimum requirements, which characterizes organic farming and other brand programs, does not cancel out the efforts and competition for a successful cost-minimization strategy. In contrast, standardization based on the guidelines prevents the establishment of valid quality production that exceeds compliance with the minimum requirements. The additional costs associated with quality production are not rewarded in organic farming either and must consequently be avoided in the existing competition for cost leadership. It is, therefore, not surprising that higher product qualities are generally not achieved with products from organic production (Tauscher et al. 2003).

The conclusions for organic production derived from the outlined interrelationships, which apply similarly to other brand programs, can be summarized as follows: If the organic economy does not want to risk its credibility negligently, there are two options for action. The first is to emphasize certified compliance with the standards as a quality factor and, simultaneously, to refrain from declaring specific product and process qualities that cannot be guaranteed or to prevent other stakeholders (trade, processors, politicians) from doing so. The second option is the challenge of developing organic farming so that its success can be measured and recognized in terms of a high level of food quality and ecological performance (process qualities). According to the previous **paradigm**, compliance with the guidelines per se leads to a high level of product and process qualities that can no longer be maintained. Consequently, there is a need for a shift from a process-oriented (guideline-related) to an outcome-oriented approach. High product and process qualities are assessable. In the following, various examples outline the premises under which an increased value can be attributed to food, and an increased price can be justified to an enlightened consumer clientele.

Real versus imagined quality

Many consumers associate compliance with higher minimum requirements with the expectation that the higher requirements profile is not limited to individual features but also includes a higher level of other features that are not explicitly mentioned and, therefore, results in a generally higher level of quality. The (inductive) conclusion from individual aspects to the superordinate whole corresponds to a transfer effect, which the advertising industry uses in many ways. The aim is to avoid the deterrent effect of too much information ("information overload") and to evoke as much attention as possible from potential buyers with as little information as possible about the characteristics of a product to influence the purchase decision in this way. Such transfer effects are the reason for the success of branded products. If targeted advertising strategies can successfully shape the image of a **brand**, this image can also be transferred to new products. In this sense, foodstuffs produced according to organic farming guidelines are also branded articles. The image established over the years is more or less transferred to all products that bear the label.

Branded products have specific characteristics that distinguish them from products of other brands, particularly those without a brand name. Brands offer various advantages for customers (Häusel 2013). They have a recognition function as well as an information and orientation function. Customers associate specific characteristics with the brand and develop brand awareness based on subjectively assigned and positively colored attributions. From the point of view of effort and benefit of shopping, the trust placed in the brand is rewarded with a high degree of efficiency. Once a decision has been made in favor of a brand and the performance promises of a brand are trusted, this saves the time and effort of gathering information and the mental effort of weighing up different products as well as ongoing price-performance comparisons. In addition, a branded product is generally expected to protect against negative surprises. Brand owners themselves have a great interest in ensuring that the additional benefits of the brand communicated by the brand owner are fulfilled. After all, a loss of trust could lead to a corresponding reluctance to buy. Retailers also benefit from customers' loyalty to branded products. Intensively advertised brands increase awareness and thus reduce the sales risk. Due to customer loyalty to branded goods, the planning of purchase quantities is more predictable than with unbranded goods, whose characteristics and benefits cannot be estimated by customers. Since branded goods achieve higher market prices, they also increase the retailer's profit. Well-known branded goods are also well suited for promotionally effective price reduction campaigns. The associated low profits per product unit can be partly overcompensated by higher sales or additional profits on purchasing other products, providing retailers with financial benefits.

From a superordinate point of view, customer confidence in the superiority of a product advertised with individual quality attributes is only justified if an independent assessment concludes a higher qualitative level. In the case of sensitive food of animal origin, official food hygiene measures merely ensure that the products are safe to eat. In addition, consumers decide for themselves, based on their sensory perception, whether the respective products meet their expectations. What is not difficult in the case of apparent characteristics is much more difficult in the case of quality characteristics. They are not accessible to direct assessment and require additional expenditures and specific expertise. As a rule, one relies on the self-interest of the provider to guarantee compliance with quality promises since non-compliance threatens to result in a loss of confidence on the part of the customer.

If you want to be sure that you get more value for the money you pay at the checkout for a supposedly higher-quality product, you have no choice but to take a closer look at the topic of "**quality**". Quality is a very frequently used word. This does not make it any easier to deal with, especially if we assume simultaneously that each person has their own, and therefore more or less different, understanding of what quality means and how it can be grasped. As will be shown, unambiguous qualitative statements about individual characteristics can only be made with specificity, i.e., with certain restrictions in the information content about the product as a whole. On the other hand, a certain ambiguity in the conceptual use of quality characteristics runs the risk of powerful interests seizing sovereignty of interpretation, as has been practiced, for example, by the food retail sector in connection with the term "animal welfare."

In the context of quality, we are dealing, on the one hand, with the description of a considerable variation of properties of products, processes, or systems on many levels. On the other hand, these properties are subjected to a qualitative assessment from different perspectives, which can vary greatly depending on the perspective. Consequently, there is a high degree of complexity in which we cannot avoid subjecting the variation of properties and their qualitative assessment from different perspectives to a differentiating and multidimensional analysis to arrive at robust statements.

According to the applicable quality management standard (DIN EN ISO 9000:2015-11), quality is defined as the "*degree to which a set of inherent characteristics of an object meets requirements*." In other words, quality indicates the degree to which a product, process, or system meets the requirements placed on it. Characteristics that are inherent to a product unit (intrinsic) are considered. Objectively assigned descriptions such as "good" or "beautiful" are not inherent, which elude measurability. Personal judgments are, therefore, not part of a quality assessment. This is based on intersubjectively measurable characteristics such as length, width, weight, and many other specifications amenable to assessment using scientific methods. Individual quality characteristics sometimes correlate more, sometimes less closely, with different attributes of a product unit. Therefore, the question is in the foreground: to what extent do individual characteristics have sufficient explanatory content for the quality of the whole, or do they only represent an insignificant secondary aspect?

The view on quality features is interest-driven, i.e., not detached from the interests that different groups of people associate with or want to see associated with it. In general, for the producer, the relationship between effort and return, and for the customer, that between cost and possible benefit is in the foreground. However, as has been the case for centuries in the various bazaars of this world, **appreciation** is at the heart of the matter, i.e., the assessment of value from a particular perspective. In the interaction between a supplier and a potential customer, there is a service (product) and a service in return (money) to which the respective side assigns a different value. In the past, the standardization of measurement and assessment procedures (classifications) of products has contributed significantly to facilitating understanding between trading partners. The knowledge of the value of a product, process, or system is much more complicated if the exchange of a good for money is less based on concrete and verifiable facts but is influenced by mental and unconscious associations and expectations, which overlay the perception of the objects or circumstances and finally their appreciation.

Human perceptions are essentially based on pattern recognition (Duda et al. 2001). This refers to recognizing regularities or similarities in perceptual individual aspects. Typical examples are text recognition and comprehension despite a reversed order of letters or face recognition despite only a shadowy representation. Recognizing patterns brings order into the initially chaotic stream of sensory perceptions. Its most important task is identifying and classifying objects in the external world. Human perception performs this task continuously and usually effortlessly. The ability to classify is also the cornerstone of inductive reasoning, i.e., the mental inference from observed phenomena to more general knowledge or the derivation of general conclusions and judgments based on circumstantial evidence. Spontaneous categorizations (mental structures) are constantly being made in the brain, which contain information in an organized form that can be referred to when the opportunity arises. According to Hofstadter and Sander (2018) is

> *"the act of categorization ... the groping, graduated, gradual association of an entity or situation with a category given in one's consciousness."*

The division into categories does not follow a black-and-white scheme but is based on shades of grey. It summarizes numerous phenomena applicable to the living being in finding his way. This approach provides a better understanding of a situation by helping to draw conclusions and make assumptions about how to categorize the current situation (e.g., interesting, harmless, or dangerous) and how it is likely to unfold. Relentless categorization of impressions is, therefore, essential for survival.

Marketing uses the fundamental structures of perception by aiming to arouse associations and expectations with specifically designed advertising messages, thus steering perception and categorization in a desired direction. The associative conclusion of a general superiority is suggested by emphasizing specific individual features. Potential customers are more or less receptive to these messages. Customers are also interest-driven. In a generalized way, it can be assumed that from their perspective, they want to purchase attractive product offers at the lowest possible price and gain a personal advantage or satisfaction. Purchase decisions are therefore not only based on the price but also on what the buyers believe they have purchased in terms of value and benefit for the price paid. Accordingly, the value assessment (appreciation) cannot be separated from individual perceptions and categorizations.

The fact that personal and individual interests are involved in appreciating products and services does not mean appreciation is purely subjective. The relationship between supplier and buyer expresses an **intersubjectivity** and an understanding between different perspectives on the same facts. Irrespective of the personal preferences cherished towards individual characteristics of a product or circumstance, science has many methods at its disposal with which the qualitative properties of products, processes, or systems can be subjected to an intersubjectively comprehensible assessment. This does not involve an absolute, context-independent assessment, as quantitative assessments are based on precisely defined and determinable measurements (e.g., weight). Evaluating a qualitative superiority of products, processes, or systems is primarily based on a comparison. The comparison is initially based on individual characteristics that are either directly visible (e.g., size, color) or based on information (e.g., ingredients, origin, etc.) that is conveyed to the potential buyer using selected additional information.

For example, wheat is classified into quality groups relevant to payment according to protein content, impacting baking quality. Milk is differentiated into up to three quality classes (differentiated by bacterial count and somatic cell content), which entail price premiums or discounts. The fat and protein content of the milk also affects the payment. In the meat industry, the classification of carcasses according to the meat-fat ratio is part of the categorization, which is used to determine the pay-out price to a large extent. To promote market activity and free trade, the classification of carcasses is regulated throughout Europe utilizing a statutory regulation. However, the classification of carcasses according to commercial classes does not provide reliable information on the sensory quality of the meat (Monteils et al. 2017). Eggs are marketed in quality classes A and B (whereby only A is permitted in the retail trade) and weight classes. In addition, the farming method of the hens is marked. If eggs are advertised as "free-range," they are distinguished from other eggs, e.g., from caged or barn eggs, based on a legally regulated labeling of the farming method. This gives them the attribute of being more attractive and of higher value. The burden of proof that these eggs come from the appropriate farming methods lies with the suppliers. They can be prosecuted for attempted deception based on the legal regulation against unfair competition. Not regulated, however, are statements on taste or positive and negative services concerning the level of animal health and welfare or environmental protection associated with the production processes.

The prominent characteristic of this type of husbandry evokes specific associations in many customers, depending on their experiences and prejudices. Inductively, connections between the kind of farming and other effects (e.g., effects on the welfare of the animals) are assumed. However, these do not necessarily have to be in a direct and verifiable connection with the type of farming. The suppliers of the products are spared the burden of providing an increased additional benefit. Although eggs from different farming systems may have the same chemical-analytical composition and do not necessarily differ regarding animal welfare and environmental protection, they are valued differently because of the reference to the farming system. This appreciation results in a different **willingness to pay,** which can be several times higher than a conventional comparative product. The valuation is based on a coherent (holistic) understanding of free-range husbandry, which is valued more highly as a form of husbandry than other forms. This higher valuation is transferred to the products originating from this husbandry system. This means that customers' appreciation and willingness to pay are based on a mental categorization without any validation of the presumption of superiority.

The price that customers have to pay for supposedly higher-quality products at the point of sale is also not directly related to the quality level of the products. The pricing results from the sales strategists' considerations, in which various variables are considered. These include estimates of the availability of the corresponding products, the expected willingness of customers to pay, and the targeted sales volumes that are believed to be achievable with appropriate pricing. Even if agricultural economists repeatedly emphasize that prices are determined by supply and demand, this should not obscure the fact that the supply of goods and end consumer prices are influenced to a large extent by market strategists who know how to induce transmission effects among customers and to exploit these for their interests. The comprehensive knowledge gained by the food retailers themselves or with the support of publicly funded scientists not only has an effect at the point of sale but also influences the profile of requirements that the food retailers place on the products and the agricultural production processes. Thus, marketing strategies do not only affect the demand and the price of products but also the production processes. Supply and demand, therefore, do not follow the free play of market forces but are subject to the manipulative efforts of actors with the appropriate means and knowledge to influence supply and demand in their interests. Those make the market with the most significant power to enforce their interests.

Based on what has been said so far, it becomes clear that it is no easy undertaking to justify "quality production" as a counter-model to the dominant cost minimization strategy and to provide it with a scientifically sound basis that is not corrupted by the subjective aspects of value-setting and particular interests. Just as letters become words with meaning through their specific juxtaposition, products are composed of diverse chemical components and characteristics. Just as terms form a statement through the context of the sentence, the particular constellation of various individual components produces a qualitative property of a product. Like the meaning content of a sentence, a qualitative statement acquires its meaning through the context in which it is made. Finally, context-dependent statements about the meaning-contents of qualitative properties are interpreted by persons with different perspectives and specific interests in each case. Suppose quality production as the basis of a higher-priced production and marketing strategy is to be deprived of arbitrary, misleading, deceptive, or even fraudulent use. In that case, one cannot avoid attributing meaning to all the levels outlined. It is necessary to reflect and present them comprehensibly, leading to scientifically reliable statements, i.e., that are based on evidence. This approach will be explained and specified by various examples.

9.1 Quality Assessment

The term **"Quality"** is derived from the word '*qualitas,*' which can be translated with the terms: 'properties,' 'characteristics' or 'condition.' The properties of a product, process, or system are characterized by very different partial aspects. These include externally perceptible forms, changes, and information that can be determined by a systematic procedure (e.g., analysis) and allow statements about areas that are not immediately visible, such as the composition of the content. A qualitative assessment corresponds to a **value judgement** about the characteristics of a product, process, or system, which refers to different appraisal systems. In general, we can distinguish between three appraisal systems, encompassing generally accepted scientific yardsticks, self-referential individual standards, and system-inherent scales, which refer to the meaning of the functionality and survivability of a system. Only the first and third appraisal systems can claim to comply with the requirements of objectivity and validity, whereas the individual standards are highly subjective. On the other hand, they are the most decisive factor in many consumers' purchasing decisions.

The appraiser and the appraisal object are involved in the appraisal process. The appraiser is a natural person who records and differentiates situations from similar ones. In doing so, the object of assessment is classified on a graduated scale (assessment dimension). The prerequisite for an appraisal is that products, processes, or systems are sufficiently distinguishable from others to be associated with quality characteristics as separate units. It must also be clarified in advance whether we are dealing with a homogeneous population in which the assessment of an object is representative of the other objects. This is the case, for example, with technical devices such as household appliances. Or is there a product that can be expected to have a considerable variation in composition, as in the case of meat from individual animals? In the case of products with a high degree of variation, it is impossible to make generalizable statements. The situation differs for product batches where raw materials from many animals have been mixed and processed into a uniform product with standardized ingredients (e.g., milk with 3.5% fat content). At the beginning, therefore, there is the question of how representative the assessment objects are subjected to a quality assessment.

9.1.1 Assessment of Quality Characteristics

A quality feature (also characteristic) is a recognizable property that distinguishes one thing or abstract context from another. A product of animal origin, such as a piece of **meat** that a customer buys in a butcher's shop, has various characteristics such as color, size, weight, and content of individual ingredients. These describe partial aspects but not the product as a whole. The content of a specific group of substances (e.g., fat content) in a product is always related to the content of other groups of substances (e.g., protein and water content). An increase in one substance group produces a proportional reduction in other substances. In addition, the relations of substance groups such as fat and water content influence characteristics such as tenderness, juiciness, or the taste of a piece of meat. The interactions between quantitative parameters and sensory characteristics are rarely linear. For example, the higher the meat's intramuscular fat content (IMF), the higher its **enjoyment value** (see Sect. 9.2.3). However, the gain in terms of enjoyment value does not increase in parallel with increasing IMF content. Not until the IMF content is above a threshold value (<2.0–2.5%) does the effect on the enjoyment value significantly increase for most consumers, but this applies not to all of them. The situation is similar for dairy products. Most consumers are likely to like a yogurt with a fat content of 6% better than one with 1.5% fat. However, this does not mean that the higher fat content is rated higher by a factor of 4. It is merely ranked higher on a scale.

To grasp the characteristic whole of an object or circumstance to be assessed, it is necessary to include various individual features that only characterize the totality in their specific combination. Separate characteristics are interrelated in many ways, so assessing one characteristic does not permit any statement about others. Also, the qualitative effect of the totality of all features goes substantially beyond listing individual features. This leads us to the first of three interlocking quality concepts: "***Quality is the totality of all properties of a product, process, or system***."

9.1.2 Assessment of the Value

The differentiation and scaling of comparable products, processes, or systems based on meaningful assessment criteria pave the way for quality assessment. Beyond this, however, other aspects require clarification and reflection. For example, an essential prerequisite for a reliable evaluation is that the person making the assessment has sufficient perceptive ability and professional competence to assess the object of the evaluation comprehensively. Since products, processes, or systems have several characteristics that distinguish them from others, questions arise about selecting the individual characteristics given priority and their weighting with one another. Weighting is not self-explanatory but requires justification. Evaluations always contain a subjective aspect because **objectivity** in the scientific sense can never exist in the assessment by and of persons. When people judge facts, they always do so based on personal patterns of perception. Even the selection of the assessment criteria contains subjective elements. This applies all the more to the weighting of individual characteristics.

At best, "relative objectivity" in the assessment can be ensured by the fact that it does not depend on the personal assessment criteria of the assessor but is based on criteria agreed upon by the members of an institution (scientific discipline or members of parliament when passing a labeling regulation). To include the subjective part of the perceptual judgment and still ensure a high degree of objectivity as the aim of the assessment, Kleber introduced the principle of controlled **subjectivity** (Kleber 1992). Controlled subjectivity means that even knowledge of perception and assessment errors and the factors that lead to them can raise awareness of the problem and thus help to prevent mistakes.

Another challenge is to combine partial assessments into an overall evaluation. Scientific disciplines are skilled in analysis, i.e., breaking down a product, process, or system into individual components and sub-segments. Once something has been analyzed (cut up or dissected), it can no longer be classified factually but only mentally in a superordinate context. The mental classification of analyzed sub-segments into an overall context is done by **induction**, i.e., in the conclusion from the individual case to the general or from the sub-segment to the whole. Even though the inductive approach is fraught with the risk of inductive fallacies, we depend on generalizations in our everyday dealings. We cannot get to the bottom of the detailed properties of products, processes, or systems in each case. As long as we know the danger of mental fallacies and formulate the mental conclusions as an assumption (**hypothesis**), we recognize that we are not in a position to generalize. A hypothesis implies that closer examination (**validation**) is required to see whether the conjecture or hypothetical statement is more likely to be proven correct or incorrect in hindsight.

Scientific disciplines naturally find it difficult to make qualitative assessments. Wherever possible, quantitative parameters are used to exclude subjective aspects of the evaluation. Quantification utilizing measuring instruments takes account of the demand for an objective (observer-independent) assessment. At the same time, however, many other (possibly also non-measurable) characteristics are not considered, although they can be very relevant for a qualitative assessment. For example, assessing the so-called "carcass quality" is based on quantifying the ratios of muscle meat to fat cover. Originally performed visually, this differentiation has been replaced in large slaughterhouses by increasingly sophisticated techniques that record carcass composition with greater precision. A higher percentage of muscle meat is equated with a higher "carcass quality."

However, a higher percentage of muscle meat only increases the utility value for the processing industry but not the enjoyment value for consumers. Through a differentiating price mask, incentives are set for primary producers to increase the proportion of value-added cuts and follow the processing industry's specifications. Due to a more or less pronounced characteristic antagonism between muscle fullness and sensory quality characteristics, the interests of processors in a high meat yield were prioritized over the interests of consumers in a tasty piece of meat (see also Sect. 9.2.3).

The prevailing assessment practice thus contradicts a quality assessment considering the totality of quality-promoting and quality-reducing characteristics. Quantitative aspects are emphasized, and the evaluation of semi-quantitative traits and the result of the interaction of the individual factors are usually neglected. This leads to a bias of the assessment in favor of specific criteria at the expense of others. At this point, it becomes clear that the scientific approach to quality assessment cannot be the measure of all things. It offers no possibilities for dealing with the ambivalence of interactions between variable individual characteristics. Thus, it fails to deal with the complexity inherent in the quality of products, processes, and systems. Although an observer-independent recording of traits is carried out, an interest-driven criteria selection is made in advance. However, based on a few quantifiable characteristics, no statements about the quality of the whole can be derived. Thus, an observer-independent assessment is feigned, and at the same time, an interest-driven approach is concealed.

The various individual features and overall impressions of products, processes, or systems have different meanings for different groups of people and stakeholders and, therefore, experience divergent assessments. The scientific approach suggests excluding the interest-driven approach and self-centeredness of the evaluation. However, one comes closer to reality in a multi-perspective view if these interrelationships are considered when attempting an intersubjective evaluation. Quality assessment does not occur in isolation from those who make it. The extent to which this happens and how it can be better dealt with was recently discussed by the psychologist and winner of the Nobel Prize in economics Kahneman et al. (2021) together with his co-authors. It is characteristic of human thinking to draw hasty conclusions from a few impressions or individual pieces of information and then look for further evidence that confirms the initial impression.

Therefore, for a comprehensive assessment involving many perspectives to emerge, independent opinions are first needed to identify the dispersion that arises from different perspectives and to examine their variability. In addition, there is a need for second-order observation and judgment. The second-order observer is anyone who observes the first observer. One person observes another person in how they arrived at their decision. The second-order observer need not always be another person. Thus, the first-order observer can also observe or reflect on their previous judgments and may conclude that the earlier decision was wrong or not entirely correct. First, there must be a change of perspective and a distanced observation of the interactions between an object or circumstance to be assessed and a quality assessment by the first-order assessor. This brings up the question: Why does who determine what and how? In this way, the subjective parts of an appraisal do not fade out but are included in the appraisal process. The assessment becomes more transparent and understandable. The differences between different perspectives enable a new dimension of knowledge that goes far beyond adding several other individual aspects (Alrøe and Noe 2010).

Taking these interdependencies into account, the definition of **quality** in the 2nd order is extended as follows: "***Quality is the value of the totality of all properties of a product, process or system attributed by someone.***" The term "**value**" (= finding something good, less good, or bad), which is often used in everyday language as a synonym for quality, expresses the relationship between the object of assessment and the person attaching certain values to specific characteristics of a product, process, or system. Accordingly, the evaluation of quality is partly made based on generally accepted minimum standards, which all products have to comply with, as well as preferences and standards on which individual persons or interest groups base their judgment and which enclose the fulfillment of previously determined features that exceed the usual or the average, respectively.

Legal requirements generally prescribe minimum standards to be met before products can be classified as saleable. The main requirements are freedom of public health and safety deficiencies specified by hygienic and toxicological indicators (level of environmental impurities, mycotoxins, bacteria, parasites, etc.). In Europe, the main food hazards are mycotoxins in nuts, pathogenic microorganisms in poultry meat and fish, pesticide residues in fruits and vegetables, and heavy metals in fish. Since 2002, an increase in the number of notifications has been observed (Pigłowski 2020). Minimum standards are designed not only for the properties and conditions of products but also for the production processes. Some of the purposes of minimum standards are to ensure equal primary conditions for all operators and to reduce unfair competition.

Product properties and production processes above the minimum requirements vary to a high degree and thus range from a very poor to a very high level. There is a general understanding that within a comparison framework, the term "quality" is only justified when the respective level exceeds the average. Besides consumers' individual preferences, other stakeholder groups are also involved in the attribution of value. Marketing is particularly relevant, as it is interested in what consumers think about food products and processes, how the willingness to pay premium prices can be estimated, and how the knowledge can be used for marketing purposes. For example, an added value is required to emphasize a specific characteristic of a food product, often based on emotional criteria. A particular brand, promise, or shopping experience can represent this. Labeling food as superior quality usually identifies the products as deriving from a production method defined by enhanced minimum standards. In organic livestock production, for instance, these include specifications for the conversion process, such as the ban of mineral nitrogen fertilizer and pesticides, housing conditions, animal nutrition and care, disease prevention, and veterinary treatment. The widespread sympathy for organic agriculture seems to stem from its value-based approach. Many consumers directly associate organic farming with enhanced product quality and production processes. However, the word "organic" has many meanings to consumers, and organic food consumers are neither homogeneous in demographics nor in beliefs. Thus, they hold various motivations, perceptions, and attitudes regarding organic food, their appreciation, and consumption. Finally, price is a prominent attribute of products related to quality based on the notion of value. Demand tends to depend more on the price differential concerning other products than the actual price. Hence, consumers are neither a uniform group covering common interests nor experts who can decide how to evaluate the quality of a product, let alone the quality of production processes.

Also, the various actors in the food production process chain have very different ideas about the qualitative differentiation of products, processes, and systems. Böcker et al. (2004) name six types of quality concepts about foodstuffs, which show the broad spectrum of different basic understandings of quality: (a) absolute quality concept: quality is equated with outstanding characteristics; (b) product-oriented quality concept: quality is a precisely measurable variable of bundles of properties; (c) customer-oriented quality concept: quality is oriented towards the satisfaction of individual customer needs; (d) production-oriented quality concept: quality is the compliance with specified standards according to (b) or (c); (e) "value"-oriented quality concept: quality is assessed in terms of a price-performance ratio; (f) labor-value-oriented quality concept: quality is the sum of the effort expended in production. All these quality concepts have emerged from various viewpoints and interests. They lead to different strategies to realize improved value creation through quality production.

In this context, it must be considered that quality assessors act self-referentially, i.e., they make a perceptual grid and evaluation scheme internalized in their patterns of thought as the starting and reference point of a quality assessment. What is generally said about taste: "*de gustibus non est disputandum,*" cannot be said about quality because: "As is well known, quality can be argued about." Regarding taste, it becomes clear that a subjective sensory perception eludes intersubjective examination and the dispute about its subjective evaluation. This idiom expresses that it cannot be rationally proven that a particular preference is correct. However, a scientific and intersubjective discussion (dispute) about the assessment parameters and their respective degrees is possible for quality assessment. This does not affect the final purchase decision; it's up to each person. Here, the question is primarily about whether the value of a product, process, or system can be assessed intersubjectively and comprehensibly. To dismiss quality assessment as "subjective" and thus as a subject area outside the sciences excludes a significant part of reality. While most scientific disciplines largely ignore the evaluation of facts by different people, this connection is focused on agricultural marketing. It is in assessing the value and benefit that a product, process, or system appears to promise for different groups of people that the willingness to pay for the purchase of goods is decided, irrespective of whether the goods deliver what the advertising promises.

In common parlance, if products or services meet their expectations, they are said to be of "good" quality. From the seller's perspective, "quality is what the consumer wants." (i.e., finds suitable). The seller sees himself confirmed, "when the customer comes back and not the goods". However, a primarily customer-oriented understanding of quality is fraught with various problems. Customer satisfaction and reactions to possible discrepancies between expectations and perceived quality levels of the purchased goods can vary significantly from one individual to another. In addition, customer satisfaction is not only influenced by the quality of the purchased goods or services and the specific preferences of the buyers but also by the price. This represents a decisive quantification in evaluating qualitative characteristics of products, processes, or systems, to which the potential buyer reacts with an "associative activation" (Kahneman 2011). Often, the price below the threshold of consciousness is linked to ideas about what one expects for a price and whether this price corresponds to the value or quality for which one is prepared to pay the price. Prices have a so-called "priming effect" for associations, thus steering in a specific direction and influencing the purchase decision. The great importance that food retailers attach to low-price offers in advertising indicates the effectiveness of price quotations.

In supermarkets, specific characteristics of a food product are highlighted, for example, with a catchy slogan that arouses associations in the buyer, like the price. What ultimately dominates in the buyer's mind, whether the price, the associated value in relation to the cost, or the simple power of habit in buying processes once initiated, is a central question that drives all advertising strategists of this world. On the other hand, whether the food products are worth the price in relation to different products or can justify the price through comparative judgments often remains in the dark. This opens the door to purchase decisions based on false assumptions and overpricing of buyers. Since buyers are generally unable to judge the value of a product or process properly, many believe they can do little wrong by orienting themselves to a low-price level, thus minimizing the risk of being overcharged. However, this assumption can also prove to be a fallacy if the purchase has negative consequences for the buyer himself (e.g., increased health risks), for other living beings (e.g., farm animals), or other protected goods (e.g., nature, environment, and climate protection).

In contrast to price-conscious buying behavior, a quality-conscious buying behavior, in which the buyer claims for himself a particular competence in quality assessment, priming brings the associated value to the fore. Using priming, the attention of potential buyers is drawn to features with a high associative potential for the targeted group of buyers and a high recognition value (e.g., organic products). Through associative attributions, specific characteristics are given a prominent meaning and differentiation. The advertising machinery surrounding us is geared solely to drawing our attention with its advertising messages to particular features that stand out from a rather indifferent perception. Attention is the central currency of the advertising industry. This is why visible characteristics also have such a prominent meaning for animal-sourced food products, as expressed, for example, by pictures of farm animals kept in free range.

The weighting of features, their (interpretation) and qualitative classification (categorization) are oriented towards the already internalized individual patterns of perception and explanatory approaches based on analogical reasoning (Hofstadter and Sander 2018).

Complexity reduction is indispensable because of the manifold features that characterize the totality of a product, process, or system. The brain is confronted at every moment with a vast amount of nerve impulses from within the body and with impulses (information) from the living environment conveyed via the sensorium. The brain can only deal with this abundance of information using reduction by continuously breaking it down into simplifying categories. The conclusion from individual sensory impressions to superordinate connections (inductive procedure) enables the individual to orientate himself and to be able to behave and act despite a general overload of stimuli and information.

In quality assessment, the inductive approach is helpful if there are highly correlative relationships between the individual characteristics and the overall quality. In the case of machine-manufactured products, for example, it can be expected that the high quality of parts goes hand in hand with the high quality of the overall product. Furthermore, the presumption that the characteristics inherent in a machine-produced product are also found in all products of the same production batch is usually true. Consequently, the result of the assessment of a product can be extrapolated to the products manufactured in the same way. By analogy, in the case of automobiles of the same series production, defined inputs in the form of fuels lead to the expectation of defined outputs in the form of driving performance and car exhaust emissions. However, even machines produced in a standardized manner exhibit a specific scattering of performance and emissions. In contrast, for complex biological processes, the predictability of the output of these processes is not given. The extent of variations in the initial and boundary conditions and the hardly manageable interrelationships between system-inherent influencing variables make inductive conclusions from partial aspects to the overall quality of pure speculation and, therefore, tempt inductive false conclusions. In the case of biological processes, qualitative statements are reduced to assumptions regarding the probability of occurrence and, thus, require verification (**validation**).

In the past, there has been no lack of effort and expense on the part of market research to determine which consumer groups are sensitive and responsive to which initial impressions and preliminary information so that an increased willingness to pay can be inferred. Among other things, the simplification of facts and a certain vagueness in the definition of advertising terms draw the attention of as many potential buyers as possible to specific individual features of products. The vaguer and more general the terms, the more people feel addressed and react with their associations to the advertising message. As shown in a review by Grethe (2017), many agricultural scientists have also participated in various studies to determine consumers' sensitivity and responsiveness to impressions and partial information that are effective in advertising. While agricultural scientists work in this way for the agricultural industry so that the latter can benefit from this through specific agricultural marketing, only a few studies can be attributed to the goal of consumer education. Thus, there is a lack of investigations and reflections on the interactions between the customers, the partial information of the advertising industry, and the overall quality from the second-order observer position to bring light into the darkness of interest-driven processes and to carry out enlightenment in the original sense.

The heterogeneity and confusion of the concept of quality have probably contributed to the fact that quality production of foodstuffs has so far only been able to develop in a rudimentary form. As long as **primary producers** are exposed to global undercutting competition, which they try to counter with efforts to reduce production costs, no impetus for a reorientation of production processes towards quality production can be expected. The proposals made so far by agricultural policy-makers and the agricultural economists advising them have also been limited to setting new **"Quality standards"** by raising minimum standards. This is intended to provide an alternative offer to the section of consumers who express a desire for qualitative improvements and criticize existing conditions. At the same time, most farms are encouraged to continue with their existing production processes, which are geared towards productivity growth, to survive on the global markets. This is to be done regardless of the immense damaging effects that have already occurred and that are associated with the prevailing production processes, as outlined in Sect. 5.3.

Causal relationships for harmful effects at the operational level are largely ignored. This also applies to the ambivalence that is not adequately realized concerning the generation of quality. Quality is not generated by emphasizing individual characteristics but by balancing positive and negative aspects and organizing a balance between synergistic and antagonistic effective processes into an overall quality. This requires penetrating complex issues and developing conceptual approaches for dealing with complexity. Furthermore, the generation of quality requires that the result of the processes in the interdependency of products, processes, or systems is accessible to verification.

Despite the failures, quality production remains a crucial instrument for improving value creation for those primary producers with the necessary prerequisites. These producers can generate a higher sales price through the proven superiority of their products and, in this way, escape the downward spiral of undercutting the competition. An elementary prerequisite for this is that a qualitative and thus differentiating statement about an issue is intersubjectively comprehensible and resilient, i.e., evidence-based. Qualitative statements should not only be valid between seller and buyer or trading partners, but as in case of disputes in court, they should also be justiciable.

9.1.3 Quality Assessment Based on External Benchmarks and Targets

Increased value creation through higher-quality and higher-priced sales products is one thing; improving production processes to take account of the public welfare interests is something else. Both sets of interests can be related to the interests of consumers who, by buying high-quality products, want to do something "good" for themselves on the one hand and for the common good on the other. In addition, they can acquire the additional benefit of the satisfaction of supporting primary producers in implementing a method of production that serves the common good. A reciprocal conditionality characterizes the relationships between the interests addressed in Fig. 9.1. Public welfare interests can only be realized if appropriate consideration is given to the particular interests of both

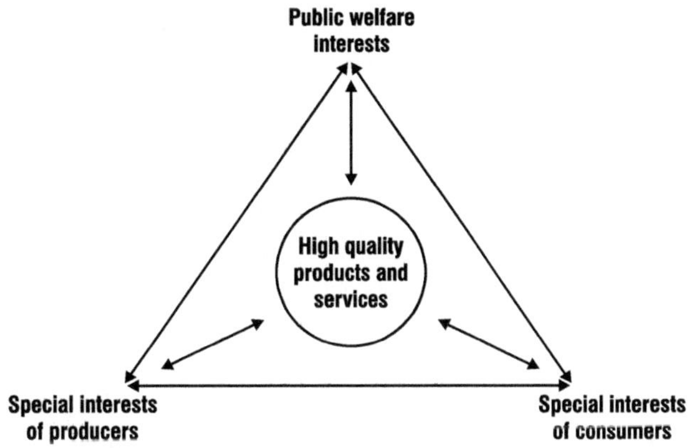

Fig. 9.1 Network of relationships between different interests about the assessment of high-quality food (own illustration)

primary producers and consumers. Primary producers will only be able to reduce the potentially harmful effects of production processes to a minimum if they are supported by doing so and provided with the necessary resources through cost-covering market prices. In the long term, consumers will only be prepared to pay the higher market prices if they are offered something of higher value in return, which is distinguishable from inferior goods. It should consider personal preferences and convey to the buyers of higher-quality goods that they are part of a community in which they participate by purchasing corresponding products and receiving appropriate remuneration for the services rendered. Linking different interests results in synergy effects and additional benefits that cannot be tapped if the focus continues on particular interests. It is also true that public welfare interests can hardly be realized without the synergistic effects and the associated additional services.

The attentive reader will not have failed to notice that the interest groups of the processors of raw goods and the food retailers are missing from this triangular constellation. While they have occupied mainly the definition and interpretation of sovereignty for themselves in the previous market events, this role is not granted to them in the context of public welfare interests. Here, processing and trade enterprises should concentrate on their very own task, namely, to act as intermediaries between primary production and consumers. To this end, just like the other interest groups not explicitly mentioned, they should orient themselves to overriding standards and align their respective business models accordingly. This includes, among other things, sounding out the appropriate recycling options for the different quality grades and the relevant prices concerning supply and demand.

High-quality products and services are at the center of the triangle. The higher quality must be accessible to an assessment based on criteria not shaped by the assessors' perspectives. Instead, the evaluation should be free of the self-referentiality of particular interests and enable assessments about higher-level objectives. An independent, intersubjectively valid, and evidence-based quality assessment is indispensable for standing up to the still-existing self-interests. Scaling products and services in a spectrum from very low to very high could provide orientation for all stakeholders. **Quality** that meets these requirements is defined as follows: "***A third-order quality is the totality of char-***

acteristics of products, processes or systems measured against external yardsticks and regarding overriding objectives."

The explicit **objectives** of the public welfare interests include consumer, animal, and environmental protection. Their realization presupposes that one emancipates oneself from an anthropocentric view and approach that focuses on the interests of individual groups of people. Anthropocentric views cannot be dissolved; they determine our view of the world. If, however, not only the interests of individual groups of people but also the interests of the public welfare interests act as a yardstick for an assessment, the self-interests are not negated but relativized by including a higher perspective, i.e., integrated into a network of relationships. The assessment of products and services concerning their contribution to the objectives of the common good should focus on the extent to which these protection objectives are achieved in food production. For example, consumer protection aims to protect consumers from potential hazards arising from food consumption. **Animal protection** aims to protect farm animals from adverse effects involving pain, suffering, damage, or even death. Environmental protection aims to keep the discharge of pollutants into the environment as low as possible in relation to the desired sales products. The extent to which this is achieved in different systemic contexts is open to assessment (see Sect. 9.4.3). Which external benchmarks are suitable for differentiation in the respective contexts and which overriding objectives should be met in the process are explained in more detail below.

Derivations

There is considerable variation in the quality of products and in the public welfare-related services associated with production processes. A valid quality assessment cannot be made in isolation from the **context** in which the products and services are embedded. The context, in turn, is shaped by different interests and perspectives that go with them. However, there is too much at stake to avoid the challenges of an **intersubjective** assessment and the scientifically based objections to the subjectively colored parts of an evaluation. Because agricultural sciences have left this subject area mainly to lobbyists, a grey area has developed here, opening the door to interest-driven manipulation. Thus, clarification is urgently needed in the form of a distinction between three approaches of quality: an **analytical**, a **preferential,** and a **functional quality approach**.

The analytical approach encompasses all values that corresponding analytical methods can gain. This leads to the 1st order quality concept, which defines quality as the totality of all properties of a product, process, or system. On this level, an intersubjective understanding of the evaluation and weighting of complex facts is always possible if the participants agree on a standard methodological procedure. The 2nd quality concept defines quality as the value of the totality of all properties of a product, process, or system attributed by someone. The preferential quality approach makes the participants' appreciation the object of consideration and allows for a 2nd order quality assessment. Finally, public welfare interests can claim a superior position, which has to be prioritized towards particular interests. The prerequisite for a valid 3rd order quality assessment is an independent and intersubjectively valid procedure about the degree by which specific public welfare interests are realized. The totality of characteristics of products, processes, or systems measured are assessed against external yardsticks and regarding overriding objectives. Only in this way can it be prevented that public interests are only addressed in appearance, but in reality, are instrumentalized to pursue self-interest. Scaling product and process qualities in a spectrum from very low to very high can be realized in all three quality concepts and is essential to providing orientation for all stakeholders involved.

The current market conditions widely ignore the large variability in quality traits and their impacts on public welfare interests. This promotes unfair competition when enabling equal prices for very different product and process qualities. For example, farmers who gain economic benefits by selling organic products at premium prices but only providing a low level of animal health undermine the efforts of other farmers attempting to maintain a high level of animal health and welfare, consumer protection, and product quality. In general, striving for a high-quality level has applied additional labor and cost-intensive efforts, contributing to a considerable increase in production costs while at the same time providing an essential competitive disadvantage in markets that do not make substantial differences concerning quality production.

9.2 Differentiation and Contextualization of Product Quality

The quality of a product marketed as an **aliment** is determined by aspects as diverse as **nutritional value**, **utility value,** or **enjoyment value**. These value-giving factors have in common that they can be recorded, analyzed, and assessed directly on the product. When discussing a food product, it would have to be clarified beforehand whether we are talking about a product in nominal terms, i.e., whether its possible characteristics can be distinguished but do not have a natural ranking. Is it a homogeneous or heterogeneous batch of products of the exact origin or a specific product intended for consumption, the quality of which is to be assessed? There is also the question of how well food products can be distinguished from one another.

Thus, an egg is distinguishable from other eggs by its shell; consequently, it can be evaluated as a singular entity. However, one egg is not like another. **Eggs** of the same species differ in size and weight, among other things. However, eggs of the same weight class are not identical in other characteristics (e.g., yolk color). To judge the color of the yolk, the eggshell must be removed. This makes the egg no longer usable as food except for immediate use. Therefore, breaking only a few eggs makes sense to conclude the entire batch from their assessment results. To make a statement about the color of the yolk of eggs from a barn unit with a specific weight class, it must be decided how many eggs from a batch must be examined to extrapolate the examination result to the quality characteristics of a batch. Consequently, the question of representativeness arises, i.e., the estimated degree of agreement between a sample to be assessed and a population or a category of products.

With **meat,** the classification of quality statements is even more difficult. A piece of meat intended for human consumption is a muscle or muscle group separated from its context (a carcass). The cutting method determines the cut's size ratio in relation to the whole carcass. As a rule, a carcass is cut into many cuts, each with a different texture and quality characteristics. The individual muscles of a slaughtered animal (more than 200 in number) fulfill very distinct functions in the organism. They are, therefore, of different compositions to be as valuable as possible for the organism. The quality assessment of a single carcass raises the question of which part of a carcass is suitable to conclude the overall condition of one part to the condition of other parts of the same carcass. Further questions of informative value arise when statements about the quality of meat products of an animal population (e.g., of a breed) are to be derived from the assessment results of a carcass to the condition of other carcasses of an animal group or several carcasses. The question of the generalizability of qualitative statements about products, which are never identical but always exhibit an undetermined degree of variation, always resonates.

A carcass is not homogeneous, but it is a definable product unit. The situation is different with **milk** and the products derived from it. Milk intended for human consumption is a liquid made homogeneous by mixing milk produced in the mammary glands of individual animals. Unambiguous attributions to the original context, i.e., the udder of a dairy cow, goat, or sheep are cut. As a mixture, raw milk from different farms is filled into milk tanks and treated in various ways before it is used as a homogenized raw product as a starting point for further processing steps (e.g., butter, cheese, yogurt). The sum of all the milk of the individual animals produces the total composition of the respective raw milk batch. The batch size corresponds to the filling quantity of the milk tanks used in the dairies. Consequently, the quality of milk is represented by a mixture of raw milk of different sizes. The extent to which milk can still be subjected to a qualitative assessment and whether the term "quality milk" still has its justification will be discussed below.

9.2.1 Nutritional Value

The content of nutritionally relevant ingredients characterizes the nutritional value of food. These include the macronutrients: proteins, fats, carbohydrates, and the micronutrients: minerals, trace elements, and vitamins. In addition, the contents of secondary plant substances or dietary fibers and the presence or absence of undesirable substances such as residues or antinutritive factors are essential. Despite an unimpaired nutritional value, unwanted substances can drastically change the overall quality of the food. The legally prescribed minimum requirements for product quality are laid down in the applicable legislation, particularly in the food law regulations, including the Food and Feed Code and other European and national regulations. The minimum requirements are intended to prevent the sale of products with severe defects or that contain harmful substances. Foodstuffs whose consumption poses a risk to consumers may not be marketed as food. However, they are still used in other ways, e.g., as raw materials for industrial products. However, where no comprehensive and effective control measures have been established, the temptation is great to sell raw goods that, strictly speaking, do not meet the legal requirement profiles but are nevertheless suited as raw materials for producing foodstuffs to limit financial losses. Depending on the depth and density of controls, there is a more or less wide grey area. For example, milk from dairy cows suffering from udder disease is milked into the tank and mixed with milk from udder-healthy animals. However, according to the regulation, an altered milk secretion should not be placed on the market. The nutritional value of milk from udder diseased animals is reduced compared to unaffected animals. Mixing conceals the differences in the nutritional value and enjoyment value of the milk. At the same time, primary producers obtain the same price for the inferior milk as for the milk from animals with a healthy udder.

There is a significant variation in the ingredients and, thus, in the nutritional value of all raw materials of animal origin. The variation collides with the requirements that the technological processes of further processing place on the raw materials. Accordingly, raw materials are categorized in terms of composition and size ratios as far as possible before further processing. In the case of raw milk, mixing different kinds of milk already leads to an adjustment of the composition. In the dairies, the milk is then separated into its nutritional substrates and recomposed for various product groups to a previously defined standard into sales products with uniform ingredients. A well-known example is drinking milk, offered to consumers with different but standardized fat contents.

The starting point for assessing the nutritional value of edible products is the National Consumption Recommendations, which are based on the requirements of an average person. This means that standardization occurs not only in a product's ingredients but also in the evaluation yardstick. The smaller the values deviate from a standardized consumer's nutritionally determined requirement values, the higher the nutrient level of a product is classified. However, consumers are not the same; they have very different nutrient requirements depending on live weight, age, gender, and physical activity, among other factors. They can also react very differently to the same ingredients due to their age (e.g., infants) or specific intolerances (e.g., lactose intolerance). General supply recommendations are, at best, orientation values for individual people. Not the defined standards but the personal needs and preferences of the consumer are decisive for assessing the nutritional-physiological quality of foodstuffs. The individual consumer is, therefore, the **reference system** that determines whether the entirety of the specific composition of the product is suitable for meeting the nutrient and energy requirements of the respective consumer. The **value** of a food product thus depends on the inherent composition and the specific requirement situation as well as on the composition of the menu of the individual consumer, of which the respective products cover a different part of the total supply. If the context of food consumption changes, the same product's nutritional quality changes. Food that is still judged to be beneficial for one person may prove to be problematic for another. While the nutritional value of foodstuffs is generally reduced to the standardized and analytically verified ingredients, the qualitative value of food concerning the intended purpose of nutrient supply can only be assessed context-dependent.

9.2.2 Utility Value

The utility value describes the usefulness of foodstuffs concerning a specific application. In this context, the chemical-physical properties of raw goods play a decisive role. They are more or less suitable for processing and cooking, depending on their composition. Since raw materials make up a more or less large proportion of a processed product, it should be possible to combine them well with other additives to meet the requirement profile of a standardized sales product. For example, different carcass cuts are usable to varying degrees for preparing specific sales products. The variation ranges from differently composed muscles of animals of various species (cattle, pigs, sheep, goats, etc.) and different types of production (fattening animals, parent animals, etc.) to animal-specific differences. In the case of cattle, the age of slaughtered animals can range from calves a few weeks old to cows 15 years old. Even meat from animals of the same breed and age, which have been fed the same feed, shows a wide range of variation. The categorization of raw goods merely results in a specific reduction in the range of variation without eliminating it. The reference system for assessing the utility value of raw materials is their suitability for producing specific sales products. The less the raw materials vary in ingredients and chemical-physical properties, the more efficiently the processing procedures can be designed.

The meat-to-fat ratio has emerged as a decisive evaluation criterion for carcasses. The higher the lean meat percentage and the lower the fat percentage, the higher the carcass is rated. Although the criterion only addresses the quantitative size of the meat-fat ratio of the carcasses, the term "carcass quality" has become established for it. The meat-fat ratio varies considerably between carcasses. The variation can be narrowed down by the feeding regime, the genetic origin, the sex, and the age of the animals. The categorization (classification) of carcasses aims to reduce the variation to a manageable and tradable level and standardize carcasses' commercial value. A uniform grading scheme allows for the comparability of commercial goods and a certain degree of market transparency. Furthermore, in combination with a differentiated payment system, the production processes can be steered in the direction desired by the trading partners.

In the European Union, the same commercial classes (E U R O P) apply to the classification of carcasses of cattle and pigs in all member countries. However, the grading system applicable in Europe differs from grading systems in other countries, such as the USDA system (U.S. Department of Agriculture) or the MSA system (Meat Standard Australia). The EUROP grading system was introduced by regulation in 1983 and is still used today in a slightly modified form. The aim was to harmonize the European meat market and to ensure a uniform application of the standard market organization throughout the EU, free from distortions of competition. It serves the meat industry as a basis for transparent payment of meat producers and forms the starting point for the price quotations of the meat trade. In most European countries, the persons (classifiers) who supervise the classification of carcasses in the classification system are employed by the operators of slaughterhouses. Classifications are subject to regular review to ensure that standards are met. Since 2000, it has also been permitted to classify carcasses based on video-image analysis. Here, checks are carried out on an ongoing basis to ensure that the classification is correct; after all, the payment of primary producers also depends on this.

9.2.3 Enjoyment Value

Sensory quality

The enjoyment value is also called a food's sensory value or quality. The human senses perceive the enjoyment value. Accordingly, appearance, smell, taste, ripeness, freshness, and consistency significantly differentiate between products regarding enjoyment value. The perception of food via a person's sensorium is naturally subjective. Each person has their idea of how a food should look, smell, and taste. However, individual preferences are not innate but are shaped by the prevailing food culture and the socialization processes and experiences during childhood and adolescence (see Sect. 6.5). This explains why system catering, among others, spares no effort or trouble in helping to condition taste. At the same time, it is ensured that the products of system gastronomy always leave the same taste impression, regardless of the place where they are prepared and offered, and thus independent of context. The example of the global one-size-fits-all nature of system catering is not necessarily at odds with the continued wide variability in consumer product preferences. After all, most people do not feed on a one-size-fits-all product daily but are generally also eager for variety. Notwithstanding the formative effects of prior taste experiences, there is also a wide variation in the weight given to individual sensory impressions. Thus, the sensory quality of food results from the interaction of the characteristics mentioned above. Consumers' specific patterns of perception decide whether the sensory impressions emanating from a product are interpreted in a positive, indifferent, or even derogatory manner. The variation in consumer preferences is reflected, among other things, in the wide range of product offerings, which are based on the same raw goods but are caused by different processing steps and additives.

Consumers' perceptual patterns determine which product offerings they prefer and purchase. Once consumers are conditioned to a specific taste nuance of a product, they expect that taste nuance to be recognized in all products marketed under the same **brand**. Consumers' expectation of the particular flavor nuance of the product they purchase creates the need for products labeled as branded goods to be consistent regardless of the context of production, processing, and 'point of sale.' Branded goods thus provide a pleasurable value with which various consumers can make friends and remain more or less loyal to the branded product. For the suppliers of branded goods, an adequate introduction of branded goods is therefore essential to winning over as many consumers as possible to the respective branded goods.

For consumers, branded goods thus set the anchor and reference point for the pattern of perception by which other products are compared and by which the superior or inferior enjoyment value of a food product is determined. The **subjectivity** of the assessment is more or less fixed and is only accessible to a limited extent to relative objectivity that can withstand an intersubjective evaluation independently of the pre-formed patterns of perception of the individuals and independently of the interest-led possibilities of influence of marketing. To circumvent the individual pre-determined patterns of perception, an intersubjectively valid assessment of the enjoyment value of a product requires testing by an independent institution, such as that provided by specially trained sensory panels. However, even specialists in food sensory analysis are not immune to being deceived and hoodwinked by technical refinements in processing and specific additives, especially in the case of processed products.

For example, in a sensory panel of the German Agricultural Society (DLG) to assess **sausage products**, a meat sausage was awarded a silver medal whose raw materials were of inferior quality, using 46% mechanically separated meat, 27% water, and 18% **meat** and bacon with various additives. The staged deception reveals how little the assessment of processed products depends on the quality of the raw materials and how far the art of faking high quality has already advanced in processed products. Regardless of its susceptibility to deception, in the retail world, the DLG seal is still considered a definitive criterion for listing processed goods in supermarkets. It is almost logical that this seal is now emblazoned almost inflationary on nearly every pack of sausages. Even with fresh meat products, there is no lack of possibilities for influencing sensory impressions. The meat's color is particularly important, significantly influencing consumers' purchasing decisions. Consequently, the processing industry has developed specific methods of preservation and sophisticated lighting techniques at the meat counter, with a particular coloring of the products simulated when consumers examine the products to influence the purchase decision in this way.

In the case of **meat,** the degree of marbling (distribution of the fatty tissue in the meat, reminiscent of the structure of marble) as a visible characteristic of fresh meat or the intramuscular fat content (IMF content) as the corresponding analytically determinable variable of the fat content, belong to the most relevant criteria of enjoyment value. The degree of marbling or the IMF content is accessible to objective determination, and their prominent explanatory power for the sensory quality of meat applies to beef, pork, and poultry meat. What makes this characteristic stand out from the multitude of product characteristics is the very close relationship between the degree of marbling and the aroma, tenderness, and juiciness of meat. The degree of marbling increases the likelihood that a piece of meat will receive a positive intersubjective assessment regarding sensory quality (Emerson et al. 2013). Due to the outstanding relevance, carcass classifications of cattle have been established that focus on this trait as the most significant evaluation criterion in various countries outside Europe (USA, Canada, Australia, South Korea, and Japan) (Polkinghorne and Thompson 2010). In the USA, a system of differentiation and classification of marbling was established as early as 1927. The objective was to promote **beef** marketing on this basis and to provide the trade and the meat processing industry with a standardized evaluation scheme. Today, beef produced in the USA and the countries mentioned above differs significantly from Central European conditions regarding sensory quality. For example, intramuscular fat (IMF) content of well over 10% is achieved in the USA. In contrast, the IMF contents of beef carcasses produced in Germany are between 1 and 4% (Geuder et al. 2012). The degree of marbling varies considerably between cattle breeds and between animals of the same genetic origin. In this respect, Kobe cattle occupy a particular position. As a result of centuries of breeding, they have particularly intense marbling and, thanks to their high quality, achieve top prices that can reach 400–600 € per kilogram.

For long periods, a trained classifier performed the carcass marbling grade assessment. Today, image analysis is used in many countries to record the degree of marbling. In contrast, European slaughterhouse operators are still very reluctant to introduce this measurement method. Although the scientific community has regarded the EUROP classification system as insufficient to meet the demands of the market for qualitative differentiation based on sensory **meat quality** characteristics to become fair (Monteils et al. 2017), the market sticks to the previous assessment. This is because they expect a more significant economic advantage from a quantitative evaluation of the carcasses than a focus on sensory quality. This does not change even though carcasses with a high proportion of lean meat often show disadvantages in sensory meat quality. Several, although highly variable, antagonisms between growth performance, slaughter yield, and sensory meat quality have been identified (Hocquette et al. 2010).

Even though it has long been known in top gastronomy and among many meat lovers that the degree of marbling represents the characteristic with the highest informative value about sensory meat quality, most consumers are deprived of higher-quality meat products. Instead, the economic interests of slaughterhouse operators and food retailers set the context for how the heterogeneity of carcasses is dealt with. An alternative contextualization would be to take as a basis, as in other non-European countries, the variation in the available degrees of marbling of slaughtered cattle and pigs. The recording of the degree of marbling allows an interval-scaled ranking of the carcasses. The degree of marbling thus offers a valid assessment standard that could guide both primary producers in the design of production processes and consumers in their purchasing decisions. Irrespective of this, each consumer could continue to decide, based on self-referential patterns of perception, whether to follow this intersubjective assessment standard or make their purchasing decision independently.

However, differentiation of carcasses based on the degree of marbling and classification, i.e., by **contextualization**) of the raw good into the respective available quality assortment is incompatible with the predominant approach, which is oriented towards utility value and economic self-interest. A further aspect is likely to promote resistance to reassessing meat quality. Differentiating carcasses based on the degree of marbling would mean that even branded programs could not avoid being geared to the availability of high-quality meat lots. High-quality carcasses, however, would be in limited supply given the expected frequency distribution. Food retailers and other suppliers would have to react to this form of shortage and the supposed oversupply of low-quality goods with corresponding price adjustments. These would counter the current situation in which, faced with an oversupply of undifferentiated raw goods, food retailers have become accustomed to dictating prices to primary producers and consumers at their discretion.

Associative quality

Further differentiation of products in connection with the enjoyment value concerns the mental connection to the context in which the source goods or the farm animals were integrated as product suppliers during their lifetime. Consumers consider animal health and welfare an attribute of the food quality concept with increasing importance over other attributes (Gellynck et al. 2006). Health and welfare conditions are used as a 'value-added' component or distinctive selling point for food products, brands, or even particular manufacturers and retailers (Buller and Roe 2012). They can be briefly summed up by the slogan "Healthy food from healthy animals" (Sundrum 2012a). What is considered harmful for the animals is ultimately bad for consumers, particularly concerning disease control (Clark et al. 2016). The complexity of 'health' can be grasped and operationalized by the freedom of severe health problems equivalent to the state of physical health without disregarding that it is a necessary but not sufficient condition for the state of mental health. A life worth living and a good life are statements about an animal's quality of life, implying a judgment made by a man acting on behalf of the animal's interests (Farm Animal Welfare Council 2009). The term 'worth' is a measure of value. The same applies to the goal of livestock farmers to make a living by producing ASF in a way that neither in excess compromises the lives of farm animals nor endangers the lives of consumers.

It is undisputed that the products of healthy animals, which, in addition to being free of health impairments, were kept under living conditions that allowed them to exercise their natural behavior largely undisturbed, are classified as being of higher value than products of diseased animals. Healthy animals are spared the extent of pain, suffering, and damage that affects diseased animals to a greater or lesser extent. Given the choice, most consumers would probably opt for products from previously healthy animals. Knowing about consuming products of animal origin, where the animals had a good life, involves a higher associative pleasure value than learning about the unhealthy context of living conditions when consuming them. The fact that consumers are widely left ignorant here deprives them of the opportunity to make a conscious choice in favor of higher quality products and contribute to improving farm animals' living conditions by purchasing appropriate products. Being sensitive to the suffering of other sentient creatures but largely unaware of the specific background from which the products of animal origin come, younger people, in particular, decide against consuming animal products to distance themselves from general grievances in the context of farm animal husbandry, and reach for substitute products where there is no concern about potential harm to farm animals.

To become more specific, the issue is explained in further detail. For example, high-quality raw milk can be defined as (1) compositionally complete (e.g., protein and fat levels within the norm); (2) free from off-flavors and odors; (3) free from detectable drug residues, added water, or other contaminants; (4) having low total bacteria counts; and (5) having low somatic cell counts (SCC), well below the regulatory public health-based limits (Murphy et al. 2016). At higher levels, somatic cells and bacteria are associated with increased activity of enzymes that damage milk components and potentially result in product defects. The ability of the enzymes related to increased SCC or bacteria counts to influence the quality of processed dairy products depends on several factors, including enzyme level, specificity, heat stability, pH, moisture, and inhibitors and activators. Thus, the potential effect will vary depending on the enzyme, the product, and the conditions. The differences between the quality of raw milk can be seen, among other things, in the suitability value (e.g., cheese suitability and shelf life) and the sensory quality. Many consumers demand and expect dairy products to be manufactured from high-quality raw materials, even if using higher-quality raw materials does not necessarily translate into improved functionality or sensory perception of the finished products.

In contrast to the expectations of many consumers, raw milk milked from the individual udder quarters of a cow is characterized by a wide variation in material composition and qualitative characteristics. The processing procedures level out the significant differences in the initial substrates. This means that milk from cows, in which inflammatory processes occur in the udder tissue due to the penetration of mainly bacterial pathogens, is used for further processing, like milk from dairy cows with healthy udders. In the dairy, the germs and the SCC, released as immune defense cells via the blood into the udder tissue and thus into the milk, are removed by centrifugation. In addition, brief heating ensures that the processed milk does not pose a risk to consumers. The individual nutrient fractions of the milk are separated and remixed in appropriate proportions to form standardized sales products. Dairy products that we, as consumers, consume are consequently foodstuffs of a high degree of uniformity that have been freed from "unhealthy" inflammatory and harmful substances. However, from a qualitative point of view, milk from a healthy udder is nevertheless of higher quality than milk from an inflamed udder. In the original context of the mother-cow-calf relationship, milk is seen as a means of enabling the calf to live, thus preserving the species. For a calf, milk from an infected udder is inferior and jeopardizes the offspring's health and survival. It can also harm the health of humans who consume raw milk. Therefore, raw milk should be heated before consumption. For processed milk products, a hazard can essentially be ruled out.

Another example is the muscle tissue of an animal transformed into a food product, which was initially integrated into the overall structure of the organism and contributed to the animal's self-preservation by ensuring specific functions. From the animal's perspective, a state unaffected by health disorders is of higher value than a state associated with pain, suffering, and damage. A piece of meat as part of the whole organism, which comes from an animal that was healthy during its lifetime, is a higher value piece for those consumers who attach particular importance to the welfare of farm animals than meat from diseased animals. In this value assessment, on the one hand, the context of origin is used as a relevant quality criterion for differentiation. It is assessed as an associative quality without necessarily translating into a sensory differentiation or an increased risk for consumer health protection. Nevertheless, it is not only an associative differentiation. Finally, it makes a relevant factual difference, particularly for the animals concerned, but also for the farm context and part of the consumers and the public welfare interests.

While most consumers expect animal-sourced food to derive from healthy animals, the reality is different. The animal health and welfare problems that derive from the intensification processes of livestock production have already been explained comprehensively in Sect. 4.1. Table 9.1 reveals the extent of the health and welfare problems in different animal species by showing some results from selected scientific surveys. The surveys reveal a desolate situation in livestock production with inapparent animal health and welfare problems, which generally become apparent only after specific diagnostic efforts. Therefore, they are widely disregarded by livestock farmers, processors, food retailers, and state authorities and not transferred by the media into the public sphere.

Table 9.1 Degree of the prevalence of inapparent animal health disorders and pathological findings at the abattoir in selected scientific studies from different animal species

Poultry			
Key bone fractures in laying hens	Foot dermatitis in turkey	Breast buttons and blisters in broiler	Liver inflammation in turkey
(Thøfner et al. 2021)	(Freihold et al. 2019)	(Huang and Ahn 2018)	(Freihold et al. 2021)
50—98%	**97%**	**49%**	**90%**
Pigs			
Stomach ulcers	Pulmonitis	Pleuritis	Liver lesions
(Swaby and Gregory 2012)	(Pallarés et al. 2021)	(Merialdi et al. 2012)	(Vismarra et al. 2023)
79%	**66%**	**48%**	**4 – 98%**
Dairy cows			
Subclinical mastitis	Pathological findings	Abomasal ulcer	Liver lesions
(Krieger et al. 2017)	(Veldhuis et al. 2021)	(Munch et al. 2019)	(Valkova et al. 2023)
51%	**55%**	**84%**	**46%**

The extent to which consumers think about the connections between products of animal origin and the state of health and welfare of the animals from which they originate is left to their discretion. While for some consumers, an associative link between the product and the context of origin represents a momentum of belonging to a larger whole and seeing themselves as having a certain co-responsibility, other consumers show no interest in the existing connections. Through a qualitative differentiation between products from healthy and diseased animals, market potentials could be used to ensure a qualitatively improved product range utilizing price differentiation and to set market-based incentives for improvements in animal health and welfare in primary production. Concerning farm animals and their living conditions on farms, on the other hand, state bodies have a constitutional mandate to protect farm animals from harm associated with pain, suffering, and damage. A reliable and intersubjectively valid assessment of the state of health of slaughtered animals would be an essential prerequisite to fulfill this mandate. At the same time, such an assessment would provide a basis for differentiating and ranking farms of origin, which could be used by government advisory and inspection bodies and for marketing products of different values.

As late as the 1990s, meat from diseased animals was marketed as so-called "free-bench meat." On so-called "free benches," named after the custom of setting up these benches apart from the regular sale of meat, the meat was sold as inferior but not hazardous to consumers' health (Zrenner and Haffner 1999). At the meat inspection, it was stamped as "conditionally fit." The veterinary examinations were more thorough and considerably costlier than routine slaughtering. On the other hand, retail prices to consumers were consistently lower than at other retail establishments. In a time of shortage, the "free benches" utilized as much animal product as possible. With an increasing oversupply of meat sold at relatively low prices and their production costs having to compete in global markets, the "free-bench meat" lost its economic base. However, instead of discontinuing the marketing of low-quality meat and thereby reducing the oversupply of meat products on the markets, the EU has refrained from differentiating carcasses according to the state of health of the slaughtered animals. The reasons for marketing meat undifferentiated are mainly due to the global competition for cost leadership. The consequences for farm animals were and are devastating.

While marketing attempts to attach positive values and transfer effects to products of animal origin to increase willingness to pay, the fact that a large proportion of the products come from diseased animals that were exposed to considerable impairments in the form of pain, suffering, and damage during their lifetime is simultaneously ignored. Consumers are generally unaware of the extent of the health impairments and, thus, the extent of inferior products they purchase. The low level of interest in the background of food production suggests that many consumers do not want to know in more detail what undesirable side effects the food has been produced. This makes it easy for agricultural marketing experts to disconnect consumers' associations with food from the production context and use appropriate advertising messages to draw attention to individual aspects that suggest superiority without providing robust evidence.

9.2.4 Evidence-Based Assessment of Product Quality

The nutritional value, utility value, and enjoyment value of products can be differentiated and compared based on scientifically recognized parameters within the framework of different **contextualization**. As a rule, individual features stand out from the context and are used in an evaluation system to differentiate between various products. Extracting individual characteristics from their original context and placing them in an evaluation system is central to quality assessment. Depending on the interests at stake, selected individual features of products are singled out and contextualized differently without this transformation process being consciously reflected upon. As a result, sometimes quite different quality statements are made. For example, the specific nutritional value of a product only acquires a qualitative value through the context of the menu in which it is embedded, as well as through the exact dietary and physiological needs of the respective consumer. The same applies to the utility value of raw materials as a function of the requirements that further technical processing places on the raw materials concerning producing a standardized processed product. The **enjoyment value** of a food product is determined by the interaction of partial components to overall quality and the context in which the product can develop pleasure. This also includes the ability of a consumer's sensorium to judge the particularity of the respective product and its differentiation from other products.

However, the quality of a product is not determined by individual characteristics, even if many product suppliers try to give this impression to present the sales product in a favorable light. Instead, the quality assessment is always a mixture of many individual characteristics, each at a different level, which come together in a product and emerge various possible combinations. No weighting of the partial aspects concerning the quality of the whole can be derived from the determination of individual characteristics. The situation is further complicated by the synergistic and antagonistic interactions between single traits, which can influence the overall properties of a product. Consequently, knowledge of partial aspects does not allow an inductive conclusion to be drawn about the totality of the characteristics, i.e., the quality of the whole.

Unlike physical and technical processes, biological processes do not follow context-independent laws and regularities. Accordingly, the results of the processes cannot be predicted precisely. Instead, the overall outcome of the processes is determined by the interactions between the processes that occur within living organisms and the external boundary conditions that affect them. Consequently, the numerous influencing factors and their interactions along the process chain result in a wide variation in the properties of a product. For example, the composition of carcasses varies not only between animal breeds, sex, or age of the slaughtered animals. Even in the case of slaughtered animals in which these influencing factors are not different and have been kept and fed under the same living conditions, a considerable variation nevertheless results. This variation can, at best, be narrowed down by targeted management measures.

The range of variation in the composition of raw materials produced by farms contrasts with the expectations placed on the uniformity of raw materials and sales products by processors, retailers, and consumers. A step towards uniformity is subdividing the variation spectrum of specific raw material characteristics into categories. The characteristics selected are usually based on the utility value attributed to them by the processing industry. Examples are the differentiation of bread grain according to raw protein content, eggs according to weight classes, or carcasses according to muscle meat content. In the processes of further processing, the different categories of the raw goods are then preferably used for the processes with which the requirement profile of a standardized sales product can be best served.

Section 6.4 already discussed standardization as a retail strategy for dealing with variation. Branded goods, which follow their logic in marketing, further develop standardization (Schenk 2001). A distinction is made between manufacturer brands and private labels. Branded goods have specific characteristics that distinguish them from products of other brands. Suppose the consumer recognizes a brand mark on a product, for example, from advertising or a previous purchase, and if positive characteristics are associated with the branded goods. In that case, the **brand** fulfills a recognition function and, at the same time, an information and orientation function. However, this presupposes that the products are offered uniformly, irrespective of the context in which they are sold. Branded goods have a utility value for consumers and an added value. This added value consists in the certainty that the brand owner has made every effort to ensure that the branded product, in addition to the standard generic characteristics and benefits, also fulfills the particular promises of the branded product communicated by the brand owner. On the part of the trademark owner, the consumer is supposed to trust that the goods are constantly available everywhere with the known characteristics, at the known price level, and the known points of sale. Branded goods thus fulfill a trust protection function, reducing the consumer's risk of making a mistake when purchasing a product. Some branded goods, such as organic products, are not only associated with an intersubjectively verifiable benefit but also convey a subjectively perceived additional benefit that goes beyond the functional value of the goods in question because they are held in higher esteem by the public.

Retailers define the specific characteristics of private labels. Often, the introduction of branded goods is preceded by extensive soundings in the market to test the attractiveness of such brands and the willingness of potential customers to pay for the branded goods. With private labels, retailers aim to achieve a higher price or profit margin than conventional comparable products. In doing so, a qualitative superiority of the branded goods is suggested to the customers, which justifies a higher price. The higher quality is usually limited to specific aspects of a branded product that distinguish it from competing products, while there is not necessarily a difference in other relevant characteristics. For foodstuffs, for example, the salient feature may relate to individual ingredients, such as a higher content of omega-3 polyunsaturated fatty acids. The organism itself cannot produce these but must be supplied with food. They are not only vital for the organism but are also associated with interested parties with a variety of positive health effects. From a scientific point of view, however, the health effects are classified as relatively low (Abdelhamid et al. 2018). Positive effects are especially evident when there is an undersupply, i.e., when the respective organism requires a specific supplementation.

Therefore, the **reference system** for assessing food quality is the respective consumer with the specific supply status. People who already consume a varied diet are unlikely to benefit from fortified foods. Since the supply status is seldom known to consumers in detail, it is primarily a question of the consumer's self-perception and need for security. The sales prices of specific branded products are often significantly higher than those of comparable products. These are based less on the additional costs incurred in processing and marketing than on the consumer's willingness to pay and the turnover that can be expected when the price is set. To resist the manipulative approach of the food industry, independent expertise would be needed to provide orientation in the interplay of numerous influencing factors in a very complex matter. In the case of the nutritional value of foods, a benchmark would provide orientation by showing the range of variation in nutrients that can be expected and indicating at what level the specific product falls on the scale from low to high. This would give individual consumers the information to tailor a meal plan to their needs. The various initiatives on nutrition labeling in the form of a traffic light system are moving in this direction. The "traffic light labeling" would show at a glance and in a way that everyone can understand, with the colors green (low), yellow (medium), and red (high), how much fatty acids, sugar, or salt a food contains compared to other products.

An analogous approach can be applied to the **enjoyment value** of products of animal origin, for which many consumers show great interest and, at a high level, a high willingness to pay. Rarely, however, do they have the expertise to assess the enjoyment value with any degree of reliability. A yardstick with which it is possible to differentiate and classify the enjoyment value of products on a scale from low to high based on a meaningful characteristic would help provide consumers with the necessary orientation about enjoyment value.

Concerning the enjoyment value of fresh meat, the degree of marbling or the intramuscular fat content (IMF) has emerged as the most significant quality criterion. It exerts the most influence on fresh meat's aroma, juiciness, and tenderness for beef, pork, and poultry (Hocquette et al. 2010) Accordingly, meat with a higher degree of marbling or IMF content is of higher quality than that with a lower level. The degree of marbling or IMF content does not cover the entire spectrum of meat quality characteristics. However, they act as a salient **variable**, called an "iceberg variable." The concept of iceberg variables is based on a proposal from the Farm Animal Welfare Council (1997). It aims to make a reliable statement about a complex issue based on a few variables that are as informative as possible. Just as the protruding tip of an iceberg signals its submerged bulk beneath the water's surface, iceberg variables are key variables that correlate closely with other variables that are not immediately visible, i.e., that have an effect below the surface and are therefore more representative of the overall quality of a product than others. The higher the degree of marbling, the higher the level to be expected in terms of the meat's tenderness, juiciness, and flavor. The classification into corresponding categories determines the qualitative value. Nevertheless, slaughterhouses and **food retailers** in Europe have so far shown no interest in introducing such a differentiation in the consumption value of meat.

The same applies to the sensory quality of **milk**. There is a consensus in the scientific community that the cell content of milk can be attributed to the most significant importance of the sensory quality of raw milk. Suppose one also considers that the raw milk with increased cell counts comes from an animal suffering from an inflammatory process in the udder tissue. In that case, the milk cell counts are the most significant quality criterion and an iceberg variable for assessing the consumption value. Milk from udder-healthy cows, whose milk cell count is generally below the threshold value of 100,000 cells per ml of milk, is, therefore, classified as being of higher value than milk from cows with significantly higher milk cell counts due to inflammatory processes. At the primary producers, the milk cell contents of the individual dairy cows are determined within the scope of the milk controls and reported back to the primary producers. No additional analyses would be required to perform a qualitative milk differentiation. However, differentiation based on the consumption value would involve additional expenditure for separate recording, transport, and further processing, which the primary producers, dairies, and food retailers would like to spare themselves. However, they do this at the price of making the pricing of anonymous (context-free) raw milk subject to global undercutting competition. It is also likely to be challenging to defend raw milk from dairy cows with udder disease toward quality-conscious consumers regarding arguments against the emerging competition from milk substitutes. By buying products, many consumers no longer want to be jointly responsible for production processes that involve pain, suffering, and damage to farm animals and are, therefore, increasingly turning to substitute products. The fact that substitute products are also associated with various disadvantages should only be mentioned in passing at this point. It can, therefore, be assumed that for a relevant proportion of consumers, milk from udder-healthy animals is rated higher than synthetically produced milk substitutes. Dairies and food retailers have shown no interest in differentiating raw milk based on the cell count level prevailing on the respective dairy farms.

What applies to the health status of the udder tissues of dairy cows from which the milk is obtained can be used analogously to meat products. The absence of pathological-anatomical findings in the assessment of carcasses and organs marks an iceberg variable for which it can be assumed with a very high probability that the animals were free of health impairments at the time of slaughter. This not only takes account of the requirements of consumer health protection. At the same time, the absence of disease also fulfills an elementary requirement for the welfare of farm animals. Products of animal origin taken from an animal context previously identified as essentially "free of disorders" can claim a qualitatively higher level than products from animals with pathological findings. While the superiority of animal products of demonstrably healthy origin requires no further explanation to the consumer clientele, such an apparent differentiation of carcasses contradicts the interests of the food retailers. They expect a more significant economic advantage from categorizing food according to the farming conditions, such as "pasture-raised milk" or "pasture-raised meat," than based on the health status of farm animals. However, this might not be engraved in stone. Substitute products for products of animal origin enter the market and try to displace them by highlighting their weak points in terms of enjoyment value. This makes it more critical to prove that products of animal origin can compete with substitute products in terms of enjoyment value.

Derivations

The handling of quality-relevant characteristics of foodstuffs is characterized by the requirement profiles (standards) that the processing industry and the retail trade place on the raw materials supplied by the primary producers. For the processing industry, the focus is not only on the nutritional and utility value but also on the hygienic safety of the products. After all, extensive batches of products can be contaminated by introducing harmful substances or pathogenic microorganisms into the processing chain. As a result, these are no longer fit for consumption or sale and must be destroyed. The primary producers are identified and held liable for the damage through the traceability system of raw goods back to the farm of origin. In addition, the primary producers are made to align their production processes with the requirement profiles of the buyers through corresponding price masks for raw goods.

On the other hand, the processing industry and retailers have so far attached only minor importance to the enjoyment value of raw materials. The significant variation of the raw materials concerning the characteristics mentioned above of the enjoyment value collides with the prevailing marketing concept of the agricultural sector. This is geared towards generating additional prices through standardized branded goods. This is done with the help of marketing, in which transfer effects are evoked in consumers through highlighted individual characteristics. Supposed differences between no-name products and branded goods replace product differentiation based on reliable quality criteria and justify a higher market price without providing proof of superiority. From the retailer's perspective, quality is that for which consumers show a higher willingness to pay due to a supposed superiority.

In this approach, organic products represent the highest quality level. They are based on compliance with the increased legal requirements of the EU organic regulation compared to conventional products. These are not geared to individual improvements but to many selective enhancements. Consequently, it is a brand program that significantly expands the requirements profile. However, as with other branded goods, the retail trade still fails to provide reliable proof of the qualitative superiority of organic products and the production processes concerning the individual farm context from which the products for sale originate.

This marketing strategy not only deceives consumers about potential expectations of higher quality products but also prevents those primary producers who strive for higher product quality from realizing more excellent value for the additional expenses involved. Without remuneration, there is no incentive for evidence-based quality production. Instead, the incentive is to produce raw materials at the lowest possible cost, largely irrespective of what impact this might have on quality levels. Finally, primary producers obtain the same price for lower-quality raw materials as for higher-quality ones. This means that higher quality commodities produced in the context of the animals and the farm stand out from the average level and have no chance of competing for cost leadership. The abandonment of evidence-based quality production marks the starting point for **unfair competitive conditions**. These are not only partly responsible for abuses on many farms but also block the possibilities for fundamental improvements. The latter can only be realized if higher prices are paid for higher-quality raw goods by the buyers in processing and trade. Accordingly, market prices should in the future be geared to the respective availability of products with different quality levels, which are classified by independent bodies based on reliable criteria on a scale from very low to very high.

The food retail industry not only possesses the market power that sets the prices for the primary producers but also the power to shape the market to its liking. By establishing branded goods, the retail trade also determines which standards are applied to the qualitative assessment of products and which are not. In this way, the retail trade claims interpretative sovereignty over what is supposedly high quality. They are actively supported by agricultural marketing and agricultural economics. Although scientists, as civil servants, would be obliged to the public welfare interests and not to the interests of the market, it has not yet occurred to them to enlighten the public about the behavior of the retail trade. Many agricultural economists see themselves as part of the market and may not realize that they are part of the problem from the public welfare perspective.

The relentless pursuit of cost leadership within our current economic system has already pushed numerous farms to the brink of financial collapse. Simultaneously, the market economy is inflicting significant damage on our communal resources, with the culprits evading responsibility. A profound reorientation and realignment of the financial framework is imperative to break free from the clutches of market power, which exploits primary producers and strips them of the means to secure their economic survival. This necessitates an agricultural policy that no longer exposes primary producers to merciless undercutting competition, a system that primarily benefits the agro-industry and food retailers.

9.3 Environmental Protection Services

Food production is environmentally sound if no burdens on the environment lead to permanent damage. Damage is a relative quantity. In the private sector, specially trained experts and neutral surveyors are used to assess the damage amount and take further steps to settle the claim. No official evaluation and damage settlement procedures have been established to evaluate the environmental damage caused by agricultural production processes that could be applied in this context. As long as no procedures for an intersubjectively valid assessment are implemented by an independent body, the most powerful interests and their perspectives will generally prevail. For decades, national policymakers have found it challenging to enact generally applicable rules for agricultural practice that would be suitable for limiting harmful environmental effects. Finally, immense economic interests and a potent agricultural lobby are at play, which can promote its interests.

9.3.1 Misguiding Sweeping Statements

In the past, the political focus of environmental protection was not on reducing emissions and tackling their causes but rather on **immission**, i.e., the reduction of **harmful effects** caused by emissions on other goods and interests. Immissions are detrimental environmental effects that, by their nature, extent, or duration, are likely to cause hazards, significant disadvantages, or significant nuisances for the general public or the neighborhood. The legal approach to environmental damage via immission protection directs the attention to the effects of damage to the damage caused to individual persons affected. On the other hand, the damage caused to the environment as a common good is not considered. Before state bodies have the authority to intervene against the release of emissions, it must be proven that this causes harm to others. Thus, the burden of proof for harmful effects lies on the side of the state control authorities. Because of the complexity of the processes involved in detrimental impacts, this is often only possible when the damage has already reached a high, almost undeniable level. The ambiguity of the causal relationships offers interest groups, who see themselves restricted in pursuing their interests by state intervention, various approaches for argumentative objections, and legal resistance.

A prominent example is the Fertilizer Ordinance, which is at the heart of environmental legislation for the agricultural sector. This is based on the European Directive of 1991 (91/676/EEC, Nitrates Directive), which protects **groundwater** from excessive nitrate inputs due to fertilization measures. The European Directive was to be transposed into national law, offering options for considerable resistance and delays to state regulation by the agricultural lobby. A quarter of a century after the Fertilizer Ordinance came into force, the declared goal of complying with the limit value of 50 mg of nitrate per liter of groundwater everywhere in Europe has still not been achieved. At the instigation of the EU Commission, the European Court of Justice ruled in an infringement case against several European Member States in 2018 that the Fertilizer Ordinance was insufficient to meet national obligations concerning the Nitrates Directive. A high court thus confirmed the failure of the previous strategies to protect groundwater from agricultural discharges.

While national politics has been unwilling to pursue a targeted and verifiable strategy, the EU Commission has acted as the overarching authority, urging member states to commit more to environmental protection and comply with agreements already made. However, the mills of EU justice grind very slowly. Despite national resistance, the EU Commission's initiatives in road transport are bearing fruit more quickly than in agriculture. Here, the EU Commission has learned lessons from past mistakes and no longer relies on the data provided by car manufacturers regarding the extent of emissions. With the EU Directive 2016/2284 (European Union 2016), new rules have been established to measure exhaust emissions under realistic traffic conditions. In this way, the exhaust emission values of car brands can be compared and ranked on a scale. At the same time, a limit value of 95 g CO_2 emissions per kilometer driven was set, which must be achieved by the suppliers' vehicle fleets within defined periods. In this way, the EU Commission has set concrete and verifiable **targets** for the automotive industry and shown them the path of technical developments to follow in the future. However, it should not go unmentioned that the EU Commission has helped in the automotive sector because more stringent emission requirements are also being imposed in those third countries where vehicles manufactured in Europe are primarily exported. This means costly measures to achieve low emission levels are no longer a competitive disadvantage in trade with relevant third countries.

This is fundamentally different for agricultural products. The impairment of exports is a significant reason for the policy of export-oriented EU member states to prevent or delay measures that could limit the competitiveness of national producers in global markets. With the legislative package "Fit for 55", the EU Commission is attempting, against much resistance, to make agriculture responsible for achieving the EU climate targets agreed in the Paris Climate Agreement. According to a proposed regulation on land use, land use change, and forestry, the EU agricultural sector aims to be **climate-neutral** by 2035. By 2030, greenhouse gas emissions will be reduced by at least 55% compared to the reference year 1990. The setting of **targets** is one thing, but the means and methods by which they are to be achieved are something else. To get closer to the targets, it would be necessary to clarify who would have to make what contribution, what changes would have to be made, and to what extent. Concerning the **operationalizability** of measures with which the declared goals are to be achieved, the policy still owes concrete specifications. Unlike the automotive industry, farmers thus lack a clear orientation as to what levels of emissions from farms will still be permitted in the future. As long as there are no concrete and sanction-proof legal requirements, the farms will continue to design the production processes and, thus, the emission quantities in a way that seems sensible to them for the economic viability of the farms.

Production outputs and the associated emissions of a farm system result from the operational orientation of all production processes. These cannot be assessed in isolation, as they are interdependent and influence each other in the complex interactions within a farm. While the production performance of a farm can be recorded with sufficient reliability, company managers generally do not have any reliable estimates of the extent of the emissions that escape from their farms into the environment. Unlike cars or industrial plants, farms do not have exhaust pipes, chimneys, or sewers where emissions can be analytically determined and quantified. Diffuse and large-scale releases via the air and surface waters require a completely different approach to quantification in the form of **balancing**. In the Netherlands, the method of balancing the individual farm discharges of **reactive nitrogen compounds** is already used throughout the country (Daatselaar et al. 2015). Before this approach can be established in other European Member States, considerable resistance from many stakeholders and their political representatives to quantifying emissions has to be overcome. After all, the interest groups fear relevant disadvantages for their clientele if greater importance is attached to protecting the environment than to protecting primary producers from **state regulatory interventions**. Unlike in other export-oriented sectors, production is decisive for the agricultural industry. Since higher environmental protection requirements are incompatible with the pursuit of **cost leadership** and price-undercutting competition on global markets, the real crux for implementing environmental protection measures lies in the export orientation of the agricultural sector.

The export orientation does not permit any measures that would impair the competitiveness of domestic producers on global markets by increasing production costs. However, this view ignores that there is a domestic market in addition to the world market. If the support of the domestic population and consumers is not to be jeopardized, the **environmental effects** of the production processes cannot be permanently ignored. The importance that must be attached to environmental and climate protection because of the severe environmental and climate damage that has already occurred means that domestic primary producers will have no other choice sooner or later. They must set out to produce agricultural commodities in the future in a way associated with demonstrably low emissions levels. The emphasis is on verification. While various labels, such as eco-label, already promote their sales products with supposedly environmentally friendly production methods, they have so far failed to provide sound evidence concerning comparable lower emissions per unit of product. Free riders, who pocket increased sales revenues without providing increased socially relevant services, cannot be countered simply by checking compliance with increased minimum standards. There is no way around the urgent need for evidence-based quantifying environmental performance in the long term to establish fair competition and justify consumer confidence and willingness to pay higher prices.

The vast majority of farm managers believe that they can spare themselves the effort of quantifying substance discharges into the environment as long as this is not demanded of them by the legislator under threat of sanctions. In doing so, they overlook that reducing substance discharges can also be accompanied by significant economic benefits, such as savings in production resources. Without appropriate quantification, farm management cannot assess whether the optimum for its financial success in terms of the cost-benefit ratio has already been reached (see Sect. 5.2). Primary producers also give away the opportunity to offer affinity consumers and local, regional, or national government bodies a higher level of emission reductions per unit of product, for which the primary producers are entitled to claim a corresponding additional price or subsidy. Suppose primary producers want to meet the demand of consumers and taxpayers for emission-reduced production and corresponding raw goods for the domestic market. In that case, they must also make a reliable offer to the buyer. Most primary producers, however, are still far from taking such a proactive approach.

However, companies that operate in a more environmentally compatible manner are not only dependent on achieving a higher sales price for their higher operating expenses. At the same time, they must be protected from unfair competitive conditions and, thus, from competitors who gain a competitive advantage by realizing lower production costs at the expense of public welfare interests. We are dealing here with multiple **'externalizations.'** On the one hand, large quantities of residues and waste produced during production processes are transferred into the environment, causing considerable pollution. The true costs of making and using externally purchased means of production and disposing of waste materials into the environment are shifted to others who cannot defend themselves against them. Externalization reflects unfair competitive conditions that give market advantages to those who act more ruthlessly than others. Finally, the refusal to take responsibility for one's actions is called externalization (Göpel 2020). Companies act against public welfare interests by irresponsibly taking advantage at the expense of others. Externalization also includes farms, which can continue this practice because the state's control bodies do not prevent them from doing so.

Although agricultural policy is responsible for ensuring fair competition conditions, it does not fulfill this task. Agricultural policy attaches less importance to the state's mandate to protect the community than to the interests of the agricultural lobby, which generally rejects any form of regulation. The contradictory behavior of primary producers complicates the situation. Since many of them are the primary sufferers of ruinous undercutting competition, it would be obvious if the majority of farmers were to advocate that the state establishes fair framework conditions. This would include ensuring that higher-quality production is protected from the unfair practices of free riders and low-cost suppliers from home and abroad. Even more than the dishonest practices of competitors, many farmers seem to fear state controls and regulations, which might affect their business practices.

Above all, the supposed representatives of the interests of primary producers in the various professional associations show no interest in an approach that differentiates between farms. After all, this would torpedo the general claim to represent all farmers if a distinction were made between farmers who act fairly and those who act unfairly. Instead, they claim to be a protective force against any impositions by the state. At the same time, they advocate an increase in all forms of subsidy to farmers from public funds "*so that everyone is helped and no one is left behind.*" On the other hand, they oppose the control of whether **subsidies** from public funds are used in the public interest. The prevailing attitude is one of economic liberalism, the effects of which, in retrospect, have produced comparatively few winners but many losers among farmers. **Contradictions** and **unfairness** among the members of the professional associations and vis-à-vis the community can only be covered up as long as the state does not differentiate between those farms that justify the subsidies by making a positive contribution to the public welfare interests and those that pocket subsidies but at the same time harm the interests of the common good. Representatives of associations, therefore, do not protect those companies that are disadvantaged in competition but those that use externalization to gain a competitive advantage at the expense of other competitors and to the detriment of the common good. In this way, the interests of the lobbyists run counter to the interests of many farms they claim to represent.

As in other sectors, those responsible only start thinking about reducing emissions when they are forced to do so by regulatory law or attracted by cash incentives. Under the current framework conditions, this applies even to organic farming operations. Their image among the population is mainly based on the assumption that organic farms produce in a more environmentally friendly way than conventional farms. The general renunciation of using mineral nitrogen fertilizers and pesticides suggests that this is accompanied by substantial advantages for more environmentally compatible production. There is no doubt in this respect as regards the discharge of pollutants per farm. However, productivity and product volumes per unit area are considerably lower than in conventional agriculture. As a result, discharges of carbon and reactive nitrogen compounds into the environment per unit of product sold can sometimes be higher than in conventional production (Sundrum 2012b). As is the case between conventional farms, there is also a considerable variation between organic farms regarding environmental protection performance. For this reason, general statements attesting to the general excellence of organic farming are not justified.

9.3.2 Context-Dependent Assessments

What is lacking on conventional and organic farms is an individual quantification of the actual discharge quantities of residues and waste materials in relation to the sales products. While most conventional farms have no interest in such quantification, organic farms would have good reasons to strive for it. On the one hand, this could justify the higher market prices to consumers. Consumers' willingness to pay is based not least on the confidence that the farming method is associated with a higher degree of **environmental compatibility**. If these expectations are unfounded, there is a risk of losing confidence and undermining a business model that cannot be maintained without additional price payments. For this reason alone, there should be a strong interest in acting proactively to counteract the threat of a loss of image. No less important is the fact that a demonstrably more environmentally compatible way of doing business can not only serve the interests of the common good and an affine consumer clientele. At the same time, it can also serve self-interest in terms of increased productivity and economic profit. However, before such a win–win situation can be realized and financial and public welfare interests can be reconciled, i.e., external effects can be internalized, various preconditions must be met.

First and foremost, it is essential to be aware of the **conflicts of objectives** that arise to varying degrees between the interests of the particular and those of the public welfare interests, depending on the specific farm context. Even if agricultural enterprises want to commit themselves to environmentally compatible management, they cannot afford to lose sight of the economic consequences. Different strategies are employed in pursuing the two production goals: increasing productivity and reducing emissions. Strategies for minimizing expenditures and increasing production output meet strategies for reducing emissions, which are usually accompanied by additional expenses and output limitations. Pursuing one goal is to the detriment of the other unless the two production goals can be combined into one overarching objective. Instead of an "either-or" approach, there is a need for a "both-and" approach and corresponding strategies. Management should strive for the greatest possible common overlap between the two objectives to contain conflicts. Since the starting and boundary conditions vary considerably between the farms, efforts to expand the intersection can only be a farm-specific approach that does not follow general advisory recommendations. If it is a question of linking and balancing two opposing but complementary sets of values, the square of values, based on Schulz von Thun (2007) (see Fig. 9.2), provides valuable services.

The prevailing focus on individual business success amounts to a selfish attitude. The design of production processes is subordinated to economic objectives. Competition with others can lead to excessive productivity (hyperproductivity). This approach is detrimental to the public welfare interests. However, farms are not self-sufficient; they are part of a community. They are, to a high degree, dependent on support, among other things, through the purchase of products for sale, through the subsidy payments granted, and through the maintenance of infrastructure to survive economically in the long term. Accordingly, farms cannot permanently ignore the public welfare interests. If, on the other hand, the focus is solely on ecological requirements, this would result in reduced

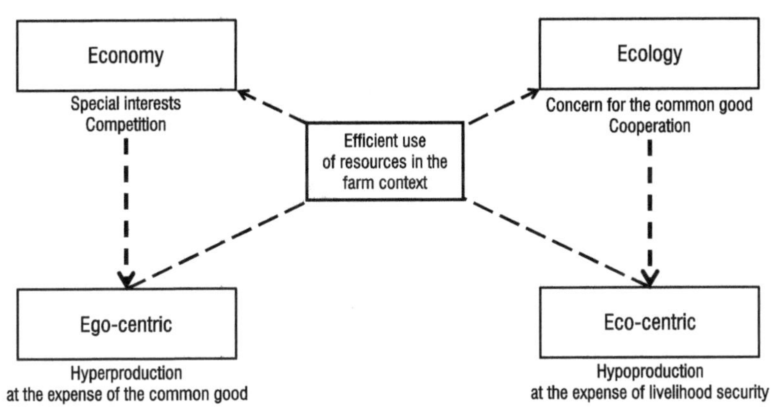

Fig. 9.2 Balance of interests between economic and ecological requirements via efficient internal use of resources (own illustration)

productivity (hypoproductivity). The monetary income from the products sold may be too low to guarantee the maintenance of the operating system. Both requirements have their justification but require a balance to contain the negative consequences of a one-sided orientation. In the square of values, the directions of development cross each other. Farms in danger of slipping to the lower left are well advised to consider eco-social values more and include them in management decisions. Those farm managers who risk over-emphasizing ecological aspects and neglecting economic requirements are advised to attach greater importance to the latter to secure their financial existence. The common intersection in avoiding the respective extreme positions is represented by **efficiency** in using internal and external resources in the farm-specific context. Efficiency corresponds to the ratio of output to input variables and, as such, is the relevant **target variable** of an economically and ecologically oriented approach. Accordingly, striving for high efficiency in the implementation processes should be the focus of individual farm resource management. However, focusing on high efficiency only in individual sub-areas is insufficient when the overall system is lost out of sight.

The prevailing agro-industrial strategy is based on intensification, specialization, and mechanization of production processes (see Sect. 5.1). This way, a high degree of efficiency is realized selectively only in sub-sectors. At the same time, consequential effects are caused in the upstream and downstream areas inside and outside the farm system, which are often ignored or externalized. Suppose the environmental impacts that arise in the external and on-farm production of fertilizers and feedstuffs and their application in the farm context are also considered. In that case, the benefits of intensification can be considerably relativized. For example, efficiency gains in meat and milk production can be caused by increased feed imports. At the same time, the resulting increase in the production of excrement reduces its usefulness as farm fertilizer. This approach is economically advantageous as long as the adverse effects of increased substance discharges into the environment are not included in the economic balance. As a rule, it is also not considered that the efficiency of nutrient conversion follows a utility function at the animal, field, or farm level. This involves the increase in benefit achieved by using a further unit of an input variable. However, the law of diminishing **marginal utility** states that the relative utility decreases with each additional unit with a concave course of the utility function (see Sect. 5.2). With further input beyond the optimum, utility becomes negative and causes considerable economic losses.

Despite its economic and ecological relevance, the marginal utility function of production processes has not yet found its way into agricultural extension practice. There are various reasons for this. First, it is due to the lack of detailed farm-specific data to display the relationship between efforts and benefits deriving thereof in the respective context. For example, managing a dairy farm requires a farm- and cow-specific assessment of the needed resources and revenues. Due to practical reasons, the various types of data from a farm, e.g., economic and biological data, commonly lack connectivity and compatibility. This means that the economic impacts associated with differences in rearing, productivity, and culling and the economic values of cow-individual service lives often remain hidden under average figures. Secondly, data acquisition efforts are usually underestimated concerning their potential usefulness to pinpoint problems, optimize herd management, and prioritize investment necessities. In contrast, the overall extent is generally underestimated, particularly the individual variation of the negative impacts and failure costs associated with exceeding gaps between the demand and supply of resources. Above all, this is because average figures cannot estimate these. The individual animal or, in the case of plant production, the single plot is the only valid reference system to assess the relationship between efforts and benefits. While continuously eager to minimize the efforts to manage production processes, farmers have become victims of their adherence to the selective economic approach. The inductive deduction of partial aspects to superordinate overall contexts is the source of severe misjudgments and, at a certain level, causes the opposite of what was initially intended. Following the selective economic approach, farmers are trapped in false thought patterns and can no longer find any constructive way out.

Considering the efficiency of all internal material conversions on the various process levels and farm areas and the inclusion (internalization) of upstream and downstream external areas requires a far-reaching reorientation. This affects the economic approach and the way of thinking and approaching things. After all, it is not only the production output and costs that are important; at the same time, the extent of material discharges into the environment is also relevant. Such a multivariable process cannot be done without a sound data basis. The operational output quantities in the form of desired sales products and undesired discharges of residual and waste materials are contrasted with the input quantities that enter the operating system as an external means of production. Accordingly, the efficiency as the ratio of output to input can be related on the one hand to the sales products and on the other hand to the unwanted outputs. Provided that no substances accumulate or are released in the operating system through organic binding to biomass, the efficiency in the production of sales products (economic efficiency) is reciprocal to the efficiency in preventing substance discharges into the environment (ecological efficiency). The higher the proportion of input variables that leave the farm system as sales products, the lower the proportion of input variables that escape from the plant in the form of residual and waste substances. Thus, the absolute discharge quantities of residual and waste substances and the ratio to the amount of sales products are highly relevant for assessing **environmental compatibility**.

Due to their high **harmful effects**, emissions of **reactive nitrogen (Nr)** are of most significant importance for the environmental compatibility of agricultural production (see Sect. 5.3). Nr is present in organically bound form (amino acids, nucleic acids, proteins) and inorganic form (ammonia, ammonium, nitrate, nitrite, nitrous oxide, nitrogen oxide). While the Nr compounds in the sales products are largely organically bound, the residual and waste materials that escape into the environment are generally inorganic Nr compounds. Internally, the Nr-compounds are either organically bound as microbial, plant, or animal biomass or are mineralized in the soil and the manure store. The particular challenge of management is to convert as high a proportion of the purchased and on-farm Nr quantities into an organic binding form as possible. These leave the farm in the form of sales products of plant or animal origin or accumulate or are recycled as microbial or plant biomass in the material cycle of the farm. At the same time, various management measures must be taken to prevent the proportion of inorganic Nr compounds from leaving the operating system via the air or dissolved in water.

Ecologically oriented **nutrient management** strategies differ fundamentally from the prevailing intensification strategies of production processes. While the latter are primarily based on increasing resource inputs, ecologically oriented ones are concerned with limiting them. Three strategies, in particular, can promote high efficiency in transforming nutrients into sales products while minimizing the discharge of residual and waste materials into the environment. (Sundrum 2019):

- **Reduction** of the nutrient input into the operating system to substitute the substances that limit the growth processes according to the minimum law and are insufficiently available internally.
- Targeted **allocation** of nutrients to the plants and animals which, according to the crop-specific and animal-specific requirement values, have the most significant requirement in the course of vegetation or production and which, according to the marginal utility function, can derive the most important benefit from a targeted supply.
- High degree of on-farm reuse (**recycling**) of nutrients from soil via plant to animal and from animals via excrement back to soil.

The more purposefully and efficiently the internal resources are used, the fewer nutrients must be supplied from outside. Nutrients bought in but not adequately used represent a loss from both an ecological and an economic point of view. However, the financial incentive of efficient use is lost if the purchased means of production are disproportionately cheap. This applies, in particular, to mineral nitrogen fertilizers concerning the environmental damage that occurs during their production and their application in the operational context. An (eco)tax on mineral nitrogen fertilizers would, therefore, be a particularly effective measure to encourage management to use them more efficiently by making them more expensive. A higher price could be more than compensated for by more efficient resource use, so there need not necessarily be an increase in production costs. In fact, given the generally excessive use of fertilizers in plant production, the result of a tax may even be a reduction in production costs due to increased resource efficiency. This way, an ecological benefit can be brought without any economic disadvantages (except for the fertilizer manufacturers). However, the unavoidable price that would have to be paid for increased resource efficiency would be the need to introduce data-based nutrient management, which most farms still believe they can do without.

This applies even more to organic farming, which pursues a much more rigorous course by generally banishing the supply of mineral nitrogen fertilizers. Organic farms depend on obtaining the growth-limiting resources of reactive nitrogen compounds through cultivating nitrogen-fixing legumes and purchasing fodder. If the soils are not to be depleted, the nutrients lost to the farm via the products sold and via discharges into the environment must be substituted by additional purchases. Consequently, the managers of organic farms have, at least in theory, an intrinsic motivation to use the compounds or reactive nitrogen as efficiently as possible to increase productivity and farm income and to prevent them from escaping into the environment. In this way, yield-increasing and thus economically profitable measures act synergistically with efforts to reduce substance discharges into the environment (win–win situation). However, even on organic farms, the available instruments for balancing and nutrient management are insufficiently used. It is, therefore, not surprising that many plots are not adequately supplied with nutrients. A recently published study found that 66% of the investigated arable fields of organic farms have yield-limiting deficiencies in nutrient management (Kolbe and Meyer 2021). The average calculated yield loss on the affected fields was 18%. In contrast, excessive nitrogen and phosphorus balances affected 25% of the arable fields, which indicates potential substance discharges into the environment. The more precisely the available nutrients are allocated to the individual plots or the individual animals according to the specific demand, the more efficiently they can be converted into sales products.

Organic farms are particularly challenged to take account of the complex requirements for the targeted allocation of nutrients. They are often structured as mixed farms with a wide variety of farm branches and, therefore, rarely exhibit a degree of specialization in conventional agriculture. In addition to the various advantages of mixed farms concerning the recycling of nutrients, there are increased demands on the management to deal appropriately with the diverse requirements and to reconcile them. Status quo surveys from organic poultry and pig farming show that there is a suboptimal and, thus, inefficient use of feed resources with insufficient production performance and increased animal losses on many farms (Blume et al. 2021b). On the other hand, the calculations showed that the respective farms have various options to improve the allocation of nutrients and the supply of animals through targeted optimization strategies and, at the same time, to increase the farm profit, in some cases significantly (Blume et al. 2021a). However, to realize significant improvements, a feeding control system on individual farms must be implemented in combination with an inter-farm comparison of the farm's resource efficiency. Otherwise, the reality of organic pig and poultry farming will not be brought more in line with the farm-specific challenges and potentials of resource efficiency.

Recycling nutrients in the inner-farm **nutrient cycle** can also promote a win–win situation. Processes designed to recycle limited groups of nutrients such as nitrogen and phosphorus increase productivity and reduce the need to substitute these nutrients by purchasing fertilizer or feed. An essential prerequisite for the reuse is the formation of **organic biomass**. This prevents the nutrients from escaping into the environment, thus being lost for reuse. The targeted promotion of the formation of organic biomass and the avoidance of emissions of inorganic compounds is not compatible with conventional housing on slatted floors, by which feces and urine are mixed to form a slurry. To counteract nutrient losses during excrement storage, it is essential to separate feces and urine.

While the reactive nitrogen in the feces is organically bound, urine contains urea as an inorganic form of nitrogen, which is split into ammonia by the enzyme urease. Ammonia is highly volatile and can very easily escape by air. The emission potential from urine in contact with feces is exacerbated by urease from specific bacteria in the feces. When urine and feces are stored separately, the nitrogen can be more easily prevented from escaping than is the case in slurry.

Carbon compounds in agriculture are part of a global carbon cycle in which carbon not only escapes from farms but can also be taken out of the air via **carbon dioxide** and can be sequestered again in biomass. Due to carbon fluxes, there is a considerable exchange of matter between the carbon reservoirs of the earth's spheres. Important parameters resulting from the material fluxes are residence time in the carbon reservoir and inflow and outflow rates. If a reservoir S1 releases more carbon per unit time to another reservoir S2 than it takes up from it, S1 is a carbon source for S2, while S2 is a carbon sink. In the context of the carbon cycle, the carbon balance is a budgetary accounting of the inflows and outflows of carbon stores. Carbon dioxide (CO_2) is the dominant carbon compound in quantity. Since the beginning of industrialization, the burning of fossil fuels has added previously long-term sequestered carbon as CO_2 to the material flows in the environment, so the concentration of carbon dioxide in the Earth's atmosphere has been rising continuously, and is thus a significant contributor to global warming and the climate crisis.

Agriculture is also involved in the carbon cycle to a relevant extent. The largest carbon store is the soil, followed by the living and dead vegetation in the biosphere. The most important interaction between the atmosphere, lithosphere, and biosphere occurs through vegetation. Through photosynthesis, plants absorb carbon dioxide from the atmosphere and release about half of this carbon back into the atmosphere through respiration. The other half is organically sequestered via biomass accretion. Much of the biomass accumulates as litter and is mineralized by soil organisms and released into the atmosphere. There is longer-term carbon storage only in the trunk of trees and the root system. A comparable small remainder is stored in the soil as **humus** (Houghton 2007).

In addition to carbon dioxide, the carbon compound **methane** (CH_4) is considered by many authors to be a significant emitter of greenhouse gases from agriculture. Methane is a colorless, odorless, and flammable gas. It has a significantly higher global warming potential than carbon dioxide but a much shorter half-life. In the earth's atmosphere, methane is oxidized to water, formaldehyde, and CO_2. In nature, it occurs as the main component of natural gas and methane hydrate bound to the ocean floor and permafrost areas. Biotic processes produce the gas in considerable quantities, either by microorganisms breaking down plant carbohydrates under anaerobic conditions or by phytoplankton, plants, and fungi aerobically. In the agricultural context, methane plays an ambivalent role. On the one hand, methane formation in biogas plants is promoted as a renewable energy source with extensive government support measures. On the other hand, methane escapes unused into the atmosphere during the storage of excrements from manure containers and rumen of ruminants. It disreputes farm animal husbandry, especially cattle, as "environmental sinners."

Due to one-sided media reporting, the release of methane from the rumen of ruminants is seen in the public perception as one of the main problems of the agricultural contribution to the greenhouse. This usually ignores that methane is part of the global carbon cycle outlined above. Methane is broken down to carbon dioxide, and as plants grow, it is organically bound again in the plant material. Up to five times more carbon is stored in the soil of permanent grassland (meadows and pastures) than under arable soil. Grassland can only be effectively used for food production by ruminants. The decisive factor for assessing the **environmental compatibility** of production processes is whether a farm, in terms of the overall carbon cycle, represents a carbon source or a carbon sink. Since there are different forms of carbon sequestration and release on a farm, this cannot be judged based on single-point sources of emissions, such as methane release from the rumen, but only by an individual farm balance sheet.

Furthermore, it is overlooked that methane in the rumen of ruminants fulfills an important nutritional-physiological function. Via methane, hydrogen ions, which are produced in the rumen during the breakdown of plant carbohydrates, are bound to carbon and removed from the rumen in gaseous form. Thus, the release of methane contributes to a relevant extent to the fact that the pH value in the rumen does not drop too much and does not impair the fermentation processes in the rumen. Above all, however, the release of methane has a positive effect on the health of the animals because it counteracts rumen acidosis, which can involve a considerable risk of disease with a wide range of subsequent impacts. Scientific studies have shown that over one-quarter of dairy cows suffer from subclinical rumen acidosis (Plaizier et al. 2008). Feeding strategies aimed at reducing the release of methane can thus contribute to an increase in animal metabolic disorders. While impairing **animal health**, metabolic disorders are jointly responsible for shortening the useful life of the animal and reducing its productivity. At the same time, more rearing animals have to be kept to replace those that leave prematurely. The additional rearing animals, in turn, release large amounts of methane themselves and can thus, in some cases, significantly exceed the potential reduction in methane releases through specific feeding strategies. The example of methane releases indicates that a focus on particular emission sources not only gives an abbreviated view of the situation but can also lead to wrong conclusions and measures when the context in which the emissions are released is not considered appropriately.

Ultimately, the methane released from the animals' rumen has a beneficial effect on the ruminants. Analogously, it is valuable when plant biomass produces methane as a regenerative form of energy production in biogas plants. However, the output of "**biogas**" is perceived differently from the release of methane from ruminants. Despite a considerable risk that methane escapes to a substantial degree from gas leaks in biogas plants uncontrolled into the atmosphere (Umweltbundesamt 2021), there is a lack of legally binding requirements for protecting the environment and ensuring the safe operation of biogas plants. While the releases of methane from the rumen of ruminants are in focus, methane releases from biogas plants and manure containers are largely ignored. To contain releases from biogas plants, regular inspections are needed to identify and address potential leaks. Methane releases from the **slurry** can best be prevented by storing manure and urine separately, thus eliminating the biochemical requirements for methane formation. Methane release from ruminants can best be reduced by reducing the number of animals and offspring. Therefore, providing optimal living conditions for the parent animals is necessary, resulting in a long and useful life. Depending on the initial and boundary conditions on the farm, different options are adequate and associated with various economic consequences. Environmentally-friendly production is not a question of single factors but the outcome of the totality of the complex metabolisms within a farm system, thus being the relevant reference system.

If as much carbon is bound in an operating system as is released during the metabolic processes, the operating system behaves in a **climate-neutral manner**. Consequently, the amounts of methane from ruminants and manure are a climate-relevant problem if the amounts released exceed those sequestered. The reference to cattle as climate polluters is unjustified if the animals are integrated into an inner-farm carbon cycle in which as much or even more carbon is organically bound than released. It is only possible to assess which subsystems have exceptionally high balances and how these can be reduced as effectively as possible with minimal effort and economic loss when based on a **carbon balance sheet**. However, this requires something that is lacking on most farms: an accounting system that would have to include not only the monetary but also the material inputs and outputs, as well as the efficiency of the conversions in the various subsystems of a farm. Instead, farms tend to be a **black box** regarding intra-farm material flows and the extent of emissions into the environment.

Managing the resources available on the farm, i.e., operating ecologically, means, in particular, keeping records of the limited resources available in the form of specific nutrients, working time, and investment funds. Knowledge of the available resources is one of the elementary prerequisites for using them efficiently in the operational context. Another prerequisite is the knowledge of the demand, i.e., when which specific subsystems need which resources in which proportions to be productive and to avoid disturbances resulting from imbalances. This is the only way to achieve an improved allocation) of the limited available resources and a more efficient material conversion for an overall objective. Since an agroecosystem does not organize itself like a natural ecosystem but requires organization by the management, knowledge of the respective availabilities, the need situation, and the efficiency of the conversions are decisive prerequisites for targeted control of nutrient flows. A targeted allocation of resources goes far beyond traditional measures of agricultural and extension practice. Comprehensive accounting is, therefore, a central instrument for adapting production processes to internal and external changes and for generating synergy effects. This is also the only way to assess the success or failure of management measures and to make any necessary corrections promptly before more serious undesirable developments occur. The same applies to any strategies to honor and regulate environmentally friendly production in an evidence-based manner.

There are many reasons why the vast majority of farms find it difficult to establish a solid accounting system for the material flows through the various areas of the farm. It would go beyond the scope of this book to reflect on this comprehensively. Apart from a low affinity to the associated desk work to be found with many farmers and the additional working time expenditures at the same time prevailing time shortage, three further aspects are likely to play an important role. First, the conversion of materials on farms is such a diverse, varied, and confusing process that the complexity makes it easy to feel overwhelmed. Accordingly, it is understandable if farmers do not want to rack their brains and would like to spare themselves the effort of mentally penetrating complex interrelationships. The standard economic analysis of decision-making holds that information is valuable only to the extent that it leads to better decisions. Among the stylized assumptions of economics, this may be one of the most unrealistic and consequential (Golman et al. 2017). Considerable research in economics, psychology, and neuroscience points to the conclusion that people derive utility from possessions, experiences, and beliefs. Most of what matters happens "inside our heads," and given that there are often multiple ways to interpret the same piece of information, how we construe information is often as important as the objective content of the information. For instance, people dislike being exposed to information that conflicts with existing beliefs (dissonance avoidance). People often avoid information when it would make them feel bad because it carries negative consequences. Furthermore, people sometimes avoid easily obtainable information about the impact of their actions as part of an excuse-seeking behavior to maintain a positive self-image. Finally, an attitude of expectation has developed about the solution potential of technical developments and the wealth of knowledge available in the advisory system and in the various specialist disciplines of agricultural science, from which many practitioners hope to find solutions while simultaneously being able to reduce efforts for information acquisition.

While minimization of efforts per product unit belongs to the primary strategy to succeed in competition, too few players have yet realized that only an overview and insight into farm-specific processes and detailed knowledge of the discrepancy between actual and target values in the various subsystems will enable them to make targeted optimization proposals. On the other hand, the price for "not wanting to know more precisely" what is going on in the operational context is generally unknown in advance. Retrospectively, the cost can be very high when performance potential is not fully exploited, and undesirable developments are not identified in time to be corrected promptly and without consequential effects. Non-obvious consequences remain unnoticed for a long time and often only become noticeable after a considerable delay. Without comprehensive data collection, many farms lack a corrective in the form of reference values with which the internal parameters can be continuously compared. They drive quasi-on-sight and rely on impressions and generalized assessments, which have become ingrained in the thought patterns of those responsible. According to these thought patterns, decisions are made without checking whether the optimum of possibilities has been achieved with the decisions.

Before the industrialization of agriculture, it was above all the traditional treasures of experience (farmers' wisdom) that farmers could orientate themselves on to extract as much food as possible from the farmland and the farm animals for the survival of a community that depended on it. During the agro-industrial revolution, the treasures of experience were replaced by the agronomic knowledge of general biological and technical laws and regularities. The significantly expanded knowledge of regularities, e.g., between the genome of plants and animals and a supply of nutrients matched to it, was jointly responsible for an overwhelming increase in productivity. For a long time, the undesired and harmful side effects of these interventions were not an adequate subject of research in agricultural science. It was and still is far more attractive to concentrate fully on the realization of productivity potentials and the economic interests of agribusiness. Research primarily focuses on laws and regularities that can claim validity regardless of the context in which the processes occur and can thus be expected to have scientific merit. In addition, the agricultural sciences generally forego external validation of the findings they produce under standardized experimental conditions. This means that it remains hidden in which operational contexts the findings lead to the desired effects, in which they do so only to a limited extent, and in which contexts the opposite effects come to light. The same applies to the undesirable consequences and side effects that the decisions and the management interventions cause when based on them. In contrast, the undesired and harmful side effects depend on the respective context and are thus only partially amenable to generalized scientific statements. Science cannot answer which measures are most likely effective in which context and, simultaneously, are associated with an excellent cost-benefit ratio. The answer to this question can only be given in the context based on continuous data collection and evaluation, i.e., sound **accounting**.

There is another weighty reason for forward-looking farms to strive for sound accounting. This lies in the need to distance themselves from blanket statements about the negative **environmental impacts** of agricultural production processes. Here, too, misguided developments in agricultural science are partly responsible for the fact that highly aggregated calculations are made based on so-called life cycle assessments, which attest to a high environmental impact, especially on livestock farming and the production of meat, and do considerable damage to the image of livestock farming. The blanket orders of magnitude ignore the significant variation between farms and the great potential of many farms to reduce the release of pollutants through targeted measures. Now that these generalizations and sweeping prejudices are out in the world, they are being used by various stakeholders (including media representatives, NGOs, and vegan product suppliers) to support their positions. Blanket rejections, such as those put forward by representatives of farmers' associations against the environmental impact of agriculture, are not very convincing and certainly not suitable to refute the sweeping accusations. The only ones who could defend themselves against sweeping prejudgments are the farmers, by presenting reliable data on the extent of substance discharges on individual farms concerning the products sold. In this way, successful farms with demonstrably above-average environmental performance can distinguish themselves from sub-optimally managed farms and counter the negative consequences of blanket accusations with solid data.

As already explained in Sect. 9.2, the need for differentiation also arises from the marketing strategies of the suppliers of substitute products, who emphasize the supposedly lower environmental impact of their production as a sales argument. The more a growing number of consumers are concerned about the production processes of the food they eat, the more studies are commissioned and carried out to shed light on the problem. However, these studies produce highly aggregated figures that counter a differentiated assessment. A highly complex situation cannot be reduced to individual parameters and general statements. Too many factors are involved in food production; their interactions produce a high degree of variation and change. Thus, reliable statements about the actual situation are impossible when based on average figures. In contrast, the contextualization of products with the totality of processes involved in the production of food in an agroecosystem over an annual period, as well as a cross-farm differentiation of the environmental services of agroecosystems concerning the products sold, allows a far more meaningful assessment.

On farms, the self-interest in the efficient use of resources is, to a certain degree, aligned with the interests of the community in reducing the discharge of residual and waste materials into the environment. However, the intersection of common interests decreases again if the degree of intensification increases excessively because the farms want to gain an economic advantage through increased input quantities of means of production (in particular fertilizers and feedstuffs) without having to pay for the associated **externalizations**. In addition to increasing the cost of means of production, charges for using the environment as a dumping ground for surplus residual and waste materials from agricultural production are, therefore, an imperative instrument for curbing the pursuit of particular interests at the expense of public welfare interests. Forgoing further intensification and measures to reduce emissions only makes sense for primary producers if they can earn economic benefits. If primary producers are to turn to a socio-politically desired, more environmentally compatible production method that also brings them an economic advantage, changes in the legal and economic framework conditions are indispensable. As long as the state does not prevent farms from gaining a competitive advantage through externalization at the expense of the community, they will continue to do so. As long as farms lack incentives to gain economic advantages by implementing emission-reducing measures, they will continue to operate in their present form. Thus, economic incentives and sanctions are indispensable instruments that the state must use in its regulatory function and as part of an overall strategy if the environmental impacts of agricultural production processes are to be significantly reduced. Community structures at the municipal, regional, and national levels would be well advised to prevent farms that cause more harm than good to the community from continuing this practice. This brings the polluter-pays principle to the fore and focuses on farms responsible for exceptionally high pollutant discharges concerning the quantities of products sold.

The recurring protests by farmers against increased requirements, for example, for the application of fertilizers or pesticides, show how difficult it is for many farm managers to subordinate themselves to socio-political necessities. On the other hand, many farmers feel left alone with their problems. Most citizens and consumers care little about what goes on in agriculture in detail. On the other hand, they lack the prerequisites for this in the form of background knowledge. It is equally difficult for many farmers to place their responsibilities appropriately in the context of society as a whole. Farmers are justifiably concerned that state regulations will put them at a competitive disadvantage vis-à-vis competitors in third countries who do not have to implement costly requirements. On the other hand, communities have a right to be protected by the state against impairments that affect not only the integrity of the citizens but also the natural basis of life and animals. Many farmers cling unwaveringly to their traditional and self-referential positions instead of addressing the legitimate concerns of the community, other perspectives, and reference systems. This makes it all the more important to explore the intersections of interests to contain the conflicting goals between particular and public welfare interests. At this level, too, only a solid data basis for individual enterprises creates the necessary preconditions.

Derivations

While many farmers and their representatives are concerned with a fundamental resistance to regulation by state bodies and with safeguarding their freedom of action, the essentials are being lost from view, namely the differentiation of environmental protection services between farms and the **contextualization** of the material discharges. The agricultural lobby has so far been able to prevent such a comparison. The Netherlands provides a positive example. Farms above a specific size record the intra-farm flows of reactive nitrogen compounds through the various subsystems of an agricultural system to contextualize the nitrogen flow within a farm system and to carry out N balancing at the level of the operating system and subsystems (Daatselaar et al. 2015). These balance sheets enable a single-farm and comparative assessment of environmental performance. Using this approach, farm-specific weaknesses can be identified and eliminated. At the same time, **benchmarking** provides farmers with an assessment of where they stand and, thus, with the orientation they need to gear management decisions to the conditions prevailing on other farms. Benchmarking provides government bodies with the necessary overview of the range of variation in substance discharges and indications of those farms that require special attention due to high pollutant discharges.

In addition to the regulatory options, incentive systems are required. One option is to link public subsidies to environmental performance. In addition, market-based instruments can be used if the market were to reward the environmental performance of companies that produce lower emissions per unit of output and thus implement higher-quality production processes. A prerequisite for this, however, is that a sufficiently large consumer clientele is prepared to contribute to environmental protection by purchasing higher-quality and higher-priced products. This is most likely to be the case if the promotion of higher quality is not left to the food retailers but is seen as a sovereign task. Only an evidence-based assessment creates the conditions to reorientate primary production and the market, which largely escapes the manipulative influence of the stakeholders involved. A change in perspective and a fundamental reorientation are required to significantly reduce environmental damage in the context of agricultural production processes. The focus should no longer be on the opportunities for domestic producers in the world markets but on the environmental protection services that the farms can provide. The output of pollutants discharged into the environment by individual farms concerning the products sold is the yardstick against which operational and inter-operational measures must be measured.

9.4 System Services of an Agroecosystem

The economic system of the "free" market economy prevailing in the Western world has also shown in the agricultural sector the extent to which the productivity of food production can be increased and production costs and market prices reduced. Meanwhile, this success story has been considerably clouded by the tremendous harmful effects of agricultural production processes. These have substantially affected the living conditions of flora and fauna in the remaining remnants of nature, as well as the lives of farm animals. At the same time, the economic existence of very many farms has been destroyed. The self-imposed goal of securing food for the world's population was also missed. Despite decades of overproduction, a large number of people continue to suffer from malnutrition or undernourishment. Last but not least, farms contribute to an existential threat to the living conditions of large sections of the population through their significant involvement in releasing climate-relevant gases.

The negative consequences of the current agricultural food production system have reached a level that calls for drastic measures to curb them. However, the agricultural economic system is mainly incapable of reconciling the particular interests of the various actors with the public welfare interests. In the competition between producers, the winners are those who can produce the products for sale at the lowest production cost. The temporary pioneers are surpassed by other producers who act even more ruthlessly towards the public welfare interests. Thus, the necessity and urgency of timely changes are evident. While large sections of the population would meanwhile agree with this description of the situation, the proposals as to what should be changed and implemented by whom, and to what extent, and above all by whom, diverge widely. The demands for change have in common that they are directed primarily at others. For example, primary producers demand more financial support from the state, higher prices, and more respect and recognition from society. Consumers and citizens demand that primary producers take a much greater account of nature, environment, climate, animal, and consumer protection. Various interest groups are more inclined to one side or the other between the primary producers and the end consumers but essentially have their particular interests in mind. The situation appears to be somewhat deadlocked because of the sometimes widely divergent interests and the resulting conflicts of objectives.

9.4.1 Prerequisites for the Provision of System Services

Considerations on possible ways out of the present misery focus on the following question: By which framework conditions can the management of agricultural enterprises be made to control and regulate the internal production processes in such a way that the enterprise can both survive in competition with other enterprises and consider the public welfare interests by minimizing harmful effects? This question implies that a reorientation of the farms' overall production goals is needed. No longer should those who produce the products at the lowest cost, thereby harming public welfare interests, prevail in competition, but those who cause the least harmful effects in food production. Consequently, it is not a question of abolishing competition between farms but of redefining the production objectives and the rules of competition that determine success in competition. What seems as evident as it is inevitable is nothing less than a move away from a "free" agricultural economy and a reorientation towards an eco-social one. Such a reorientation requires not only individual actors but all actors and members of the interest groups involved to rethink and be prepared to make severe changes to previous behavioral patterns and practices.

Before considering further implications for stakeholders, it is essential to consider the objectives to which living systems such as agroecosystems are geared. The Greek origin of the word "system" translates into "*a whole composed of several individual parts*." Biological systems such as cells, organs, or the whole organism are separated from other systems by cell walls or connective tissue skins. System boundaries separate what is *internal* (inherent to a unit) from what is external to the system (*external*). In the case of an agroecosystem, system boundaries require definition. For some, the boundary stones of the usable land belonging to an operating system mark the spatial dimension of the system. Others see the need to include usable land outside the said farm boundaries in the operating system, where, for example, fodder is produced for the farm and surplus farmyard manure is spread. Thus, which system boundaries are agreed upon is important because different system boundaries result in different qualitative statements.

Systems can be subdivided into subsystems according to structural or functional aspects. The division into subsystems can be vertically on different levels and horizontally on the same level. For example, in hierarchically arranged system levels, the living organism is divided into organ systems, the organs into cell tissues, the tissues into cells, and the cells into organelles. The subsystems are always the components of the superordinate system. The composition of the organism provides an illustrative example of hierarchically arranged subsystems that have a functionally differentiating relationship to each other. Different organisms of a population represent subsystems on the same level. The respective properties of the subsystems result from the specific relationships to the overall system and of the overall system to the subsystems, i.e., from the reciprocal relationships of the parts to the whole (Ison 2010). **Thinking in systems** describes a systematic approach that deals with the interrelationships of the parts to the whole and the whole to the parts. In a hierarchical arrangement of systems, the superordinate system represents the **context** of the respective subsystems. The vertical and horizontal structuring of systems provides access to the interrelationships between the subsystems in the respective context and the effects of the context on the subsystems. In contrast to a scientific approach, in which individual partial aspects are removed from the context and subjected to an isolated, largely context-independent observation and assessment, in a systemic approach, the objects are always considered part of a superordinate system whole.

Unlike a natural science approach, living systems are viewed as open systems in continuous exchange with the living environment. Inputs from outside (external) and received into the system can be in material form (e.g., nutrients) and as information. Outputs are released from the respective systems and released into the surrounding environment. In contrast, systems that exchange energy but not matter with the surrounding environment are called closed systems. For example, an insulated can does not exchange matter with the environment; however, the temperature gradient concerning the ambient temperature is equalized—albeit with a time delay—by heat emission or absorption. According to Bertalanffy (1968), the founder of systems theory, theories of closed systems are the domain of the scientific disciplines of physics and chemistry. Physical chemistry can accurately predict reactions, turnover rates, and the chemical equilibrium in a closed vessel. The initial conditions determine the state of equilibrium in a closed system. If the initial conditions are changed, the final state also changes. Consequently, the input has a determining effect on the output.

The processes in living systems are fundamentally different. Living systems are selective about the input and the effects that the input has on living systems. The extent to which the input of nutrients is beneficial or even detrimental to the goal of maintaining living processes depends not only on the nutrient composition but also on the needs of the living systems in terms of their ability to function and survive, as well as on the interactions and adaptation processes between the structures involved. This determinacy by the system's inherent structures and need to survive marks an essential difference between closed and open systems. It explains why a natural science approach provides limited access to living systems. The self-preservation of living systems is only possible through a finely tuned interaction of the subsystems within the system with the living conditions outside the system. Accordingly, the survivability of living systems depends on the functionality of the subsystems and on the availability of essential resources, which can be accessed externally or internally (nutrient depots) in competition with other subsystems. For example, the animal organism depends on an adequate supply of food to supply the organ systems, cell tissues, and cells with what they need for their ability to function and survive. The cells, in turn, subordinate themselves to the higher-level functional principle of the organism by contributing their specific functions to the maintenance of the organism as a whole in the context of physiological processes, insofar as this is permitted by the availability of resources and the functionality of the structures (see Sect. 8.5.2). Given the ongoing dynamic changes within and outside living systems, assessing functional states is always only a snapshot. An assessment is also complicated because the same functional state of a system can come about in very different ways (equifinal). Therefore, causal interrelationships are not generalizable but context-dependent. For this reason, forecasts about future developments can only be made to a limited extent. At best, depending on the number of systems examined in a population, more or less reliable estimates can be made of the probability of expected effects occurring.

In living organisms, life-sustaining processes are maintained through circular causal chains and feedback mechanisms. Regulations are carried out via sensors, set points, and actuators. For example, changes in ambient temperature are continuously registered. Regulating countermeasures ensures that a constant core temperature can be maintained in the body. The access of different organs to the limited resources available in the body, such as the glucose content in the blood, is controlled both decentrally in the respective tissues and centrally in the brain. The individual organ systems do not accumulate their reserves but draw on the reserves of a shared nutrient pool. Thus, a costly stockpiling of organ-specific reserves can be avoided, and a high efficiency in the use of resources can be realized for the organism as a whole. At the same time, the organism can react flexibly to changing requirements and adapt to changes within and outside the system. The process of "achieving stability through change" in different situations of demand is called **allostasis** (Sterling 2012). Despite all the differences in the adaptation processes of living systems, they have the common goal of maintaining the subsystems and the system as a whole; the regulating processes are functionally and goal-oriented (teleological) towards self-preservation.

However, the ability of organisms to adapt to changing environmental conditions is limited. For both plant and animal organisms, it is equally valid that conditions of stress and a threat to their ability to survive are present, particularly in the case of an insufficient supply of water and nutrients and the case of conflicts with pathogenic microorganisms. Whether the reactions and adaptation processes of living organisms are successful depends, above all, on the living conditions (boundary conditions) and the extent and speed of the changes. If the adaptability is overtaxed, this will result in disturbances, diseases, or even death. The ability to survive is an overall performance of the organism's self-regulation in dealing with the respective living conditions.

In self-regulation, two principles come together: **competition** and **cooperation**. Since many conditions and nutrients necessary for life are only available in limited quantities, organisms are constantly competing. This is true not only in the wild but also in agroecosystems. The omnipresent competition between organisms ensures a competitive and, thus, survival advantage for those that can cope with the adaptation services most efficiently. Competition also exists within an organism between different organs and cells. Without a well-functioning regulation of the competitive behavior of the subsystems for limited nutrients available in the nutrient pool of the animal organism, damage to the higher-level system inevitably occurs. Instead of consuming each other in competition, the organ systems subordinate themselves as subsystems to the respective requirements of the organism as a whole. Thus, for example, there is no haphazard multiplication of body cells, but cell multiplication is coordinated with the respective need for regeneration. Using the example of rampant tumorous cell growth, it is easy to understand that without containment of the competitive situation between the cells, the survival of the organism as a whole is endangered. The same is valid for allocating glucose or oxygen to the different organ systems according to the demand that results from the respective function for the maintenance of the organism as a whole, depending on the requirements of the living conditions. The **allocation** of the limited available resources is thus always at the expense of the supply of other subsystems but always to the advantage of the superordinate system.

In this way, the competitive behavior between subsystems is counterbalanced by the equally evolutionary principle of cooperation. Genes cooperate in chromosomes, chromosomes cooperate in genomes, genomes cooperate in cells, cells cooperate in tissues, tissues cooperate in organs, organs cooperate in animals, and animals cooperate in animal associations. Cooperation offers the possibility of escaping a potentially ruinous competitive situation by working together and aligning oneself with a higher purpose. Cooperation is the purposeful interaction of different subsystems directed towards the common goal of self-preservation in and through the system. While competition occurs between subsystems of the same level, cooperation is organized from the superordinate system level. For the self-preservation of the superordinate system, it is essential for survivability that the competition between the subsystems is contained by the coordination of regulatory processes so that the competing behavior is to the advantage and not the disadvantage.

Analogous to the network of interactions between cell assemblies, organs, and the organism as a whole, there is a reciprocal interdependence between plant and animal organisms and the subsystems of an agroecosystem located at different levels. Here, too, the superordinate whole depends on the functioning and purposeful regulation of the subsystems among themselves. The site conditions and soil conditions, which can hardly be influenced, as well as the size and structure of the farm, set the framework within which **management** has to cope with the diverse requirements and needs of the subsystems. Whereas organisms and natural ecosystems are capable of self-regulation, farm management is required to ensure conservation through appropriate control and regulation processes in agroecosystems. In doing so, the subsystems must be given the attention and resources they need to contribute to the well-being of the whole. At the same time, management must ensure that the subsystems are protected from disruptions, stresses, and excessive demands on their adaptive capacity to ensure economic success.

At the same time, the individual agroecosystems are part of higher-level community structures with which they interact in various ways. The infrastructures relating to upstream and downstream economic sectors, the availability and price of means of production, processing and marketing opportunities, and, last but not least, market prices significantly restrict management's room for maneuver. Management is therefore challenged not only by the internal requirements of regulating and coordinating processes. It must also keep an eye on market developments. These determine the internal options for action via the costs for the means of production and the possible revenues for the sales products. Management must also not lose sight of the behavior of competitors. After all, it has to hold its own against them on the market.

Due to insufficient cost recovery, many farms have been more or less in crisis mode for years, to which all other requirements must be subordinated. The "free" market economy only gives an economic chance of survival to those farms that can cope with market prices better than others, i.e., can produce their sales products more cost-effectively than other farms. However, the supposed "selection of the best" contains a fatal fallacy. Under the prevailing economic conditions, companies are forced to gain price advantages at the expense of the stability of internal processes, employees and competitors, and the public welfare interest. Market-based competition and the economic benefit theorem focus on the winners, while the consequential effects and costs are largely ignored, and the losers are left to fend for themselves.

While the markets are flooded with an oversupply of cheap food and, at the same time, considerable harmful effects are caused, there are massive shortages of various **resources** in inner farm situations. Shortages include high-quality means of production (e.g., feed), working time, and investment funds (e.g., technical aids, new stables). Many livestock lacks an adequate supply of nutrients and sufficient protection against infections with pathogenic agents or against the aggressiveness of conspecifics, which manifests itself, for example, in tail biting in pigs or feather pecking in poultry. Management lacks the time, money, and often also the knowledge to take measures that could specifically counteract the excessive demands on the adaptability of farm animals. The same applies to measures to reduce pollutant emissions into the environment. The management of a farm, therefore, primarily manages the shortage. The consequences show up in an excessive demand for adaptation to market conditions. Production capacities were often expanded to achieve economic economies of scale, animated by general advisory recommendations. This reduced the available working time for the respective requirements and contributed to an oversupply of raw goods in the markets, pushing market prices down even further. In short, farms are part of a rural economic system geared towards extensive exploitation of natural resources. Since this system does not provide farms with the necessary resources and framework conditions to secure economic existence and thus exploits farms, farms exploit biotic and abiotic resources while they can still access them.

There seems to be a way out of this vicious circle only for farms that are in a position to orientate the practical structure of the internal production processes towards a high level of quality production. Suppose the quality level of the products sold exceeds the average level offered on the market. In that case, they represent a higher value, which can be converted into a higher added value through higher sales proceeds. This approach has been taken on various occasions via brand programs (including organic production). However, in these programs, the proclaimed quality advantages remain somewhat vague. The existing programs lack, above all, scientifically reliable evidence that the increased minimum standards implemented in the production and processing conditions (input variables) are also associated with outstanding output variables, i.e., concerning qualitative characteristics of the sales product (product quality) and the process qualities (animal welfare and environmental protection). The willingness of consumers to pay a higher price depends mainly on their trust in the suppliers or on the extent to which the advertising messages catch on with them. However, without a comprehensive quality certification, the willingness of consumers to pay premium prices for specific products is based on a relatively insecure footing, and the relationship of trust between consumers, food retailers, and suppliers is too weak to set up a viable business model.

9.4.2 Health and Life Value of Food

Various preconditions must be fulfilled before sustainable value creation can be achieved using traceable and proven quality performance and additional price generation. This includes, among other things, establishing valid assessment systems for continuously recording overall quality performance and continuous feedback on the assessment results to the primary producers. These can use the input for the implementation of selective and structural improvements. Furthermore, financial incentives and cost-covering payment systems are required to realize the optimizations of the production processes associated with additional expenditure. Finally, the qualitative (extra) services must be communicated to consumers to promote their willingness to pay further prices. This is most likely to succeed if consumers were independently informed about the frequency distribution of quality services. These rank on a scale from very good to very poor. Here, the frequencies of occurrence of scaled quality levels generally follow a normal distribution (Gaussian bell curve). This means that high-quality products are only available on the market to a limited extent and that most products are of average or inferior quality. According to economic law, supply and demand determine price, and the supply of sales products with different quality levels has to result in various price levels.

The idea of animal health and welfare as a public good seeks to impose limits and restrictions on individual interests, placing what is good for the community above what is desirable for individuals and interest groups. At the same time, public welfare interests must accommodate at least the interests of several, but not all, stakeholders. A basic understanding of the structure of effects of sub-groups of a superordinate community of interests results in a mutual conditionality and the necessity to organize a balance between diverging interests. This can only be realized when an agreement exists on a common goal. Otherwise, the respective activities and the deployed resources cannot be used in a goal-oriented and efficient manner. The more clearly the common goal can be outlined, the more likely it is to provide all participating stakeholders with the necessary orientation and knowledge for future actions. The principal goal of animal health and welfare of farm animals as a public good is to protect them from pain, suffering, and damage that challenges their well-being and self-preservation. Other stakeholders' interests must be accommodated to safeguard farm animals' interests. The resulting collaboration must consider the interests of livestock farmers who have to protect farm animals from harm by providing appropriate living conditions for the animals to cope with. In addition, consumers must be prepared to pay the required price to produce food from healthy farm animals. The involved market participants and, last but not least, the legislators are responsible for establishing framework conditions that enable fair competition for animal-sourced food with outstanding quality in terms of animal health and welfare.

At least in theory, farm animals, livestock farmers, and consumers who adhere to a common understanding of animal health and welfare form a community defined by the common interest in the absence of severe health problems that threaten the fitness and survival of farm animals. At the same time, health problems threaten the productivity of livestock and food safety when deriving from diseased animals. Food from diseased animals threatens the expectations and associations of consumers concerning the quality and value of food. It thus reduces their willingness to pay premium prices or purchase animal-sourced food. "Health" is a precious state, representing what is highly desired by humans for themselves and also prioritized when it comes to the best wishes to others, e.g., on birthdays. While human health predominates, consumer concern (European Union 2017), disturbances of animal health are critical to purchasing choices (Buller and Roe 2012). When introducing the new term "health and life value of food," the value is not defined by what consumers attribute to their individual preferences and willingness to pay. Instead, the food value is determined by the state of physical health of the farm animals the food derives from, together with the health state of the respective farm system being able to provide a high level of animal protection service in terms of a large percentage of animals free from severe health problems. In addition, food of high health and life value has to derive from a production and marketing chain that ensures that the nutritional value of the food is not canceled out by negative external impacts imposing excessive burdens on other living systems. Thus, "health and life value of food" provides a conceptual viewpoint that allows us to perceive, recognize, and describe a complex state. It enables us to define elements according to their original characteristics, the interrelations in which they participate, the perspective of the organization in which they operate, and the perspective of the whole in which they are integrated.

Food with a high health and life value can only be produced when the respective livestock farmers can afford the required efforts and costs without jeopardizing the economic viability of the farm system. This is considerably hindered in a market economy that forces the farmers to strive for cost leadership because realizing a high animal health and welfare level requires additional efforts and resources. Thus, the key question and the bottleneck in the issue of animal health and welfare is how livestock farmers who strive for a high health and life value of food can be honored by premium prices to cover the additional efforts and costs. It goes far beyond the scope of this chapter to explain how this might be best implemented. Nevertheless, it should have become clear that the reference system for food's health and life value is the respective farm system and the achieved animal protection service. From consumers' perspective, retailers warranting that a specific brand label of animal-sourced food is based on the produce of healthy animals might be sufficient. However, this would not necessarily ensure that the state of animal health and welfare of farm animals will be improved. Instead, produce from healthy animals could just be selected from what the market offers without providing adequate incentives to farmers to invest in improving living conditions to benefit all animals in a herd.

How farms service animal protection has to be assessed on a valid basis to enable fair competition between livestock farms. Farm systems that strive for high-quality animal-sourced food cannot compete globally in a market dominated by context-invariant raw produce. Thus, evidence of animal-sourced food of superior quality is mandatory. It must come from animals free of severe health problems and livestock farms with a prevalence of mortality and morbidity clearly below average. To enable the production and marketing of such food, a new market has to be established, relying on evidence-based marketing processes and fair competition. "Health and life value of food" is a relative value, ranging between the average and the highest possible level. It refers to what farms are currently able to perform. Livestock farms must be assessed and ranked according to their animal protection service. Benchmarking enables food processors and retailers to obtain raw produce from selected farms with an outstanding level of animal protection service. A benchmarking system ensures that food of high health and life value remains a scarce commodity that can only be produced and offered if the production process is not based on the excessive exploitation of farm animals and livestock farmers. The "health and life value of food" represents a common value appreciated by various stakeholders who share a common vision of food. Such food can only be produced when economic and health requirements are balanced, and stakeholders cooperate, subordinating personal interests to benefit a common goal, as illustrated in Fig. 9.3.

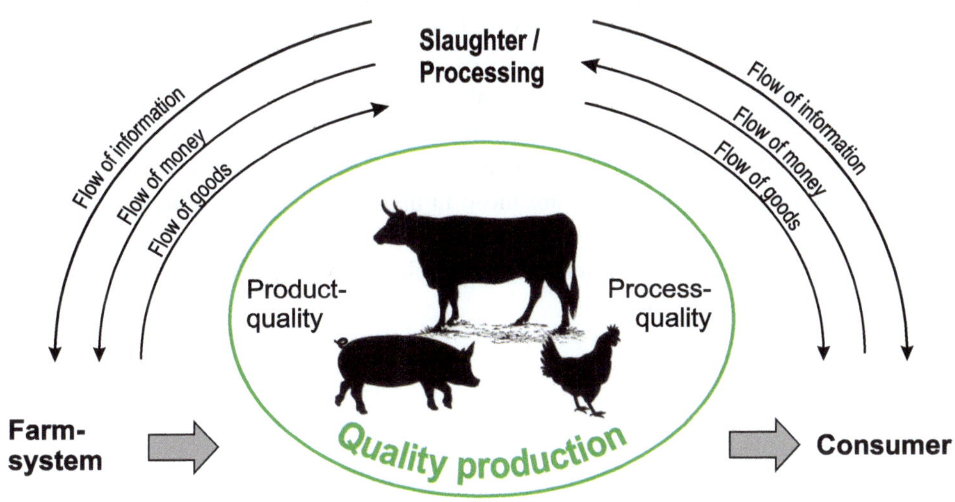

Fig. 9.3 Flows of goods, money, and information between the actors in a value chain as products move from the farm to the consumer (own illustration)

The current strategies for producing animal-sourced food are supply- and market-oriented and thus contrary to public welfare interests. So far, legislators have done too little to establish market conditions enabling qualitative competition. Whereas the development of organic agriculture has proven that it is possible to establish a separate market for food that refers to higher prices, organic livestock production is a generalizable example showing that enhanced minimum standards are inappropriate and misleading for increasing animal health and welfare (Sundrum 2014). Without an external validation of the farm's service concerning animal protection by the freedom of severe health problems, the issue is left to the self-referential estimations of the livestock farmers and the interest groups involved. Organic livestock production has proven that high-quality production cannot be realized without a clear definition of food that does not refer to the means to an end but to the outcome of production processes. The same applies to preventing unfair competition through evidence-based Quality production. This relies on external validation within market conditions that generally strive for low costs at the expense of other stakeholders.

The necessity of intervention and regulation of the agricultural sector by superordinate structures, which contain the ruinous competition by cooperative structures and the orientation towards evidence-based quality production, is evident. Suppose community structures, organized from municipalities and counties to federal states and the federal government up to the European Community, consider a public welfare-oriented food production urgently necessary. In that case, they must have the power to provide appropriate framework conditions and establish structures that promote the achievement of objectives, assess the actual conditions about superordinate target values, and intervene in a regulating manner in the case of undesirable developments. Despite the current underutilization of these possibilities, the potential for positive change is significant. However, given the far-reaching and devastating damaging effects that the agricultural system has already caused, the question arises as to why the community structures have not yet been able to regulate the farms as subsystems of the community. Why has it not yet been possible to organize food production so that those farming systems are successful in the competition which produces the highest quality and the lowest externalities in relation to the products sold? Why are those farms that cause particularly pronounced damage to the common good not prevented from doing so? As always, the reasons for failures occurring in complex systems are manifold. Some of the aspects considered to be decisive are discussed below.

9.4.3 Barriers

Concerning the current framework of the food chain, which ranges from the production of raw goods by primary producers to the consumption of food by consumers, there is no doubt about who has the say in matters. Food retailers have access to a very great variety of raw products offered on a global level. Due to a general oversupply of raw goods in relation to the demand, retailers are in a solid position to drop the purchase price wherever it seems possible. Furthermore, the availability of raw goods is not hampered by legal requirements to make a difference between the levels of product and process quality traits of raw goods. As illustrated in Fig. 9.4, food retailers are gatekeepers who use considerable variation in raw goods' properties and quality levels by levelling the variation to self-referential standards. The standards are primarily defined by minimum requirements or legal rules such as the EUROP-meat classification of carcasses. Standardization reduces the existing variation in the expression of characteristics and is a central means to facilitate free trade between countries. However, what is seldom adequately reflected is that standards enable the interchangeability of raw goods of the same standards. The interchangeability drastically increases the availability of raw goods and thus allows the retailers to put pressure on the primary producers to sell their raw goods as cheaply as possible. Otherwise, the retailers can purchase the raw goods elsewhere. In reality, however, the raw goods are very different regarding quality levels, the context in which they are produced, and the positive and negative externalities created by the production processes. Instead, the raw goods are handled, prepared, or processed and are manufactured into a large variety of food products that mostly fit the benefits of the food retailers when striving for an increase in turnover rates and to prevail over the competitors. This system, as it stands, is not sustainable and requires immediate and decisive attention and action.

The production and trade of wine represent one enlightening exception. Here, contextualization plays a significant role. Among other things, the soil, its unique characteristics, and the climate conditions make a difference, and it is called "terroir." This

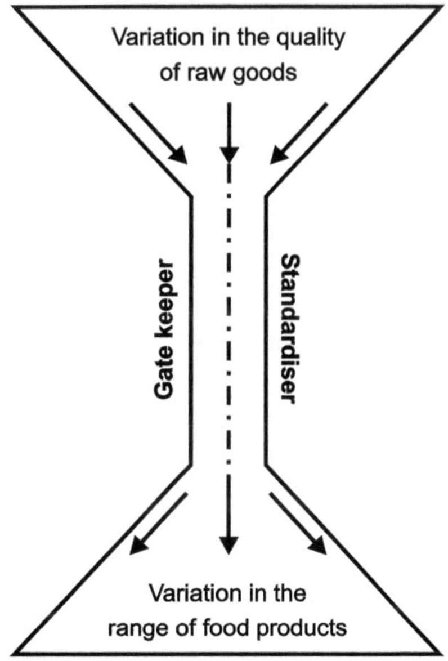

Fig. 9.4 Food Retailers as Gatekeepers, who standardize the significant variation of the level of product and process quality of the raw goods and transform them according to their benefits (own illustration)

term, 'terroir,' encapsulates the essence of a wine, making it unique and unexchangeable. Above all, the label of the wine bottles has to provide certain information, including the year, the grape variety, the bottler or vineyard, respectively, the origin, and last but not least, the quality grade. This makes individual wine lots very specific and prevents them from being exchangeable.

The trade value of raw goods is determined primarily by quantity, while product and process quality traits are widely insignificant. Nevertheless, food retailers often claim to offer products of high quality without providing transparency and evidence. At the same time, they want to increase turnover rates by offering products of supposedly high quality with comparable low prices and at the expense of the farmer's possibility of investigating substantial improvements in the quality level. No progress can be expected concerning improved product and process quality if the primary producers do not gain any benefit and profit from the market to compensate for the additional efforts and resources needed. Instead, producers are very much at the mercy of retailers and supermarkets regarding the price paid for the raw goods. In recent years, they have been chronically underpaid and left alone by the market, which does not offer substantial monetary incentives to improve product and process quality. The current market conditions widely ignore the large variability in quality traits and the impacts on public welfare interests. This promotes unfair competition when enabling equal prices for very different performances concerning product and process qualities. Primary producers who gain economic benefits by selling inferior products at the same price as high-quality products undermine the efforts of other farmers attempting to maintain a high-quality level. The latter have to apply additional labor and cost-intensive efforts, contributing to a considerable increase in production costs, thereby providing an essential competitive disadvantage.

Even though many consumers express their wish for high-quality products, the current payment and marketing systems counteract all efforts to follow consumer demands and fail to communicate adequately differentiated quality levels. The only way out of this dilemma is to make a difference between farm systems according to their respective level of product and process quality by an evidence-based approach of bonus payments to promote top quality. Only a direct assessment of the quality level on the farm level and a payment system that honors quality grades beyond average can improve the currently unsatisfying situation. This, however, requires a shift in the paradigm from a standard- and input- to an output-oriented systemic approach. Such an approach would provide the prerequisites for effectively and efficiently balancing the multiple variables and complex interactions within each farming system.

Scales are needed to assess differences in the level of quality. These are determined by the boundaries of what is currently available. In the case of raw goods, the boundaries are defined by the range between the average level of qualitative traits and the highest possible quality level. This requires an overview of the levels of qualitative characteristics available on the market and an independent actor to conduct such surveys. This is, by nature, a public authority task. By no means should it be left to the particular interests of the food retailers to establish scales for the differentiation of quality levels.

One of the reasons why the framework conditions of the agricultural economy continue to be maintained and the necessary reorientation has failed to materialize is probably because those who are the majority among the victims of the agricultural economic system show themselves to be very reserved towards state regulatory measures. Despite the large number of farms that have been unable to withstand the pressure of competition in recent decades, the farmers concerned and the supposed representatives of their interests defend the agricultural system, which is primarily responsible for this, against any criticism. The failure of many farms is seen as a quasi-natural consequence of structural change. Any thoughts of regulating market processes are resisted. This, however, does not stop the lobbyist from demanding further support and funding from the state, done without any prospect of a quid pro quo vis-à-vis the public welfare interests.

No less problematic is the fact that low market prices are misunderstood by many consumers, supported by consumer associations, media, and politicians, as an ancestral customary right. Representatives of various interest groups never get tired of warning of a rise in consumer prices, which would be detrimental to poorer populations. Agricultural policy is misunderstood here as social policy. The perception is focused on what affects their particular interests. What else might be relevant to public welfare interests is widely ignored. There has also not (yet) been sufficient awareness of the existential threat to farms and of the causal links between the harmful effects of agricultural production processes and low market prices.

There is no broad support for government intervention in the existing agricultural system, neither among the farmers nor voters. In such a constellation, it is naturally difficult for politicians to initiate changes involving considerable impositions on various interest groups. This means that the political parties and the executive lack the incentives to make their mark on the respective electorate with political proposals for solutions to the problems of the agricultural sector. In addition, critical decisions on agricultural policy are taken at the European level. National regulations can only be introduced if they exceed the minimum legal requirements at the EU level. Some Scandinavian countries have shown that national requirements significantly higher than the European minimum requirements can certainly be in the national interest. This applies even, or especially, to countries which, as in the case of pork production in Denmark, are heavily dependent on the export of their products (Danish Agriculture & Food Council 2014). In these countries, however, this is only possible if not only the low-price segment but also a high-price segment is served in the global markets. In countries where it is impossible to condense agricultural concerns and the associated consequences into an overriding issue of national and public welfare interest, there is, at best, scope for fundamental political debate. These do not lead to concrete solutions, but they have a great advantage for some stakeholders in that they distract from the increasingly explosive problems. In this way, the prevailing framework conditions can be maintained to the benefit of the previous profiteers of the economic system and to the detriment of most farmers and the public welfare interests.

The individual farmers contribute to the oversupply of raw goods to very different degrees. The extent to which farms cause damage and impairment of goods of public interest varies just as much. The more market prices have come under pressure in recent years due to overproduction, the more those farms that have been able to keep their production costs low have benefited. Low production costs have often been achieved through a high degree of externalization at the expense of the common good. However, the **trade-offs** between particular and public welfare interests vary significantly from farm to farm. However, the differences between farms are primarily ignored by the state. This also applies to market participants. Irrespective of the very different levels of animal welfare and environmental protection that the farms produce, they achieve more or less the same sales revenue for the raw materials. The extent to which the production processes on the individual farms are detrimental to the common good is, therefore, of no interest to the state, the competition authorities, or the market participants. Farms are protected mainly from regulatory state intervention compared to other economic sectors. This is true even if this is presented differently from the perspective of the representatives of the farmers' associations.

For decades, agricultural policy in the Western World has been geared towards conditions of free trade while ensuring that domestic producers participate in competition on global markets. An essential economic liberal attitude dominates, so options for implementing national regulations were primarily dispensed with in the past. At the same time, the polluter-pays principle, which applies as a matter of course in other areas of life when dealing with damage cases, e.g., in traffic or property offenses, has been largely suspended in the agricultural sector. The pollutant emissions from agriculture are an expression of irresponsibility, which has assumed such large proportions because those responsible for it are not held accountable by others. While for most areas of life, the basic rule applies that the individual actors must behave in such a way that no one else is harmed, endangered, hindered, or inconvenienced more than is unavoidable under the circumstances, this everyday rule does not seem to play a significant role in agriculture. The agricultural lobby has discredited potential extensions of existing legal requirements as part of an utterly exaggerated prohibition policy and as coercion of free enterprise. Following the motto that farmers know best what needs to be done, regulations have been and continue to be fended off, arguing with completely excessive additional bureaucratic expenditure, which would undermine the competitiveness of farms and their economic existence. If, as in the case of the revision of the Fertilizer Regulation forced by the EU Commission, an increase in legal minimum requirements is nevertheless achieved despite the resistance, the farmers' associations defy the regulation as a bureaucratic monster. At the same time, it is concealed that the alleged complicatedness of the legal requirements is also a consequence of the exemptions that the farmers' associations wrest from the legislator to demonstrate particular effectiveness to their clientele.

The shallow level of legal requirements compared to other sectors of the economy contrasts with the amount of money farmers receive in direct payments from the EU budget and additional national **subsidies**. Not only do farmers receive a lot of money from the state, but they also do not have to provide any relevant services in return, apart from committing themselves to complying with minimum legal requirements. If one compares the bureaucratic expenditure of farmers with the spending that other recipients of state benefits have to make, the privileges of agriculture become particularly clear. The aversion to bureaucratic expenditure, widespread among the population and frequently reported in the media, is often used to portray farmers as victims of state paternalism. At the same time, the public is given the impression that farmers are entitled to subsidies from the state coffers as compensation for the poor income development in agriculture compared to other sectors of the economy, as a kind of permanent alimentation. Farms in Europe are on the drip of state subsidies, without which many could no longer survive economically.

In this mixed situation, a national agricultural policy that balances social and private-sector concerns is, at best, rudimentary. The parties at the federal and state levels are too closely bound up with the interests of their respective clientele for agricultural policy-makers to be able to take up a neutral mediating position. At the same time, the mandate of export-orientation has established dependence on world market prices and the excesses of global undercutting competition. Agricultural policy is supported by an agrarian economy, which, in its belief in the market, has consistently argued that politics should stay out of market events. The undesired and harmful side effects of agricultural production processes were not considered for a long time because they were not classified as problems inherent in the system but as problems of individual farms. This begins to change when the harmful effects become ever more significant, the public protests ever louder, and the buying restraint for certain food products and production processes ever stronger. The demonstrations and the messages conveyed in this way have an increasing resonance in the media and, thus, in public perception. It is not the insight of the acting actors into the necessity of change but rather the resistance of superordinate community structures, which is calling into question the future viability of the previous business model and bringing about a process of change, albeit one that is still very much in its infancy.

Now that the need for action is growing, we are beginning to realize that there is a considerable lack of data that would allow us to assess the extent of undesirable consequences and side effects on individual farms. At best, only highly aggregated figures exist. However, the degree of **emissions** cannot be generalized; they are the results of specific production processes in the respective operational context. Measures to reduce emissions require a strategy tailored to the farm-specific situation. Accordingly, the farm management must know the sub-areas in their operations from which exceptionally high emissions emanate and how these can be countered effectively and efficiently. Furthermore, emission-reducing measures will only be implemented if they result in economic benefits.

Analogous to data relevant to environmental protection, the national governments deal with animal health and welfare-relevant data. For example, the EU regulations require that data on mortality rates of farm animals be collected on each farm. The same applies to pathological findings collected from slaughtered animals at the slaughterhouse as part of the meat inspection. These data are used for traceability in the case of epidemic emergencies and to provide government bodies with a specific overview of the respective orders of magnitude. However, only in a few European Member states is the data used to establish a nationwide ranking of farm performance regarding animal protection services to identify those farms with exceptionally high damage cases. In this context, politicians point to obstacles under data protection law. However, it is also true that politicians are not willing to assert themselves against the agricultural lobby's resistance and use the existing benchmarking instruments to identify particularly conspicuous farms.

Because of the significant variation in harmful effects or environmental and animal protection services between farms, only those measures tailored to farm-specific situations and exerting a targeted influence on them can be expected to have relevant effects on enforcing public welfare interests. While the key to tailor-made solutions to socially relevant problems lies in accessing and changing individual farm situations, agricultural policy locates it entirely elsewhere, namely in further developing generalizable scientific and technical solutions. Accordingly, agricultural policy is prepared to provide the field of agricultural sciences with extensive funding. This research funding is accompanied by the hope that the new methodological approaches, particularly in genetics and digitalization, can be made useful for the agricultural industry. It is no coincidence that the problem-solving approach of agricultural policy is reminiscent of the metaphor of the house key lost in the dark. Instead of looking for it near where it is thought to have slipped away, it is sought in the cone of light from a nearby lantern. After all, the conditions here are well-illuminated.

Various side effects are associated when those responsible for agricultural policy proclaim their belief in scientific and technical progress in linguistic unison with the representatives of the Farmers' Association, agricultural economics, and agricultural science. By raising vague hopes for future solutions, the prevailing problems no longer appear as a fateful horror scenario. However, the reference to possible scientific and technical progress provides a supposedly valid justification for the fact that nothing fundamental needs to be changed in the present circumstances. This is true as long as the scientific community does not communicate any clear and concordant statements to the contrary. It may take some time until this is the case. Suppose one considers the time before the manufactured influence on climate change was generally accepted in the environmental sciences. In that case, it becomes clear that a considerable acceleration of the discourse processes in the agricultural sciences is necessary to say goodbye to traditional thought patterns and to become compatible with the urgent challenges facing the relationship between agriculture and society.

Section 7.4 already discussed the role of agricultural sciences. The predominant orientation is one reason for their inadequate problem-solving competence (Sundrum

2024). The main concern here is to search for regularities and causalities of interrelationships that can be claimed to be valid independently of the context. Consequently, the complex and context-dependent processes on farms do not belong to the priority field of agricultural sciences. "On-farm research" is still considered a field, forest, and meadow research, with which hardly any scientific merits can be acquired. One of the consequences of a scientific orientation is that insights into and understanding of higher-level contexts are increasingly lost. To counter this loss, many scientists have turned to modelling as an instrument for dealing with complex interrelationships. However, even this approach does not provide concrete instructions for agricultural practice. On the one hand, the **indicators** on whose parameters the model calculations are based only illuminate individual aspects and are, therefore, of limited informative value. In addition, the selection of indicators is usually based on pragmatic considerations. The predominant indicators are those that can be collected relatively quickly and thus fill the data set on which the modeling is dependent. Above all, however, conflicting goals and feature antagonisms, which significantly shape events within systems and between subsystems, cannot be validly captured and mapped using modelling. Consequently, no reliable predictions can be made using modeling in living systems with a wide range of initial and boundary conditions. At best, modelling is suitable as a generator of hypotheses, the external validity of which needs to be verified in the context of individual farms.

In current affairs, agricultural policy and sciences cannot provide the impetus for reorientating the rural economy. The predominant trend towards specialization has largely brought interdisciplinary collaboration between different scientific disciplines and, thus, the epistemic principle of **cooperation** to a standstill. Even multidisciplinary research projects cannot hide the fact that there is often a lack of a common language and understanding. Interdisciplinary or even transdisciplinary cooperation that justifies this name can only get off the ground if it follows a common objective oriented less to general goals than to solving inherent problems of individual agroecosystems. Even if research projects succeed in working across disciplines on strategies for solving context-dependent problems, validating the effectiveness of measures in the farm-specific context quickly comes up against the limits of practical implementation (Sundrum et al. 2016). Many farms lack the financial resources and necessary working time to implement the recommendations adapted to the specific starting and boundary conditions. All measures associated with additional expenditures are subject to costs. Thus, they only have a chance of being implemented if they can be expected to pay off.

No less problematic is that scientists create the false impression, not only vis-à-vis politicians but also vis-à-vis the general public, that complex problems can be solved by scientific progress in matters of detail. Corresponding references to potential solutions in the form of individual measures and miracle cures then find their way into the media and are widely reported. Many examples show how media representatives take over the focus of scientists on individual topics and problem solutions, which are then generally elevated to solutions. The result is **inductive false conclusions**, which find their way into the general public and arouse false impressions and expectations. By reducing complexity to the use of individual "miracle cures," the fact is lost that environmental problems arise in the context of the entire company and can only be solved effectively and efficiently there, taking into account the specific biological, technical, structural, resource-related and economic conditions in an agroecosystem. The urge to reduce complexity is an all-too-human need for understandable explanations. It is frequently used to avoid the cognitive dissonance that threatens to arise when trying to visualize the real problem.

9.4.4 Characteristics of System Services

Instead of simplistic approaches and general recommendations, problem-solving strategies require **contextualization**, i.e., an appropriate classification (rationalization) and reflection of dysfunctions within the **system boundaries**. Often, the key to solving problems lies in an internal regulation of procedures towards an overriding goal. In the case of agroecosystems, the goal is to achieve the highest possible level of overall system performance suited to ensure economic viability. Therefore, all **subsystems** at the various process levels should be aligned in their functional sequences to the most efficient use of the resources available within the farm system. The ability of living organisms to functionally align and coordinate all subsystems and their processes with the goal of self-preservation via adaptation processes serves as an illustrative object (see Sect. 8.2). Through the synergistic interlocking of all sub-systems concerning the functioning of the whole, resource-consuming disturbance variables such as plant and animal diseases should be minimized, and externalization should be avoided in favor of a high level of animal and environmental protection, which is synonymous with the realization of public welfare interests. The protection service of the individual farm systems can be assessed and ranked on a scale from very low to very high. The ranking enables the individual farms to position themselves to other farms concerning the current level of animal and environmental protection performance and to make further decisions on reducing the discrepancy between the current and a future target level.

This type of economy can only hold its own in the face of competition if the socially relevant services are also in demand and adequately remunerated. An eco-socially oriented agricultural economy should no longer trade in anonymous foodstuffs that are decoupled from their original context and that meet arbitrary standards set by food retail marketing, but in foodstuffs that can be characterized and differentiated by the level of system performance of their farms of origin. The economic framework as the superordinate system in which the farm systems are embedded has to be organized to enable the financial viability of those farm systems that produce food without excessively compromising public welfare interests. The interdependence of supply and demand, of services and services in return, should no longer be left to a "free" market economy, which has devoted itself almost exclusively to the goal of cost and price leadership and has thus produced the consequences that now urgently require containment. If primary producers are expected to provide socially relevant services, municipal, regional, and national communities as superordinate systems should contribute by promoting appropriate economic framework conditions.

Concerning the inward-looking view, the counter-model of an economic system oriented towards system services cannot avoid focusing solely on partial areas of the operating system. Partial farm services, such as the quantity production of **cereals**, **milk**, or **meat**, should no longer be evaluated in isolation based on production costs and market prices. Valuation also includes impact assessment regarding the effects caused in the context of the operating system and terms of externalities. The value of a partial service is also measured in terms of its contribution to the economic viability of the whole operating system. For example, the contribution of a dairy cow to farm income cannot be determined solely based on its milk yield but only if the individual animal input is related to the farm benefit, including the consequential effects. High milk yields, for example, can overtax the adaptability of dairy cows and lead to premature and involuntary departure for slaughter due to severe diseases. The consequences are not only reduced **animal protection services** but also economic losses. These occur when the cow has not been able to refinance the costs of rearing and running costs with the milk produced during its shortened life, i.e., it leaves production with financial losses before the start of the profit phase.

While agroecosystems always have to face limited availability of specific resources in the form of high-quality nutrients, working time, attention, or investment funds, a one-sided **allocation** to sub-areas is always at the expense of other sub-areas. The challenge is coordinating the allocation to avoid negative marginal utility effects and expand synergistic effects. A balanced allocation of resources between the various sub-areas of an operating system cannot be achieved if they compete unregulated for resources. Synergistic effects and increased added value emerge for the operating system as a whole only when the sub-areas are understood as part of a superordinate whole and it is possible to coordinate and organize the allocation of resources. What is needed here is no longer the pure agroeconomic doctrine of competition but the biological and evolutionary principle of a regulating balance between competition and cooperation within a living system for the benefit of its preservation.

Cooperation between sub-areas is characterized by the expectation of an overall effect resulting from the individual processes in the sub-areas; a better result is achieved than if the sub-areas are optimized only for themselves. The individual sub-areas contribute their specific part to the overall success and ensure the entire system's survivability. Which structures and resource allocation strategies are effective and efficient in the respective contexts can only be assessed retrospectively based on the cost-benefit ratio and the overall results achieved compared to the previous situation. Cooperative motives for action require a break from the exclusive orientation towards selective optimality (Nida-Rümelin 2020). They can only be grasped in terms of a structure of interaction, which puts the parts into perspective about the superordinate whole, i.e., it relates them to each other. Rationality consists of an intentionality that makes the processes sufficiently coherent, i.e., coherent and thus explainable and understandable with the superordinate goal.

The **individuality** of agroecosystems is another characteristic that distinguishes them from agro-industrial production sites. Just as there are no identical natural ecosystems in the world, there are no uniform agroecosystems. Each operating system and the management responsible for it have individual strengths and weaknesses, options, and limitations. Consequently, no universally valid solution strategies are equally effective regardless of context. Which methods are suitable to what extent for the provision of system services in the respective context requires individual explorations and the verification of their effectiveness in the follow-up to their implementation. Also, the result of the impact structure with its manifold interactions and causes of variation cannot be assessed based on individual indicators but only step by step (iteratively), in which conclusions are repeatedly drawn from the whole to the parts and from the parts to the impact on the system as a whole. It is easily understandable that farm management can use this procedure to varying degrees, causing a significant variation in system performances between farm systems.

It is the primary producer who holds the key for the consumers who purchase higher-quality and higher-priced products from farms with above-average system performance. They must provide evidence of delivering an above-average performance of their agroecosystem concerning public welfare interests. System performance assessment is not a task that can be left to interest groups. Instead, it is a sovereign task that the public welfare interests must guide. Previous chapters show that animal welfare and environmental protection services are process qualities, just as product qualities are measurable. They are not achieved through the implementation of individual measures or compliance with increased minimum standards but through the totality of the performances of the individual agroecosystem. Since the farm systems differ considerably in terms of the ratio of sales products produced to animal and environmental protection performance, they represent the relevant reference system for assessment and benchmarking.

9.4.5 Political Challenges

In general discourse, it is widely accepted that the best agroecosystems are those that can produce products for sale with a high level of product quality combined with a low level of adverse health effects on livestock and low levels of pollutants released into the environment. The urgency of b**enchmarking** agroecosystems according to their system services cannot be overstated. It would provide primary producers and all other stakeholders with the guidance they have so far lacked. Not only for outsiders but also for the majority of primary producers, an agricultural enterprise is still a "**black box**." While other industrial operations are regularly tested for the pollutants released into the environment and integrated into emissions trading with CO_2 certificates, agricultural operations have remained largely undisturbed and unaffected. Things can't stay this way.

Finally, within the framework of international agreements, the national governments have committed themselves not only to general climate protection but also, e.g., at the European level, to air pollution control of substances such as ammonia and particulate matter, most of which originate from agriculture (see Sect. 5.3). Compliance with these agreements can only be achieved through efficient reduction strategies tailored to farm-specific conditions.

At the European level, the European Commission (2020) has presented a concept for a fair, healthy, and environmentally friendly **food** chain entitled "From farm to table." The so-called "Green Agreement" shows how Europe could become the first **"Climate-neutral"** continent by 2050.

> *"Consumers should be empowered to choose sustainable food, and all actors in the food chain should see this as a responsibility and an opportunity. They can use the transition to sustainability as a brand and secure the future of the EU food chain before their competitors outside the EU do so. The transition to sustainability offers a pioneering advantage to all actors in the EU food chain."*

Even if the statements made so far are still very vague and there is still a long way to go from setting targets to achieving them, they leave no doubt that in the future, there will be intensified competition between countries for the favor of consumers about sustainable solutions in the production of food. Suppose one takes the title of the EU Commission's initiative "From farm to table" literally. In that case, this already indicates that in the future, competition will be sparked not only between states but, above all, between farms in favor of consumers and socially relevant services. This is expected to have a significant impact on primary production. Policy challenges include classifying the range of different system services of agroecosystems into quality categories, assessing farm performance appropriately, and transforming it into marketable metrics. A new price structure for foodstuffs will likely develop due to the interplay between supply and demand. Countries that mainly serve the low-price segment will, in the future, find it difficult to offer appropriate products to a consumer clientele that is developing in terms of quality awareness for food. References to the regional origin of the products produced will hardly be sufficient to arouse positive associations in the long term and certainly not to satisfy quality requirements. Regionally produced products only inform about the origin. Where a farm is located, however, does not influence the level of quality that the individual farm achieves with its production processes. Unless they participate in separate marketing structures, farms are exposed to the same price pressure of global undercutting competition regardless of location.

Currently, the **subsidies** from public funds, which keep many farms afloat, are not linked to the services that the subsidy recipients provide for the community and, thus, for the donors. This means that an essential control element is missing, which could be used to influence the production processes and the extent of damage. The same applies to the production of higher-quality products. It would be, above all, the task of politics as the representative of the public welfare interests to coordinate the reintegration of farms into higher-level community structures and into a system of give and take, demand, and support. Reintegration could start from different community structures at municipal, regional, federal, national, and European levels. All reintegration measures should aim to strengthen the interrelations between the community structures and their members. This presupposes, however, that the members of community structures, in addition to their particular interests, also have the interests of the common good in mind and orient themselves accordingly.

Reconciliation between public welfare and particular interests requires clear objectives for all actors to orient themselves. Contrary to what the protagonists pursue in a "business-as-usual" approach, an appropriate objective is not in compliance with higher minimum requirements. The overriding aim is producing food in its original and comprehensive meaning, i.e., as a **means of life**. The standard for assessing foodstuffs is not limited to the content of specific ingredients and avoiding contamination with harmful substances. The decisive factor is whether they originate from an operational context that substantially contributes to promoting life processes inside and outside the agroecosystem and not to their destruction. **Food** is not only an essential means of life for those who consume it to meet their nutritional needs. At the same time, food is a means of economic survival for those who have produced these means of life and can expect a price that secures their financial existence. Also, food production processes should not destroy the livelihoods of flora and fauna in the near or far surroundings of the farms, should not pollute the environment excessively with harmful substances, and should not contribute to global warming. Also, food of animal origin can only claim to be called food if the farm animals that provide the products for humans have been able to live a good life during their lifetime.

The value of food should be measured by its contribution to sustaining life processes in the respective contexts. Unlike food derived from anonymously produced raw materials, food should not be judged in isolation but only related to its context. Food is itself the result of complex processes in an agroecosystem. The assessment of food is closely linked to the recording of farm system performance, which should be classified into quality categories through a cross-farm comparison. We should only speak of **high-quality food** if it has been produced in farm systems that rank above average regarding system performance concerning public welfare interests.

Given the outlined characteristics of system services in connection with food production, it is unsurprising that such an approach will not meet all the interest groups' approval. Resistance is expected, particularly from groups that have benefited from previous economic practices. However, it is equivalent to an elementary fallacy to confuse or equate public welfare interest with the consent of all or the most significant possible number of supporters or members of interest groups. For this reason, commissions set up by politicians are of dubious value. Here, the representatives of the participating interest groups agree on a minimum consensus, excluding the non-participating groups. It remains completely open to what extent this consensus is conducive to the common good. Just because groups articulate and demand their specific interests according to their self-referential standards, these cannot automatically be classified as justified. Likewise, particular interests are not equal among themselves; most importantly, they are not on the same hierarchical level as public welfare interests. Just as the **subsystems** in a living organism must subordinate their consumption of resources to the overriding goal of the organism's self-preservation, and just as the subsystems in an agroecosystem are contained to contribute to the economic subsistence of the farm as a whole, so too should the representatives of particular interests contribute to the welfare of the community structures of which they are members and in which they are embedded. At the very least, specific interests should not counter the public welfare interests to any relevant degree. A public debate is long overdue on what should be counted among the elementary public welfare interests concerning food production. This then serves as a yardstick against which the various groups can orient themselves concerning their particular interests or must allow themselves to be measured against it.

If agriculture is to be converted from the production of raw materials to high-quality food production, this can only be achieved through **regulatory intervention**. When the well-being and, even more so, the viability of a community is at stake, particular interests must be subordinated to those of the common good. Degrees of freedom are to be limited where the community is threatened with relevant damage. In such cases, the community has the right and duty to intervene and regulate the coexistence of the community members to benefit overriding objectives. In food production, the appropriateness of regulatory measures is measured by whether and to what extent the system services of agroecosystems can be improved for the community. Those farms performing below average and unwilling or unable to make fundamental changes should be shown ways out of agricultural production. According to this reading, farms that excessively harm the public welfare interests have forfeited the license to produce food and the right to community support. Therefore, the claim of the farmers' associations, which loudly advocate in their proclamations that no farm should be left behind, is not only unrealistic but downright harmful to the community. While in the economic system of the "free" market economy, which the farmers' associations support, farms are continually being driven to ruin and left to their fate, in the future, it must be a question in the sense of realizing the public welfare interests, that not those farms which make a positive contribution to the interests of the common good make way, but those which are not adequately able or willing to make adequate community contributions.

A fair relationship between primary producers and consumers involves placing an appropriate value on food in terms of the context in which it was produced and ensuring that this value is comprehensible to food purchasers. Consumers must be able to rely on the information they receive about the value of the food they buy. This can only be guaranteed if there is an assessment by independent bodies. In addition, the food price should correspond to the value attached to the products in terms of the public interest. A fair price is characterized by the fact that it covers at least the costs of production, is fair to all actors along the value chain and is a sign of appreciation. Where undercutting competition dominates and cheap offers flood the market, neither adequate appreciation nor fair payment can be established for services rendered for the common good. It is concluded that the prevailing market system undermines fair dealings between the actors involved. It is the market system that fails in terms of producing high-quality food as much as it fails in terms of reducing harmful effects. Fair conditions can only be enforced by a state institution. Consequently, agricultural policy fails to prioritize public welfare interests over economic interests and defends the former against a highly well-organized agricultural lobby.

To take corrective action here, the system services of agroecosystems should represent the common reference basis for the public welfare interests and those of primary producers. Financial incentive schemes must be linked to **evidence** of substantial improvements. In addition to financial support, farms need a comprehensive knowledge of orientation as well as action-guiding knowledge to better deal with the complexity of internal processes and to be able to regulate them concerning system performance. Even if technical aids and the progress of digitalization can have a supporting effect here, they are only a means to the end of improving system performance and not an end in themselves. The meaningfulness of the use of funds is measured by the increase in effectiveness and efficiency that can be achieved in system performance. Continuous internal controls and feedback mechanisms enable timely corrections. Instead of continuing to set the course for quantitative growth, the various actuators must be aligned and synchronized for qualitative growth. Qualitative growth means that resource productivity in value creation is constantly increasing, i.e., growth is achieved with ever-lower inputs of non-renewable resources and environmental consumption (Mohr 1999). It is based on the fact that material resources and physical labor are increasingly replaced by intellectual labor: Explicit knowledge replaces raw materials and energy. The use of knowledge should make more out of less. This can only be achieved if synergistic interactions are strengthened and antagonistic ones are contained, resulting in a high efficiency of resource use in the system as a whole. This approach is contrary to a continued standardization of process flows, which serves the goal of cost minimization.

Current agricultural practice results from a normative fixation on economic parameters that are geared solely to market events and essentially ignore the context or the negative consequences and damaging effects of undercutting competition. In addition, the polluter-pays principle has been ignored mainly in agriculture. Consequently, farm managers can hardly be held responsible for the damage they cause. As suppliers of anonymous raw materials for the processing industry, they are inadequately integrated into the overall social context. This is also evident from the defensive attitude expressed towards social demands for more excellent containment of harmful effects through regulatory measures. While hardly any concrete requirements are imposed on primary producers by society, they are massively harassed by the forces of an economic system that does not provide them with the resources they need to survive economically in the long term. Nevertheless, the displeasure of primary producers is not directed against those who cling to the exploitative system of the "free" market economy but against those who want to transform it into an eco-social market economy.

There is nothing wrong with pursuing one's interests as long as other people and the public welfare interests are not harmed. In the book's previous chapters, it has been shown who and what has already been harmed to what extent due to pursuing particular interests in agribusiness and which actors have profited from the conditions in a certain way. The assertion of specific interests at the expense of other groupings and the public welfare interests is only possible because superordinate community structures allow this to happen and do not take effective counter and regulatory measures. The public welfare interests are not realized by denouncing moral misconduct and appealing to the individual actors to please moderate themselves and not to overdo it in the pursuit of self-interest. To reduce harmful effects and to achieve containment of conflicts of interest, the community structures themselves must ensure that framework conditions are implemented that set limits to the actions of individual members of the community. In this sense, regulatory measures are not an end in themselves but a **means to the end** of minimizing harmful effects on the community.

Regulatory processes cannot be established in a democratically constituted system without a wide range of weighing-up processes having taken place beforehand. So far, the focus has only been on weighing up the advantages and disadvantages that result for the respective interest groups. In contrast, a change of perspective, connected with the question of which advantages and disadvantages result for the community and how the advantages and disadvantages affect one's interests, has so far been absent. They do not adequately fulfill the role assigned to the state organs as representatives of public welfare interests. Policies focus primarily on the interests of their electorate to be re-elected rather than prevail the superordinate public welfare interests against the subordinate particular interests. At best, compromise formulas are reached between different interest groups depending on the balance of power. Instead, relations between primary producers and higher-level community structures should be characterized by appropriate give-and-take. The primary supply of **primary producers** consists of raw materials. It is the result of processes in the complex interactions of an agroecosystem. Food's value for the demand side, i.e., for consumers and public welfare interests, is decided in the context of the operating system. It is measured by the ratio of production to system services. Those who offer the products on the market must prove the superiority of system services and expect a corresponding return in the form of an additional price and state subsidies. It can be assumed that the relationships between primary producers, consumers, and Community structures could be developed to mutual advantage if the offers corresponded better to the respective demand.

The local, regional, federal, national, and European Community structures offer primary producers various financial assistance and support measures. Primary producers are happy to use these because they depend on them. However, what primary producers have to offer to the community and public welfare interests in return has not been convincing. The offer of primary producers with low production costs is mainly directed at the market but not at the community and community structures. So far, public discourse has not drawn a clear line between market and community interests. This confusing mix of interests is not conducive to enlightening citizens and consumers about the genuine underlying interests and the winners and losers.

In many respects, primary producers' current supply of raw materials does not meet what many consumers want, which the community desperately needs. This applies to high-quality food and production processes with little or no negative impact on the public interest. Instead, primary producers flood the market with an oversupply of anonymous, low-quality raw goods. Under the current conditions, neither the primary producers nor the buyers of the raw goods can assess the specific harmful effects of the respective production processes. If higher prices are to be achieved for the products produced in the future and at the same time, the public welfare-oriented objectives of state support measures and subsidy payments are to be met; there is no alternative to **contextualizing** and differentiating the products sold. Only based on a realistic assessment of the socially relevant services provided can primary producers offer to consumers and community structures that justify subsidies in the form of higher market prices for higher-quality products and state support for public welfare-oriented system services.

The relationship between give and take has become very skewed in the agricultural sector in recent decades. Supply is not matched to quality-differentiated demand. Neither do primary producers make consumers a differentiated offer that corresponds to their respective wishes and needs, nor do primary producers provide services for the common good that justify the enormous sums with which farms are subsidized from public budgets. Moreover, state subsidies increasingly lose their legitimacy vis-à-vis other people in need when primary producers use the **subsidies** to serve market interests at the expense of public welfare interests. It is likely to become increasingly tricky for agricultural policy to justify promoting a rural economy from public funds that run counter to public interests because this perpetuates the production of harmful severe effects and undermines public welfare interests. It is time that subsidies from the EU budget were returned to the basic principle of a state support policy: Support of structures and processes through public money to promote public welfare interests.

Community structures have so far failed not only because of the complexity of the processes underlying the emergence of harmful effects. The lack of clarity contributes to the fact that most decision-makers do not understand the causal relationships. However, the interest in clarification has also been limited so far. Even from the ranks of the agricultural sciences, no enlightening activities are to be expected because of the advancing specialism. An agricultural system that does not attach adequate importance to the health disorders of farm animals and their relevance for consumer and animal protection and product quality is unhealthy. Instead of making a qualitative differentiation of products based on the immediate context from which the products originate, agricultural marketing advocates an associative contextualization via standards (e.g., organic or pasture-raised milk). This imagined linking of individual aspects with a supposed superiority corresponds to the transfer effects the food retailers use in many ways for their interests when establishing brand products. This disguising marketing strategy is misleading and anti-enlightening and runs counter to the efforts to improve the economic framework conditions and the pursuit of public welfare interests.

Regulatory interventions should not focus on generally valid guidelines for enhancing minimum standards of production processes but on the differences in system services between the individual agroecosystems as outcomes in the respective context. The central task of the state is to assess these based on binding methodological procedures and to classify them on a scale from very low to very high. Of no concern to the common good is how management balances economic and environmental problems within the agroecosystem. What is relevant are the outcome variables that result from the complex interactions within agroecosystems in the form of animal and environmental services. On the part of state bodies, within the framework of sovereign tasks, monitoring a reliable assessment of system services and ensuring corresponding clarity and transparency would be urgent. On this basis, it is up to the market participants to negotiate the pricing of system services by comparing supply and demand.

The market economy system misleads consumers into believing that monetary values also reflect the qualitative value of products. The 'free' market system decouples agricultural products' monetary and qualitative value. The assumption that they have something to do with each other is one of the elementary errors of thinking in agricultural economics and has significantly promoted undesirable developments. The agricultural economic system of the "free" market economy is a system that does not want to be regulated by superordinate structures but trusts that the market itself will sort it out. Now that it is evident that the market is failing to support public welfare interests, it is time to transform the "free" agricultural economy into an eco-social agricultural economy. While in the past, intensification, specialization, and mechanization have shaped the development of agricultural production processes and achieved considerable productivity gains, the diminishing marginal utility and the extent of undesirable side-effects and harmful effects leave no doubt that the strategies pursued to date have long since exceeded what is beneficial for the majority of farms, but above all for the public welfare interests.

Nevertheless, due to existing path dependencies, the belief in the market's effectiveness and technical and scientific progress is unbroken in large parts of the agricultural sector. The economic model of rational choice orthodoxy must believe in the miracle that selectively optimizing behavior over time yields a coherent overall practice and can resolve problems that have arisen elsewhere (Nida-Rümelin 2020). The miracle has so far failed to materialize. Contrary to politicians' joint assessment and the theorems of agricultural economics, the market has proved incapable of balancing particular interests with public welfare. Therefore, it is all the more important to initiate a reflection and debate within the scientific community on how lessons can be learned from the harmful effects of misguided economic theories on public welfare interests and how a reorientation of the socially relevant agricultural sciences in the direction of a systems science can succeed.

The agricultural system, which has created the problems inherent in the system, will not be able to solve them itself. A market economy geared to undercutting competition cannot free itself from the dependency relationships created but requires a regulating hand that provides framework conditions in which the market economy can continue to develop its undisputed potential. At the same time, limits must be set on the growth aspirations of the market. The damaging effects of particular interests on the community define the boundaries. It is the very task of politics to enforce the public welfare interests and to contain excessive specific interests. So far, it has been unwilling and unable to free itself from the infiltration of an overpowering agricultural lobby. Therefore, what is needed above all is educational work so that decision-makers, as well as citizens and consumers, can reorient themselves and be better able than in the past to judge which form of food production is conducive to their own and the public welfare interests. It is a central concern of this book to contribute to this.

References

Abdelhamid AS, Brown TJ, Brainard JS, Biswas P, Thorpe GC, Moore HJ, Deane KH, Al Abdulghafoor FK, Summerbell CD, Worthington HV, Song F, Hooper L (2018) Omega-3 fatty acids for the primary and secondary prevention of cardiovascular disease. Cochrane Database Syst Rev 7:CD003177. https://doi.org/10.1002/14651858.CD003177.pub3

Alrøe HF, Noe E (2010) Multiperspectival science and stakeholder involvement. In: Darnhofer I, Grötzer M (eds) Building sustainable rural futures: The added value of systems approaches in times of change and uncertainty; 9th European IFSA Symposium; 4–7 July 2010 in Vienna, Austria; proceedings. BOKU, Wien, pp 527–533

von Bertalanffy L (1968) General system theory: Foundations, development, applications. George Braziller, New York

Blume L, Hoischen-Taubner S, Over C, Möller D, Sundrum A (2021a) Status quo der nutritiven und ökonomischen Situation sowie Potentiale des Einsatzes heimischer Proteinträger auf ökologisch wirtschaftenden Geflügel- und Schweinebetrieben: Teil 2: Innerbetriebliche Wertschöpfungspotentiale des Fütterungsmanagements und des Einsatzes einheimischer Proteinträger. Ber Landwirtsch 99. https://doi.org/10.12767/buel.v99i2.350

Blume L, Hoischen-Taubner S, Möller D, Sundrum A (2021b) Status quo der nutritiven und ökonomischen Situation sowie Potentiale des Einsatzes heimischer Proteinträger auf ökologisch wirtschaftenden Geflügel-und Schweinebetrieben: Teil 1: Bedarfsgerechte Nährstoffversorgung, Tierverluste, Ressourceneffizienz und Wirtschaftlichkeit. Ber Landwirtsch 99. https://doi.org/10.12767/buel.v99i2.349

Böcker A, Herrmann R, Gast M, Seidemann J (2004) Qualität von Nahrungsmitteln: Grundkonzepte, Kriterien, Handlungsmöglichkeiten. Schriften zur internationalen Entwicklungs- und Umweltforschung, vol 8. Peter Lang, Internationaler Verlag der Wissenschaften, Frankfurt am Main

Buller H, Roe E (2012) Commodifying animal welfare. Anim Welf 21:131–135. https://doi.org/10.7120/096272812X13345905674042

Clark B, Stewart GB, Panzone LA, Kyriazakis I, Frewer LJ (2016) A systematic review of public attitudes, perceptions and behaviours towards production diseases associated with farm animal welfare. J Agric Environ Ethics 29:455–478. https://doi.org/10.1007/s10806-016-9615-x

Daatselaar CH, Reijs JR, Oenema J, Doornewaard GJ, Aarts HFM (2015) Variation in nitrogen use efficiencies on Dutch dairy farms. J Sci Food Agric 95:3055–3058. https://doi.org/10.1002/jsfa.7250

Danish Agriculture & Food Council (2014) Fact file: Danish pig production. www.agricultureandfood.dk

Duda RO, Hart PE, Stork DG (2001) Pattern classification, 2nd edn. Wiley, New York

Emerson MR, Woerner DR, Belk KE, Tatum JD (2013) Effectiveness of USDA instrument-based marbling measurements for categorizing beef carcasses according to differences in longissimus muscle sensory attributes. J Animal Sci 91:1024–1034. https://doi.org/10.2527/jas.2012-5514

Union E (2016) Richtlinie (EU) 2016/2284 des Europäischen Parlaments und des Rates vom 14. Dezember: über die Reduktion der nationalen Emissionen bestimmter Luftschadstoffe, zur Änderung der Richtlinie 2003/35/EG und zur Aufhebung der Richtlinie 2001/81/EG. Amtsblatt der Europäischen Union 59:1–31

Farm Animal Welfare Council (2009) Farm Animal Welfare in Great Britain: Past, Present and Future, London, UK

Food and Agriculture Organization of, International Fund for Agricultural, United Nations Children's Fund, World Food Programme, World Health Organization (2018) The state of food security and nutrition in the world: Building climate resilience for food security and nutrition. The state of the world, vol 2018. FAO, Rome

Freihold D, Bartels T, Bergmann S, Berk J, Deerberg F, Dressel A, Erhard MH, Ermakow O, Huchler M, Krautwald-Junghanns M-E, Spindler B, Thieme S, Hafez HM (2019) Investigation of the prevalence and severity of foot pad dermatitis at the slaughterhouse in fattening turkeys reared in organic production systems in Germany. Poult Sci 98:1559–1567. https://doi.org/10.3382/ps/pey473

Freihold D, Bartels T, Bergmann S, Berk J, Deerberg F, Dressel A, Erhard MH, Ermakow O, Huchler M, Spindler B, Thieme S, Krautwald-Junghanns M-E, Hafez HM (2021) Investigation of the occurrence of pathological carcass alterations at the processing plant in meat turkeys reared in organic production systems in Germany. J Appl Poultry Res 30:100145. https://doi.org/10.1016/j.japr.2021.100145

Gellynck X, Verbeke W, Vermeire B (2006) Pathways to increase consumer trust in meat as a safe and wholesome food. Meat Sci 74:161–171. https://doi.org/10.1016/j.meatsci.2006.04.013

Geuder U, Pickl M, Scheidler M, Schuster M, Götz K-U (2012) Mast-, Schlachtleistung und Fleischqualität bayerischer Rinderrassen. Züchtungskunde 84:485–499

Golman R, Hagmann D, Loewenstein G (2017) Information avoidance. J Econ Lit 55:96–135. https://doi.org/10.1257/jel.20151245

Göpel M (2020) Unsere Welt neu denken: Eine Einladung. Ullstein, Berlin

Grethe H (2017) The economics of farm animal welfare. Ann Rev Resour Econ 9:75–94. https://doi.org/10.1146/annurev-resource-100516-053419

Häusel H-G (2013) Emotional Boosting English Version, 1st edn. Haufe Fachbuch, vol 1639. Haufe, Stuttgart

Hocquette JF, Gondret F, Baéza E, Médale F, Jurie C, Pethick DW (2010) Intramuscular fat content in meat-producing animals: development, genetic and nutritional control, and identification of putative markers. Animal 4:303–319. https://doi.org/10.1017/S1751731109991091

Hofstadter DR, Sander E (2018) Die Analogie: Das Herz des Denkens. Klett-Cotta, Stuttgart

Houghton RA (2007) Balancing the global carbon budget. Annu Rev Earth Planet Sci 35:313–347. https://doi.org/10.1146/annurev.earth.35.031306.140057

Huang X, Ahn DU (2018) The incidence of muscle abnormalities in broiler breast meat—a review. Korean J Food Sci Anim Resour 38:835–850. https://doi.org/10.5851/kosfa.2018.e2

Ison R (2010) Systems practice: how to act in a climate change world, 1st edn. Springer International Publishing AG, London, Cham, Springer, London

Kahneman D (2011) Schnelles Denken, langsames Denken, 11th edn. Siedler, München

Kahneman D, Sibony O, Sunstein CR (2021) Noise: Was unsere Entscheidungen verzerrt – und wie wir sie verbessern können, 1st edn. Siedler, München

Kleber EW (1992) Diagnostik in pädagogischen Handlungsfeldern: Einführung in Bewertung, Beurteilung. Diagnose und Evaluation. Grundlagentexte Pädagogik. Juventa-Verl, Weinheim

Kolbe H, Meyer D (2021) Schlaggenaue Analyse von 32 Betrieben des ökologischen Landbaus im Freistaat Sachsen: Nährstoff- und Humusmanagement. Ber Landwirtsch 99. https://doi.org/10.12767/buel.v99i2.315

Krieger M, Sjöström K, Blanco-Penedo I, Madouasse A, Duval JE, Bareille N, Fourichon C, Sundrum A, Emanuelson U (2017) Prevalence of production disease related indicators in organic dairy herds in four European countries. Livest Sci 198:104–108. https://doi.org/10.1016/j.livsci.2017.02.015

Merialdi G, Dottori M, Bonilauri P, Luppi A, Gozio S, Pozzi P, Spaggiari B, Martelli P (2012) Survey of pleuritis and pulmonary lesions in pigs at abattoir with a focus on the extent of the condition and herd risk factors. Vet J 193:234–239. https://doi.org/10.1016/j.tvjl.2011.11.009

Mohr H (1999) Wissen – Prinzip und Ressource, 1st edn. Springer, Berlin Heidelberg, Berlin, Heidelberg

Monteils V, Sibra C, Ellies-Oury M-P, Botreau R, La Torre A de, Laurent C, Monteils V, Sibra C, Ellies-Oury M-P, Botreau R, Laurent C (2017) A set of indicators to better characterize beef carcasses at the slaughterhouse level in addition to the EUROP system. Livest Sci 202:44–51.https://doi.org/10.1016/j.livsci.2017.05.017

Munch SL, Nielsen SS, Krogh MA, Capion N (2019) Prevalence of abomasal lesions in Danish Holstein cows at the time of slaughter. J Dairy Sci 102:5403–5409. https://doi.org/10.3168/jds.2018-15757

Murphy SC, Martin NH, Barbano DM, Wiedmann M (2016) Influence of raw milk quality on processed dairy products: how do raw milk quality test results relate to product quality and yield? J Dairy Sci 99:10128–10149. https://doi.org/10.3168/jds.2016-11172

Nida-Rümelin J (2020) Eine Theorie praktischer Vernunft, 1st edn. de Gruyter, Berlin/Boston

Pallarés FJ, Añón JA, Rodríguez-Gómez IM, Gómez-Laguna J, Fabré R, Sánchez-Carvajal JM, Ruedas-Torres I, Carrasco L (2021) Prevalence of mycoplasma-like lung lesions in pigs from commercial farms from Spain and Portugal. Porcine Health Manage 7:26. https://doi.org/10.1186/s40813-021-00204-3

Pigłowski M (2020) Food hazards on the European Union market: the data analysis of the rapid alert system for food and feed. Food Sci Nutr 8:1603–1627. https://doi.org/10.1002/fsn3.1448

Plaizier JC, Krause DO, Gozho GN, McBride BW (2008) Subacute ruminal acidosis in dairy cows: the physiological causes, incidence and consequences. Vet J 176:21–31. https://doi.org/10.1016/j.tvjl.2007.12.016

Polkinghorne RJ, Thompson JM (2010) Meat standards and grading: a world view. Meat Sci 86:227–235. https://doi.org/10.1016/j.meatsci.2010.05.010

Schenk H-O (2001) Funktionen, Erfolgsbedingungen und Psychostrategie von Handels- und Gattungsmarken. In: Bruhn M (ed) Handelsmarken: Zukunftsperspektiven der Handelsmarkenpolitik, 3rd edn. Schäffer-Poeschel, Stuttgart, pp 71–98

Schulz von Thun F (2007) Miteinander reden: Fragen und Antworten. Rororo, vol 61963. Rowohlt Taschenbuch Verl., Reinbek bei Hamburg

Stengel O (2021) Vom Ende der Landwirtschaft: Wie wir die Menschheit ernähren und die Wildnis zurückkehren lassen. oekom verlag, München

Sterling P (2012) Allostasis: a model of predictive regulation. Physiol Behav 106:5–15

Sundrum A (2012a) "Healthy food" from healthy cows. In: Konvalina P (ed) Organic farming and food production. InTech

Sundrum A (2012b) Optimierung der Nährstoffeffizienz in der Milchviehhaltung. In: Verband Deutscher Landwirtschaftlicher Untersuchungs- und Forschungsanstalten (ed) Kongressband

2012 Passau: Vorträge zum Generalthema: Nachhaltigkeitsindikatoren für die Landwirtschaft: Bestimmung und Eignung. VDLUFA-Verl., Darmstadt, pp 122–129

Sundrum A (2014) Organic livestock production. In: van Alfen NK (ed) Encyclopedia of agriculture and food systems: 5-volume set, 2nd edn. Elsevier Science, Burlington, pp 287–303

Sundrum A (2019) Real-farming emissions of reactive nitrogen—necessities and challenges. J Environ Manage 240:9–18. https://doi.org/10.1016/j.jenvman.2019.03.080

Sundrum A (2024) Why has animal science not led to improved farm animal health and welfare? Front Anim Sci 4:1214889

Sundrum A, Emanuelson U, Fourichon C, Hogeveen H, Tranter R, Velarde A (2016) Farm centric and equifinal approach to reduce production diseases on dairy farms. In: ICPD Scientific committee (ed) p 186

Swaby H, Gregory NG (2012) A note on the frequency of gastric ulcers detected during postmortem examination at a pig abattoir. Meat Sci 90:269–271. https://doi.org/10.1016/j.meatsci.2011.06.015

Thøfner ICN, Dahl J, Christensen JP (2021) Keel bone fractures in Danish laying hens: prevalence and risk factors. PLoS ONE 16:e0256105. https://doi.org/10.1371/journal.pone.0256105

Umweltbundesamt (2021) Biogasanlagen. https://www.umweltbundesamt.de/themen/wirtschaft-konsum/industriebranchen/biogasanlagen#einfuhrung. Accessed 8 Sept 2021

Valkova L, Voslarova E, Nincakova S, Passantino A, Vecerek V (2023) The incidence of liver damage found during postmortem examination at the slaughterhouse. Animals (Basel) 13. https://doi.org/10.3390/ani13050839

Veldhuis AMB, Smits D, Bouwknegt M, Worm H, van Schaik G (2021) Added value of meat inspection data for monitoring of dairy cattle health in the Netherlands. Front Vet Sci 8:661459. https://doi.org/10.3389/fvets.2021.661459

Vismarra A, Lenti A, Genchi M, Kramer L, Geldhof P (2023) Seroprevalence of Ascaris suum compared to milk spot prevalence at slaughter in Italian fattening pigs. Vet Parasitol Reg Stud Reports 37:100828. https://doi.org/10.1016/j.vprsr.2022.100828

Zrenner KM, Haffner R (1999) Lehrbuch für Fleischkontrolleure: Mit 3 Tabellen. Enke, Stuttgart

Glossary

Abstraction Abstraction is the process of grasping a general rule based on a set of examples by deducing joint features from concrete facts, real cases, or specific examples.

Action-guiding knowledge Action-guiding knowledge describes what someone needs to know to solve a problem or task and behave competently in a situation. It aims at planning, deciding, and implementing concrete actions and includes the ability to verify the positive and negative consequences of the implementation. In addition to the knowledge of details, the overview is required to grasp a given situation in its → **complexity** and to derive working hypotheses regarding the likelihood of changes occurring. See also → **Orientation knowledge** and → **Dispositional knowledge**.

Agroecosystem An agroecosystem comprises the subsystems "soil," "plants," and "animals," as well as humans in their dual role as part of and shapers of the overall system. As control variables, humans influence inputs and outputs as well as interactions within the → **system**. Through humans as an interface, the fields of technology, economy, and society affect the system as a whole.

Aliments Aliments are any substance or product intended to be ingested by humans in a processed, partially processed, or unprocessed state to nourish the body, whether or not the production process involves a high degree of harmfulness and destruction.

Analogy An analogy is a rhetorical figure in which a relationship between things and properties or between them or their evaluation is explained through known, similar, or partly identical relationships. Just as there can be no thinking without → **concepts,** there can be no concepts without analogies. They fuel our thinking and provide an inexhaustible resource for human creativity.

Animal protection services Animal protection is a systemic service the whole farm provides. The rationale and the reference system for demonstrating successful animal protection efforts are the individual animals and the operating system in which the animals live. The percentage of animals in a herd that has been successfully protected from pain, suffering, and harm, i.e., largely unharmed, over a defined period, or the percentage of animals exposed to varying degrees of distress, reflects the protection service of a farm.

A. Sundrum, *Public Welfare-Oriented Production of Food,*
https://doi.org/10.1007/978-3-662-69040-6

Animal protection Animal welfare can be subsumed under all human activities that aim to meet the individual needs of animals by providing them with appropriate food, care, and housing (§ 2 TierSchG) and by avoiding overtaxing their adaptive capacities to protect them from pain, suffering, and harm. Animal welfare is equivalent to protecting animals from harm, suffering, and damage.

Animal welfare The term "animal welfare" is based on the definition of the World Organization for Animal Health (OIE, 2008): "*Animal welfare means how an animal is coping with the conditions in which it lives. An animal is in a good state of welfare if it is healthy, comfortable, well-nourished, safe, able to express innate behavior, and if it is not suffering from unpleasant states such as pain, fear, and distress. Good animal welfare requires disease prevention and veterinary treatment, appropriate shelter, management and nutrition, humane handling, and humane slaughter or killing.*"

Anthropocentrism Anthropocentrism is a philosophical view in which man and all his expressions of life are the center and purpose of the whole world. Anthropocentrism means that humans see themselves as the center of worldly reality and judge it according to self-referential evaluation systems.

Anthropomorphism Anthropomorphism is the (inadmissible) projection of human structures onto non-human realms; in this case, the attribution of human characteristics to animals, where human characteristics are recognized or assumed in both form and behavior.

Associative quality Associative quality is a mental link with the process of creating products. It is mainly used in marketing branded goods and → **brand programs**, using specific terms and images that consumers associate with positive memories and emotions, thus influencing their willingness to pay. Associative quality is associated with pleasure value.

Benchmarking Benchmarking compares business processes and performance metrics to best practices from other companies. It measures performance using a specific indicator, resulting in a performance metric compared to others.

Big Data Big data refers to huge amounts of data generated in one context and analyzed for others. Data are often incomplete and sometimes inaccurate and are often compiled from different sources.

Biodiversity Biodiversity is the variability of living organisms and the ecological complexes they form. It comprises the following three levels: (1) the diversity of → **ecosystems** or communities, habitats, and landscapes, (2) the diversity of species, and (3) the genetic diversity within species.

Brand program Brand programs are synonymous with seals of approval intended to describe a product's quality. The purpose of these programs, usually private sector schemes, is to provide consumers with positive indications of a product's quality or characteristics and to identify the manufacturer as an exceptionally trustworthy supplier. They are characterized by the producers maintaining a higher quality level than the minimum legal requirements in specific sub-areas, thereby setting new production standards.

Category Categories are ways of thinking and, therefore, the basis of all experience. Hofstadter and Sander (2018) understand a category as a mental structure that emerges over time, containing information in an organized form that can be accessed given the appropriate circumstances. The act of categorization is the groping, gradual, gradual linking of an entity or situation to a category predetermined in one's consciousness.

Causality Causality refers to a cause-effect relationship. It concerns the sequence of events and states of affairs related to each other. Accordingly, A is the cause of the effect B if B is brought about by A.

***Ceteris paribus* assumption** The Latin phrase **"Ceteris paribus"** means "other things being equal." Experiments are carried out under *ceteris paribus* conditions to test the effect of an influencing factor and to exclude influences of other aspects from the experiment. The *ceteris paribus* assumption means that the results of experiments are valid under the defined boundary conditions and that no conclusion can be drawn about the effect of known or unknown factors on the claimed relationship.

Coherent A system of ideas (a theory) is coherent if its parts are meaningfully connected.

Commons Commons are the cultural and natural resources accessible to all members of society, including natural materials such as air, water, and a habitable earth. These resources are held in common even when owned privately or publicly. Commons can also be understood as natural resources that groups of people (communities, user groups) manage for individual and collective benefit.

Common good Common good refers to the good which, for social reasons, should benefit as many community members as possible.

Complexity Complexity refers to the behavior of a system, a whole made up of individual interrelated parts whose components may interact with each other in different, usually non-linear, ways. The behavior of a complex system cannot be understood from knowledge of the behavior of individual components. The rules that govern the interactions are largely unknown.

Consistent A system of ideas (a theory) is consistent if there is no explicit contradiction, nor can a contradiction be deduced from it.

Cooperation Cooperation is voluntary collaboration for mutual benefit or to achieve common goals. On the other hand, if the mutual interaction between the actors is not intentional or purposeful, it is called interaction. Cooperation and interaction are essential features of human work in social systems.

Cost leadership A strategy to compete with other suppliers by keeping production costs low and prices the lowest. In the context of this book, it is distinct from → **quality leadership**.

Deduction A deduction infers from given premises to a special case or from the general to the particular. Deduction is used as a counter term to → **induction.** In natural sciences, predictions obtained by deduction must be empirically testable. If the observations do not agree with the predictions, the theory must be rejected.

Derivation A derivation is a logical inference of something new from given statements.

Digitization Digitization refers to the process of making → **information** computer-processable. If the information consists of numbers, it must be stored in a computer-readable form. If the information consists of → **categories** or types of relationships, numbers must be assigned to these categories and types so that they can be stored in the computer.

Dispositional knowledge Dispositional knowledge is a knowledge of causes, effects, and means. It is the knowledge that science and technology make available for given purposes through various publications. See also → **Action guiding knowledge** and → **Orientation knowledge**.

Ecosystem An ecosystem refers to the components of a delimited or specific type of natural area (e.g., a nutrient-poor watercourse) and their interactions. The term can refer to different spatial scales (local, regional) and covers both natural (e.g., undisturbed raised bogs) and semi-natural (e.g., calcareous grasslands) ecosystems, as well as ecosystems that are strongly influenced by human activities (e.g., agroecosystems).

Ecosystem services Ecosystem services refer to the direct and indirect contributions of ecosystems to human well-being, i.e., services and goods that provide direct or indirect economic, material, health, or psychological benefits to humans. The term is associated with an anthropocentric perspective and an ecosystem benefit to humans.

Emergence Emergence refers to creating one or more new properties from a complex whole that cannot be deduced from its constituent parts. Accordingly, properties that occur at a higher level, such as in the organism, cannot be explained or predicted from properties at a lower level, such as the organs. Thus, life is an emergent property of the cell but not of its molecules and organelles. The concept of emergence is the classical antithesis of the idea of reduction. "**Weak emergence**" is when novel properties at a higher level can be explained by the interaction of constituents alone, and "**strong emergence**" is when novel properties at higher levels can be neither attributed to nor predicted from the interaction of constituents. A "**system emergence**" occurs when novel properties at higher levels interact causally with properties at lower levels. This top-down and bottom-up → **causality** is often part of a systemic approach, which, unlike the reductionist approach, sees each component as an interdependent part of the whole.

Empiricism Empiricism holds that all we can ever know about reality is an interpretation of information our senses provide.

Enjoyment value Enjoyment value is commonly referred to as the sensory value or quality of → **food.** Enjoyment value also includes the → the **associative quality**. Enjoyment value is perceived through the human senses and associations. The perception of food through a person's sensory system, including the brain, is inherently subjective. Nevertheless, the characteristics relevant to product and process quality can be assessed intersubjectively to draw valid conclusions.

Entropy Entropy measures the disorder or disorganization of the components. The lower the entropy, the higher the components' organization level.

Environmental protection services Like animal protection services, environmental protection services are services provided by the farm as a whole. They require an active effort to minimize the discharge of pollutants from the operating system, thereby avoiding externalities. Environmental protection services are equivalent to → **public goods services**. They are related not only to the operating system but also to its production performance. The pollutant discharges in relation to the quantities of products sold can be ranked on a scale from very low to very high, thus allowing comparisons between individual farm performances.

Epistemology Epistemology is a philosophical sub-discipline concerned in particular with the question of what (human) cognition is, how exactly we can determine its scope, and what we can know or cognize.

External validation External validation is concerned with the question of the generalizability of the results of a study. According to the classical notion, statements or conclusions drawn in empirical studies have a high degree of external validity if (a) the results can be generalized to the population for which the study was designed and (b) they can be transferred beyond the specific setting of the study to other designs, instruments, places, times and situations, i.e., they are general or generalizable. External validity increases with each successful replication of the findings under different experimental conditions.

Externalization Externalization in business and economics is passing on costs to the community or future generations. It expresses unfair competitive conditions that give market advantages to those who act more ruthlessly and self-centeredly than others while refusing to take responsibility for one's actions.

Food Food is a means of life. In this sense, food differs from aliments in that it not only provides nutrients to the consumer but also considers the life processes of many other living things in the upstream and downstream context of production. In other words, → **food** is food only if its output is not excessively detrimental to the flora and fauna of the environment or farmed animals, nor primary producers and other people, i.e., it does not run counter to public welfare interests.

Forecast A forecast is a statement about future events, conditions, or developments. It differs from other statements about the future (e.g., prophecies) in that it is based on → **scientific methods** and knowledge.

Hypothesis A hypothesis is an assumption about a cause and effect whose validity has not been proven or verified. It represents conjectural knowledge that should be testable. Therefore, its provisional, untested character is usually emphasized, as it is only considered proven after successful empirical testing. When it is tested, it is either proved, verified, or disproved. When formulating a hypothesis, it is usual to state the conditions under which it should be valid.

Induction Induction means the abstract inference from observed phenomena to a more general knowledge, such as a general → **concept** or a → **law**. The term is used as an antonym for → **deduction.** Deduction infers a particular case from given premises, whereas induction is the opposite. In an inductive inference, the uncertainty remains of coming up with a case that does not confirm the general proposition. When this occurs, the inference is an inductive fallacy.

Information Information belongs to a cognitive system because of its relation to context. It is collected and structured for a specific purpose and has a contextual meaning. Information is the knowledge a sender conveys to a receiver through an information channel. Information can take the form of signals or code, and the recipient gains knowledge from the information. Information is valued by the recipient's interpretation of the whole at different levels. The sender or recipient can be not only a person/human being but also an animal or an artificial system (such as a computer).

Intersubjectivity Intersubjectivity means that a (complex) subject is equally recognizable and understandable to several observers. For example, one can agree on how one perceives something, classify it, or clarify its meaning. Intersubjective is distinguishable from "subjective" (see → **subjectivity**). On the other hand, intersubjectivity is also distinguished from → **objectivity**: objective facts are ideally provable, independently of conditions related to individual observers.

Justification Justification is an essential part of knowledge. A justified statement is characterized by convincing and adequate reasons supporting the statement.

Knowledge Knowledge involves personalized information about an object or state of affairs gained through experience, reasoning, intuition, or education. Knowledge is when a statement is true and justified (distinct from opinion).

Knowledge transfer The term is often used to describe transferring knowledge from science to practical application. However, due to the → **subjectivity of** knowledge, it is → the **information** passed on and only becomes knowledge through personal interpretation, understanding, and experience. Ignoring the subjective context is one of several criticisms of knowledge transfer.

Law In the natural sciences, a law is a statement valid indefinitely in space and time and empirically proven, i.e., successfully withstood previous attempts at falsification. For other sciences, however, laws are gradual statements because a definition too narrow as a spatially and temporally unlimited all-statement does not do justice to social reality.

Marginal utility Marginal utility indicates how the level of utility changes when the effort required to increase utility changes. It is generally assumed that the more effort is expended, the more the total utility can be improved. However, the law of diminishing marginal utility applies: the relation between effort and utility depends on the initial level of the target. At a high level, a higher level of additional effort is generally required to produce the same additional benefit as at a low level. The benefit of further effort reaches a limit. If this limit is exceeded, the marginal benefit becomes negative.

Model A model is a simplified representation of reality or a streamlined version of a *theory*, which is used to represent a state of affairs appropriately so that it is possible to describe, explain, and predict developments in the associated issues.

Neuromarketing Neuromarketing is a field of marketing that uses insights from brain research and psychology to optimize advertising. Neuromarketing experts use various methods to investigate which processes in the consumer's brain influence purchasing decisions, and companies use this knowledge to further their marketing interests.

Normative Normative indicates how something should be, i.e., how something should be evaluated. The adjective also describes any evaluative or evaluative legal terms that appear in a legal rule. Normative is also the legislative activity of parliaments. Normative provisions in company law are statutory provisions. Normative elements require a legal or social evaluation. Their meaning derives from an evaluative interpretation.

Objectivity Objectivity is the independence of judgment or description of a thing, event, or state of affairs from the observer. Since the possibility of a neutral point of view, which would make absolute objectivity possible, is generally not given, objectivity is an ideal of the sciences. Since it can be assumed that every point of view is subjective, scientific results are measured by specific recognized scientific methods.

Operationalization Operationalization is the process of making a concept measurable. It defines how a theoretical construct (e.g., welfare) will be observable and quantifiable. It is of great importance in all empirical sciences because it is the basis for being able to make measurements.

Orientation knowledge Orientation knowledge is the knowledge of standards of action and justified purposes and goals. The possession of orientations becomes orientation knowledge as soon as the orientations appear to be rationally justified. See also → **action-guiding knowledge** and → **dispositional knowledge**.

Paradigm Paradigm is a central concept in the philosophy of science. It refers to the set of basic views constituting a scientific discipline in a historical period, which prescribe what questions are scientifically admissible and what can be considered scientifically satisfactory.

Parameter In science, a parameter is generally one of several measurable factors, such as temperature or pressure, that characterize a → **system** and influence its behavior. In experiments, one parameter is often varied while others are held constant.

Problem A problem is a task that an actor wants to solve to achieve a specific goal: the solution.

Process quality Process quality of food production is mainly determined by how crops are grown and livestock reared, taking into account animal welfare, environmental protection, and sustainability considerations. It is linked to the extent to which the production processes have positive and negative consequences. The characteristics cannot be directly measured, assessed, documented, and controlled by product traits but must assess the production process.

Product quality The quality of a product (food) is determined by the totality of its characteristics. Food quality is mainly determined by nutritional value, utility, and enjoyment value. These are properties that can be controlled directly by the product's traits.

Public goods Public goods can be used by different people simultaneously without excluding others from their use. Examples are the public road network, internal security, and clean air.

Public services The voluntary nature of service provision implies that it can only be considered of general interest if the services provided exceed the scope and quality required by the regulatory framework.

Quality The term **"Quality"** can be described by the terms properties, character, or condition. Here, the properties of a product, process, or system are classified into three hierarchical levels of order:

1. First-order quality is determined by the **totality of all the properties** of a product, process, or system (analytical quality approach).
2. The second-order quality is the value of the totality of all properties of a product, process, or system attributed by someone. Value is measured by the extent to which a person considers something good. This corresponds to consumer-oriented quality as the extent to which an offering meets customer requirements (preferential quality approach).
3. A third-order quality is the set of properties of products, processes, or systems measured against intersubjectively valid **external yardsticks** and concerning overriding objectives (functional quality approach).

Quality leadership Quality leadership is the ambition to be at the forefront of market competition in terms of the level of quality produced. It conflicts with the quest for → **cost leadership**.

"Quality standard" Quality standards describe the minimum level of quality that must be achieved to meet defined requirements and, in the case of products or services, serve as a marketing strategy to attract and retain customers. In food production, → **standards** are defined based on individual aspects, while many other elements are not considered. This contrasts with the definition of quality as the totality of all characteristics. Because of this contradiction and the intentionality of using the term, the term is put in quotation marks.

Rationality Actions of actors are rational if they help to achieve the objectives of actors and if they are done in such a way that the available means are used in the best possible way according to the preferences, i.e., efficiently. An action is rational if embedded in a higher-order whole of a well-reasoned action structure.

Reduction Reduction is the possibility of attributes of higher levels (e.g., organizations) being reduced to attributes of lower levels (e.g., individuals).

Reliability Reliability refers to the trustworthiness of statements derived, for example, based on a method. It is a necessary condition of validity.

Reproducibility In science, reproducibility means the repeatability of empirical scientific research results, with the intention of controlling and verifying the reported research results.

Standard A standard is a comparatively uniform or unified way of describing, producing, or performing something. It is used as a guideline, particularly in the fields of technology, methodology, and marketing.

Standardization Standardization is the standardization of objects and procedures. It reduces the existing variation in the expression of characteristics to a manageable number of standards that can be handled, i.e., by the trade.

Subjectivity Subjectivity is that which is accessible only to the individual and for which no universality is claimed. See also → **intersubjectivity** and → **objectivity**.

Subsidiarity principle The principle of subsidiarity states that state institutions should always and ideally only intervene in a regulatory capacity when the capacity of individual actors, groupings, or state institutions alone is insufficient to solve a community-oriented task. The principle of subsidiarity is an essential concept of federated states such as the European Union.

Subsistence Subsistence is a philosophical principle of self-preservation based primarily on human interaction with the environment to secure existence and satisfy basic needs. It refers to everything materially and socially necessary for everyday survival: food, clothing, shelter, and care and companionship.

System A system is generally defined as a definable natural or artificial "entity" made up of different components with different characteristics, which, because of certain ordered relationships between them, can be regarded as a common whole.

Theory A theory is a system of propositions for explaining a group of phenomena. The statements are content-related and have been confirmed by independent experiments or observations. Theories are used to achieve the goals of research: understanding, description, explanation, prediction, and design.

Validity Validity is a concept that attempts to provide a qualitative judgment of the truthfulness of statements. Validity is a property of conclusions drawn based on methods and their results and is intended to provide information about the quality of statements.

Value judgment A value judgment is a → **normative** statement. It is to be distinguished from statements of facts because observation cannot justify them.

Variable Variables are symbols that represent the characteristic values of the property under investigation.

Willingness to pay Willingness to pay is a personalized property. It results from the sum of the factors that make individuals willing to pay more for a product than others. Estimates of willingness to pay are based on survey data, using terms that evoke specific associations in respondents and are correlated with willingness to pay.

Index

A. Sundrum, *Public Welfare-Oriented Production of Food*,
https://doi.org/10.1007/978-3-662-69040-6

MIX
Papier aus verantwortungsvollen Quellen
Paper from responsible sources
FSC® C105338

If you have any concerns about our products,
you can contact us on
ProductSafety@springernature.com

In case Publisher is established outside the EU,
the EU authorized representative is:
Springer Nature Customer Service Center GmbH
Europaplatz 3, 69115 Heidelberg, Germany

Printed by Libri Plureos GmbH
in Hamburg, Germany